T0261184

Forest Products and Wood Science
An Introduction

Seventh Edition

Forest Products and Wood Science
An Introduction

Seventh Edition

Rubin Shmulsky

P. David Jones

WILEY Blackwell

This edition first published 2019
© 2019 John Wiley & Sons Ltd

Forest Products and Wood Science: An Introduction, 6th Edition, Rubin Shmulsky and P. David Jones,
9780813820743, May 2011 http://eu.wiley.com/WileyCDA/WileyTitle/productCd-081382074X.html
5th Edition, James L. Bowyer, Rubin Shmulsky, John G. Haygreen, 9780813820361, May 2007
4th Edition, James L. Bowyer, Rubin Shmulsky, John G. Haygreen, 9780813826547, Dec 2002
3rd Edition, Jim L. Bowyer, John G. Haygreen, 9780813822563, Feb 1996

All rights reserved. No part of this publication may be reproduced, stored in a retrieval system, or transmitted, in any form or by any means, electronic, mechanical, photocopying, recording or otherwise, except as permitted by law. Advice on how to obtain permission to reuse material from this title is available at http://www.wiley.com/go/permissions.

The rights of Prof Rubin Shmulsky and P. David Jones to be identified as the authors of this work have been asserted in accordance with law.

Registered Office(s)
John Wiley & Sons, Inc., 111 River Street, Hoboken, NJ 07030, USA
John Wiley & Sons Ltd, The Atrium, Southern Gate, Chichester, West Sussex, PO19 8SQ, UK

Editorial Office
The Atrium, Southern Gate, Chichester, West Sussex, PO19 8SQ, UK

For details of our global editorial offices, customer services, and more information about Wiley products visit us at www.wiley.com.

Wiley also publishes its books in a variety of electronic formats and by print-on-demand. Some content that appears in standard print versions of this book may not be available in other formats.

Limit of Liability/Disclaimer of Warranty
While the publisher and authors have used their best efforts in preparing this work, they make no representations or warranties with respect to the accuracy or completeness of the contents of this work and specifically disclaim all warranties, including without limitation any implied warranties of merchantability or fitness for a particular purpose. No warranty may be created or extended by sales representatives, written sales materials or promotional statements for this work. The fact that an organization, website, or product is referred to in this work as a citation and/or potential source of further information does not mean that the publisher and authors endorse the information or services the organization, website, or product may provide or recommendations it may make. This work is sold with the understanding that the publisher is not engaged in rendering professional services. The advice and strategies contained herein may not be suitable for your situation. You should consult with a specialist where appropriate. Further, readers should be aware that websites listed in this work may have changed or disappeared between when this work was written and when it is read. Neither the publisher nor authors shall be liable for any loss of profit or any other commercial damages, including but not limited to special, incidental, consequential, or other damages.

Library of Congress Cataloging-in-Publication Data

Names: Shmulsky, Rubin, author. | Jones, P. David (Paul David), 1976– author.
Title: Forest products and wood science : an introduction / Prof. Rubin Shmulsky, P. David Jones.
Description: Seventh edition. | Hoboken, NJ : John Wiley & Sons, Inc., [2019] |
  Includes bibliographical references and index. |
Identifiers: LCCN 2018049393 (print) | LCCN 2018055002 (ebook) | ISBN 9781119426363
  (Adobe PDF) | ISBN 9781119426370 (ePub) | ISBN 9781119426431 (hardcover)
Subjects: LCSH: Wood. | Lumber.
Classification: LCC TA419 (ebook) | LCC TA419 .S423 2019 (print) | DDC 674–dc23
LC record available at https://lccn.loc.gov/2018049393

Cover Design: Wiley
Cover Images: Log of tree and tree leaves courtesy of Janiece Pigg, Land of shrubs, Ticks of tree, and Wood cutter machine courtesy of P. David Jones

Set in 9.5/12pt Sabon by SPi Global, Pondicherry, India

10 9 8 7 6 5 4 3 2

CONTENTS

Appendices, review questions, and additional readings are located online at www.wiley.com/go/shmulsky

This textbook was initially written as an introduction to the anatomical and physical nature of wood and the relationship of these characteristics to its use as an industrial raw material. Over its nearly four decades of service, the book has been expanded in places to discuss and address other nonwood bio-based products and building materials, but wood remains the focus. Wood-based materials discussed include solid wood products, engineered composites, energy, and paper and fiber products. Also discussed are the basic processes involved in the production of the principal wood and other bio-based products and how their properties affect their use and performance. Related issues examined herein include the use of wood for energy and chemicals, environmental implications of wood-based materials, and the global wood supply picture.

College-level students in Forestry, Natural Resources, and Wood Science are the primary audience for this text. However, given the high degree to which wood and wood-based products are embedded in our culture, students of architecture, structural engineering, building construction, and related disciplines should develop an appreciation of the many potential forms and uses for wood and other bio-based materials. Society in North America and around the world depends on wood and other bio-based products, namely from sustainable forests, for basic shelter, sanitation, heat, and numerous other daily and basic needs. Wood products are ubiquitous. Thus, a sound understanding of the keys to the proper selection of raw materials for any wood-based product, and of the type of wood product best suited for a given application, is extremely important.

The term bio-based is somewhat novel. It includes materials and products from wood as well as agricultural residues and other lignocellulosic, nonfood sources. In the United States, this broader term often seems more environmentally benign. To date, wood or products from the forest remain the largest component under the bioproducts umbrella. It is known that regardless of any name change, wood and forest-derived woody materials will continue to account for the vast majority of all bio-based materials. Processing costs, storage issues, and environmental footprints lead to this situation. To paraphrase Shakespeare: a wood product by any other name is still a wood product.

This text is divided into four parts. The first part (Chapters 1–6) introduces the growth and structure of hardwoods and softwoods, chemical and structural characteristics, and inherent variability. The second part (Chapters 7–11) examines the physical properties of wood, relates these properties to wood's chemical and structural character-istics, and explores how forest-related factors impact wood quality and properties. The third part (Chapters 12–15) discusses the major wood-based products, the basic manufacturing processes associated with each, and how raw material selection affects product properties. The final part (Chapters 16 and 17) focuses on the global raw materi-als picture, the increasing use of wood as a source of energy and chemicals, and environ-mental implications of the use of wood. Products from other nonwood bio-based sources are sprinkled in throughout.

The increasing sophistication of scientific tools that allow the nanoscale measure-ment, visualization, and manipulation of material building blocks continues to provide new insights and possibilities across almost every field, including tree physiology and wood and materials science. Thus, even in such fields of inquiry as tree growth and wood structure – topics that have been extensively researched and reported over a long period of time – knowledge of fundamental processes and structures continues to change. Similarly, advancements relative to new product development, product enhancement, and production technology are ongoing. Nanotechnology continues to slowly develop and to show promise with respect to bringing positive change across society and through-out the wood-based industries. That said, it is traditional wood products such as lumber, plywood, other composites, and pulp and paper that continue to hold the key to deriving economic, social, and environmental benefits from vast forest land areas.

This, the Seventh Edition, is a significant revision of the Sixth Edition, which was published in 2011. In particular, discussion is given to cross-laminated timber (CLT), a structural composite which is novel to North America and which has the potential to significantly enhance wood products markets. Although targeted for North America, efforts continue to broaden the text's international appeal in response to the exponential growth of worldwide distribution. The first edition of this text was initially published more than three decades ago in 1982. Through time, the authors have sought to maintain the basic fundamental concepts while revising the time-sensitive applications and tech-nologies. In general, the fundamental laws of physics and nature, along with basic prin-ciples of biological and physical science and engineering, are unchanging. Production methods, technologies, and statistics, however, require continual updating and revision.

The authors also express their sincere thanks to the previous authors, Drs John Haygreen (posthumously) and Jim Bowyer, as well as the reviewers who played a vital role in this or earlier editions. These include but are not limited to: Kent T. Adair, Terry L. Amburgey, Donald G. Arganbright, James P. Armstrong, Stravros Avramidis, H. Michael Barnes, Thomas E. Batey, Jr, Frank C. Beall, B. Alan Bendtsen, Dwight W. Bensend, Evalgelos J. Biblis, Brian H. Bond, Arthur B. Brauner, Charles C. Brunner, Ben S. Bryant, Honorio F. Carino, Vincent L. Chiang, Poo Chow, Gilbert L. Comstock, Terry E. Connors, Harold A. Core, John B. Crist, Bruce E. Cutter, Robert W. Erickson, Richard L. Folk, Charles E. Frazier, David W. French, Douglas J. Gardner, Barbara L. Gartner, Roland O. Gertjejansen, Irving S. Goldstein, Barry S. Goodell, Thomas M. Gorman, Robert L. Govett, Hans M. Gregersen, Richard F. Helm, R. Bruce Hoadley, Peter J. Ince, Judson G. Isebrands, John J. Janowiak, Fred A. Kamke, Philip O. Larson, Joseph R. Loferski, E. A. McGinnes, Thomas M. Maloney, Timothy A. Martin, David J. Moorhead, Kenneth J. Muehlenfeld, Wayne K. Murphey, Helmuth Resch, Irving B. Sachs, Simo Sarkanen, John F. Senft, Craig E. Schuler, Tor P. Schultz, Audrey Zink Sharp, John R. Shelly, Todd F. Shupe, Richard A. Skok, Douglas D. Stokke, Otto

Suchsland, Edward I. Sucoff, Richard J. Thomas, Ulrike Tschirner, Elisabeth A. Wheeler, Edwin H. White, Ed Williston, James B. Wilson, K. C. Yang, Raymond A. Young, John I. Zerbe, and Steven C. Zylkowski. Special thanks go to Donald L. Buckner, John B. Crist, Frank C. Owens, Tor P. Schultz, Shane Kitchens, Ron Teclaw, John P. Limbach, Elisabeth A. Wheeler, Mississippi State University, North Carolina State University, Southern Pine Inspection Bureau, Timber Products Inspection, Inc., Trus Joist/ Weyerhaeuser, and the University of Minnesota for their photographic contributions.

Rubin Shmulsky has an MS degree in Forest Products and a PhD degree in Forest Resources from Mississippi State University. He has a BS in Building Materials Management and Wood Technology from the University of Massachusetts. He is professor and department head of Sustainable Bioproducts (formerly Forest Products) at Mississippi State University. Previously he served on the faculty of Wood and Paper Science at the University of Minnesota. He is a board member of the Society of Wood Science and Technology and an active member of the Forest Products Society and American Society of Civil Engineering.

P. David Jones has a PhD in Forest Resources from University of Georgia, an MSF degree from Stephen F. Austin State University, a BS in Forest Resource Management from Clemson and an Associate's degree from Abraham Baldwin Agricultural College. Currently he provides outreach and consulting services internationally to the Forest Products industry. Previously he served on the faculty of Forest Products at Mississippi State University after serving as a Post Doctoral Research Associate at Virginia Polytechnic and State University. He has provided service to a variety of professional and trade organizations over the years.

INTRODUCTION

At the start of 2018, it was estimated that the population of the world was approximately 7.5 billion, a 212% increase from just 50 years earlier. Annualized, this growth rate is on the order of 1.5% per year. This dramatic and continual increase continues to create problems and challenges for leaders, including those in the political, religious, industrial, and natural-resource-related arenas. The consumption of goods in many areas of the world is today growing at a rate greater than the rate of population increase. Population and economic growth worldwide has resulted in substantial increases in the consumption of food and of raw materials needed to provide shelter, fuel, paper, packaging, and durable and nondurable goods of all kinds. Higher economic standards bring better healthcare, nutrition, and longer lifespans. Feeding 7.5 billion people is a daunting task. It is humbling to think that at any given time, should farming cease, there is enough food in the world to last for only 40 days. Higher standards of living in North America lead to bigger houses which consume greater forest resources per capita.

The long-term availability of crude oil has been a focus of concern since the 1970s. Concerns were magnified in the 2003–2008 time frame, when the per barrel price tripled from approximately $35 to over $100. Since then it has declined but the potential for great variability remains. Energy consumption and allocation concerns are likely to grow in the decades ahead. In addition, the availability of non-energy raw materials will likely gain more attention going forward as consumption expands and as, in the case of nonrenewable minerals, supplies of the highest grades and most accessible ores diminish. The use of food as energy, such as the case of corn-based ethanol, remains an issue of great contention. Supplies of wood, a renewable resource, are likely to continue to expand, the result of a long-term trend of growth exceeding removals in natural forests and tree plantations. The drop in paper consumption associated with an electronic and "paperless" society feeds back to individual and corporate land owners in the form of reduced timber value. Conversely, concerns about forest amenities other than timber are likely to continue to adversely impact the availability of timber from natural forests.

As shown in Tables I.1 and I.2, wood is a major raw material globally and within the United States. Globally in 2015, almost as much wood was produced on a mass basis for industrial products (non-fuel) as steel. In the United States, more wood was used on a mass basis in 2015 than any other basic material, and more than all metals and all plastics combined. Yet in most high school and college curricula, wood receives minimal attention.

The use of wood for energy is increasingly important. Wood has long been the principal source of energy for heating and cooking in less-developed countries. Worldwide, a little over half of all wood that is harvested is used for home heating and cooking, a reality that is revealed in the difference between the roundwood and industrial roundwood numbers in Table I.1; the remainder is used in manufacturing an array of wood products. In the United States, a much greater proportion of wood is used as a raw material for wood products manufacture, ultimately providing shelter for citizens. Moreover, in the United States about 85% of the wood consumed for energy is used in the forest industries; the other 15% is used for other heating needs, such as heating homes in rural areas. The extensive use of wood energy by the forest products industries indicates that these interests have already made the leap to green energy. Since the 1970s, forest industries have been increasing the utilization of wood residues to supply the energy required to manufacture wood products. Residues such as bark, sawdust, product trim, and pulping liquors can now be almost completely utilized. Currently, the primary wood products industries of the United States are about 60–70% energy self-sufficient, meaning that only 30–40% of the energy used for the manufacture of wood-based materials is purchased from energy providers or produced from fossil fuels. This is a very significant advantage of wood as compared with cement, steel, plastics, and aluminum. The use of wood for the production of commercial energy, particularly electricity, has grown with

TABLE I.1.   Annual world consumption of various raw materials, 2015.

	Billion metric tons	Billion cubic meters
Roundwood	2.55	3.71
Industrial roundwood	1.27	1.85
Cement	4.1	2.72
Steel	1.6	0.203
Plastics	0.322	0.269
Aluminum	0.058	0.022

Source: Data for wood from Food and Agriculture Organization of the United Nations (2015). Data for cement, steel and aluminum from USGS (2017). Data for plastics from Association of Plastics Manufacturers in Europe (2016).

TABLE I.2.   Annual US consumption of various raw materials, 2015.

	Million metric tons	Million cubic meters
Roundwood (2013)	276	402
Industrial roundwood (2013)	246	358
Cement (2015)	93	61.6
Steel (2015)	110	13.9
Plastics (2015)	50.9	42.4
Aluminum (2015)	5.39	2.07

Source: Data for wood from Food and Agriculture Organization of the United Nations (2015). Data for cement, steel and aluminum from USGS (2017). Data for plastics from American Chemistry Council (2017).

the desire for alternative energy. Current policies and "green power" mandates make it economically feasible to pelletize sawdust, other residuals, and small-diameter trees in the USA, ship the pellets across the Atlantic Ocean, and use them for electrical power generation in parts of Europe. In the future, wood is certain to play a larger role in production of electricity and may provide a source of fuels such as cellulosic ethanol, pyrolysis oil, torrified wood, and liquid hydrogen as well.

It should be apparent that wood is a vital and versatile raw material. In addition to the economic and energy benefits of wood production, trees enhance the esthetic and environmental character of every region. Managed forests hold the key to mitigating climate change while also meeting basic societal needs, in perpetuity. The renewability of forests and the assurance that with wise forest management a supply of wood will be available indefinitely favor the use of wood as a basic raw material.

The use of wood for heat and as a material for shelter, dyes, tannin, and containers dates back to the early days of human history. Historical transportation via wood boats and ships, warfare that included long bows and catapults, wood-handled farming implements, and so on have shaped human culture for thousands of years. But the use of industrial wood in significant volume goes back only about 150 years.

In colonial America, wood was the foundation on which society was built. Buildings and furniture, spinning wheels and looms, dishes and pails, wagons and carriages, dinghies and ships, bridges and sidewalks, railroad ties, plows and hay rakes, milling machinery and sawmills, and products of every kind and shape were made from wood. Wood was also a major fuel source, used for heating and cooking and as the principal fuel of industry. There remains much basis for the axiom "communities develop where forests grow."

As an early English colony, America was prized for its abundance of tall pine trees, from which ship masts could be fashioned. As the colonies gave way to rapidly expanding cities, and as populations expanded, wood abundance in many areas turned to scarcity as unrestrained wood use, combined with land clearing for agriculture, resulted in greatly diminished forests. But as wooden wagon trains carried homesteaders steadily westward, new forests were encountered and the clearing of forests continued. Wood for fencing of pastures alone required enormous volumes of timber, with some 3.2 million miles of such fencing estimated to have been in existence in the mid-1800s. Development of the steam engine led to the need for great quantities of additional wood – for steamboat fuel and for railroad ties and trestles – and provided a means of moving large volumes of wood and other materials to population centers.

One of the early drivers of inquiry into whether things might be done to increase the efficiency of wood use was the tendency of wood to rot. The huge volumes used for ships, marine pilings, fencing, ties, trestles, bridges, and telegraph line poles required replacement after only a few years of use owing to natural deterioration. As noted by MacCleery (1993), just replacing railroad ties on a sustained basis required from 15 to 20 million acres of forest land in 1900. Interest in finding a way to preserve wood to eliminate or slow decay processes provided an impetus to an early field of inquiry in what would later become known as the field of wood science.

As the population of the United States grew – from an estimated 3 million in 1785 to 77 million in 1900 – wood consumption grew rapidly (Figure I.1); primary uses of wood were lumber and fuelwood. Rapid growth in wood use continued for about another decade but then, despite ongoing increases in population, a dramatic shift in wood use occurred.

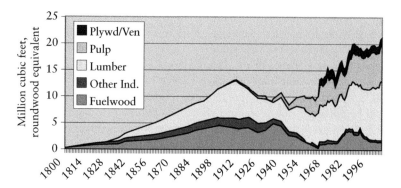

FIGURE I.1.   US consumption of wood and wood products, 1800–2005. Source: Frederick and Sedjo (1991), Howard (2006).

First, lumber consumption declined almost as fast as it had increased. The causes of the decline were many, including substitution of non-wood materials for many applications, increased efficiency of wood use, and the development of new technologies. Development of wood preservatives and preservative treatments alone resulted in a substantial reduction in the quantity of wood needed for the replacement of ties, marine pilings, poles, fencing, and similar products. Another development – the invention of barbed wire – meant that, as the 3.2 million miles of wooden fencing estimated to have existed in the mid-1800s began to deteriorate, far smaller quantities of wood were needed for their replacement. In addition to declining lumber consumption, growth in the use of wood as a source of energy leveled off at the turn of the century and then began to decline as fossil fuels became increasingly more important. Wood energy rebounded during the Great Depression of the 1930s, but then began a steep decline that continued through the early 1970s. By 1945, overall consumption of wood in the United States had fallen to a level similar to that of 1880 despite an almost threefold increase in population during the intervening period.

But then World War II ended, and the postwar boom that ensued is the stuff of legend. New homes were built at a stunning pace and with them the production of durable and nondurable goods of all kinds skyrocketed. Industries of all kinds grew at a breakneck pace, fueling the growth of communication and with it the demand for paper. As the economy grew, wood use rebounded, reaching record levels by the late 1960s and with new records set almost every year thereafter. The oil shocks of the 1970s triggered new interest in wood as a fuel, and growth of wood use for energy rose rapidly through the 1980s, helping to push wood use to ever higher levels.

Growth of wood use in the 1960 and 1970s closely matched the growth in population, meaning that wood use per capita remained relatively constant during this period. However, in the economic boom years from the late 1970s through the mid-1980s wood use grew more quickly than population numbers, and wood use on a per capita basis rose substantially. Subsequently, per capita consumption has grown more slowly, and as part of the economic recession of 2009–2011, it has declined significantly (Figure I.2). This change is in large part due the increase in multifamily housing units which require fewer board feet of lumber and building products per square foot of living space. Recently, as the economy has rebounded, per capita consumption appears to be rebounding as well.

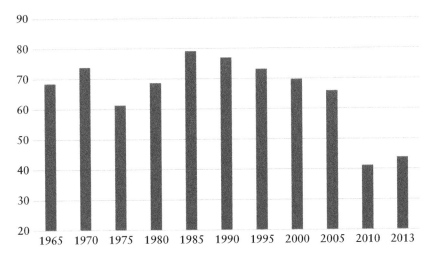

FIGURE I.2. Annual per capita consumption (cubic feet) of wood products – United States (1965–2013). Source: Howard and Jones (2016), United States Department of Agriculture, Forest Service, Forest Products Laboratory.

As rapidly as wood consumption rose in the postwar years, the rise would have been far more spectacular were it not for innovation relative to both processes and new products. For instance, in the 25 years between 1948 and 1973 the yield of lumber from a given quantity of logs doubled, while the quantity of useful products obtained quadrupled. New products brought into production during this period include particleboard, hardboard, and waferboard. Significant increases in paper-making efficiency were also achieved during this period.

New families of products made from fibers, particles, and flakes served to greatly expand options in wood products manufacturing and to increase the yield of final products. Subsequently, innovation brought to the market more new composite products. Structural composites such as laminated veneer lumber, parallel strand lumber, and wood composite I-beams allowed the use of less wood for a given application, thereby further stretching wood supplies. At the same time, improvements in recycling technology greatly increased wastepaper recovery and reuse rates, with these numbers up by 50–65% in the last 20 years alone. New products, such as crosslaminated timber offer greater architectural freedom, the possibility to develop new markets for wood fiber, and ultimately the opportunity for enhanced forest management.

Overall, increased consumption efficiencies are reflected across all products. These efficiencies help manage timber consumption and contribute to sustainability as the population continues to grow. An indication of the significance of technology gains is provided in a position statement of the Society of Wood Science and Technology (SWST 2005). Included in this statement is the following summary:

Recent assessments of the state of the nation's forests have revealed a more-than-fifty-consecutive-year record of net forest growth in excess of forest removals despite steadily increasing demand for fiber and chemicals supplied by these forests and corresponding increases in harvest levels. New technology developed by wood science and forest products research and development programs, coupled with successes in forest management, made it possible for the nation's forests to supply substantial quantities of critical raw materials while providing clean water, recreation, biodiversity, and a myriad of other forest values. To continue to maintain all of these values and also keep pace with the nation's growing need for wood and fiber will require proactive action on several fronts, including advancement of knowledge and technology to improve wood utilization, improved technologies for dealing with plantation-grown wood and other bio-fiber, ongoing improvement of environmental performance of manufacturing technologies, and increased use of recycled wood-based materials.

Those involved in the growing of wood or in the manufacture and use of forest products need to continue the record of advancement begun by their predecessors. To succeed, they need to understand the physical and chemical nature of wood and the variation of wood properties and characteristics between species and different growing conditions. Furthermore, knowledge of the processes for converting wood into products and of how wood properties affect these processes is important to the continued improvement in resource productivity and forest sustainability. This textbook is intended to further such knowledge and understanding.

References and Supplemental Readings

American Chemistry Council. (2017). U.S. Resin Production and Sales. Washington, DC. https://plastics.americanchemistry.com/Plastics-Statistics/ACC-PIPS-Year-End-2017-Resin-Stats-vs-2016.pdf.

An Analysis of Plastics Production. (n.d.) Demand and Recovery in Europe (Summer). https://www.plasticseurope.org/application/files/5715/1717/4180/Plastics_the_facts_2017_FINAL_for_website_one_page.pdf.

Association of Plastics Manufacturers in Europe. (2016). World Plastics Production 1950–2015. https://committee.iso.org/files/live/sites/tc61/files/The%20Plastic%20Industry%20Berlin%20Aug%202016%20-%20Copy.pdf.

Clawson, M. (1979). Forests in the long sweep of American history. *Science* 204: 1168–1174.

Food and Agriculture Organization of the United Nations (FAO). (2006). Forestry statistical database. http://faostat.fao.org/faostat/collections?version=ext&hasbulk=0&subset=forestry.

Food and Agriculture Organization of the United Nations (FAO). (2015). Yearbook of Forest Products 2015. http://www.fao.org/3/a-i7304m.pdf.

Frederick, K.D. and Sedjo, R.A. (eds.) (1991). *America's Renewable Resources: Historical Trends and Current Challenges*. Washington, DC: Resources For The Future.

Haygreen, J.G., Gregersen, H., Hyun, A., and Ince, P. (1985). Innovation and productivity change in the panel industry. *For. Prod. J.* 35 (10): 32–38.

Howard, J.L. (2006). U.S. Timber Production, Trade, Consumption, and Price Statistics 1965–2005 (prepublication data from personal correspondence with author). USDA For. Serv.

Howard, J.L. and K.C. Jones. (2016). U.S. Timber Production, Trade, Consumption, and Price Statistics 1965–2013. USDA For. Serv. FPL-RP-679.

MacCleery, D.W. (1993). *American Forests: A History of Resiliency and Recovery*, Forest History Society Issues Series. Washington, DC: USDA.

Moore, P. (2000). Trees are the answer. *For. Prod. J.* 50 (10): 12–19.

National Commission on Materials Policy (1975). *Material Needs and the Environment Today and Tomorrow*. Washington, DC.

Schultz, H. (1993). The development of wood utilization in the 19th, 20th, and 21st centuries. *For. Chron.* 69 (4): 413–418.

Snellgrove, T.A. (2000). The role of technology in conservation and sustainability. *Renewable Res. J.* 18 (3): 6–10.

Society of Wood Science and Technology (SWST) (2005). *National Need in Capacity for Forest Products Research and Development*. Madison, WI: Society of Wood Science and Technology, March.

USGS (2017). Commodity Statistics and Information for 2015. US Geological Survey. https:// minerals.usgs.gov/minerals/pubs/commodity (Accessed 13 July 2017).

Youngquist, W.G. and Fleischer, H.O. (1977). *Wood in American Life*. Madison, WI: For. Prod. Res. Soc.

Youngquist, J.A. and Hamilton, T.E. (1999). Wood product utilization – a call for reflection and innovation. *For. Prod. J.* 49 (11/12): 18–27.

Don't forget to visit the companion website for this book:

www.wiley.com/go/shmulsky

There you will find valuable material designed to enhance your learning, including:

- Datasets

Scan this QR code to visit the companion website

1

Tree Growth and Production of Woody Tissue

Trees are complex organisms. Originating through vegetative propagation or from sexually fertilized eggs that become tiny seed-encased embryos, trees grow to be one of nature's largest living organisms.

Like humans, trees are delicate when young and typically grow vigorously when given proper nutrition and a suitable environment. As juveniles, they form tissues that differ from those in mature trees. They respire. They require a balanced intake of minerals to maintain health. They metabolize food, but unlike humans they also synthesize their own foods. If wounded, they react quickly to effect healing. As age progresses, vigor is maintained for a lengthy period but then begins to wane. Life processes eventually slow to the point that the tree has difficulty healing wounds and warding off disease. Finally, the tree dies.

The focus of this book is not on the growth process but on an important product of growth: wood. However, a brief study of the process of wood formation provides a useful basis for a study of wood itself.

Wood is formed by a variety of plants, including many that do not attain tree stature. A tree is generally defined as a woody plant 4–6 m (15–20 ft) or more in height and characterized by a single trunk rather than several stems. Plants of smaller size are called shrubs or bushes. Species that normally grow to tree size may occasionally develop as shrubs, especially where growth conditions are adverse. Because of the size attained, wood produced by plants of tree stature is useful for a wider range of products than wood from shrubs and bushes. For this reason, wood produced by trees is emphasized.

Forest Products and Wood Science: An Introduction, Seventh Edition. Rubin Shmulsky and P. David Jones.
© 2019 John Wiley & Sons Ltd. Published 2019 by John Wiley & Sons Ltd.
Companion website: www.wiley.com/go/shmulsky

Classification of Woody Plants

Woods, and the trees that produce them, are divided into two categories: hardwoods and softwoods. Hardwood and softwood trees are botanically quite different. Both are included in the botanical division spermatophytes (Table 1.1), meaning that they produce seeds. They are, however, in different botanical subdivisions. Hardwoods are in the subdivision angiospermae and softwoods are in the gymnospermae subdivision. *Angiosperms* are characterized by the production of seeds within ovaries, whereas *gymnosperms* produce seeds that lack a covering layer.

Needle-like leaves characterize softwood trees. Such trees are commonly known as evergreens, because most remain green the year around, annually losing only a portion of their needles. Most softwoods also bear scaly cones (inside which seeds are produced) and are therefore often referred to as conifers. Included in the softwood group in the Northern Hemisphere are the genera *Pinus* (pine), *Picea* (spruce), *Larix* (larch), *Abies* (fir), *Tsuga* (hemlock), *Sequoia* (redwood), *Taxus* (yew), *Taxodium* (cypress), *Pseudotsuga* (Douglas-fir), and the genera of those woods known commonly as cedars (*Juniperus*, *Thuja*, *Chamaecyparis*, and *Calocedrus*).

In contrast to softwoods, *hardwoods* are angiosperms that bear broad leaves (which generally change color and drop in the autumn in temperate zones) and produce seeds within acorns, pods, or other fruiting bodies. Referring again to Table 1.1, note that angiosperms are subdivided into monocotyledons and dicotyledons. Hardwood-producing species fall within the dicotyledon class. Hardwood genera of the Northern Hemisphere include *Quercus* (oak), *Fraxinus* (ash), *Ulmus* (elm), *Acer* (maple), *Betula* (birch), *Fagus* (beech), *Populus* (cottonwood, aspen) and others. Included in the mono-cotyledon class are the palms and yuccas. Many of the roughly 2500 species of palms produce relatively large-diameter fibrous stems, which are strong if left in the round condition but tend to fall apart when cut into lumber; some species of palm, however, produce stems suitable for the production of local-use construction "lumber." Composite panels and flooring can be made from partially refined stems and paper can be made from the fiber.

TABLE 1.1.   Trees in the plant kingdom.

Divisions:	Thallophytes	Bryophytes	Pteridophytes	Spermatophytes
	Algae	Mosses	Ferns	(seed plants)
	Fungi	Liverworts	Horsetails	
			Rushes	
Subdivisions:		Gymnosperms	Angiosperms	
		(naked seed)	(seed in fruit)	
Orders:	Cycadales	Ginkgoales	Gnetales	Coniferales
Classes:	Monocots	Dicots		Yucca
	(palm-like)	(rare)		
Families:	Cupressaceae	Taxaceae	Pinaceae	Taxodiaceae
	Cedar	Yew	Fir	Redwood
	Juniper	Larch	Hemlock	Baldcypress
	Cypress		Pine	
			Spruce	

25 families in the United States.

Not only do hardwood and softwood trees differ in external appearance, but the wood formed by them differs structurally or morphologically. The types of cells, their relative numbers, and their arrangements are different, the fundamental difference being that hardwoods contain a type of cell called a *vessel element*. This cell type occurs in most hardwoods but very seldom in softwoods. Hardwoods are classified as ring-porous or diffuse-porous, depending upon the size and distribution of vessels in a cross-section. Woods that form very large-diameter vessels part of a year and smaller ones thereafter are called ring-porous. Woods that form vessels of the same size throughout the year are classified as diffuse-porous.

All hardwoods do not, incidentally, always produce hard, dense wood. Despite the implication in the names *hardwood* and *softwood*, many softwoods produce wood that is harder and more dense than that produced by some hardwoods. Balsa wood, for example, is from the hardwood species *Ochroma pyramidale*.

Distribution of Hardwoods and Softwoods

Hardwood species occur in every major region of the United States. They are predominate in the East, forming an almost unbroken forest from the Appalachians westward to the Great Plains. Across the Plains, the trees that line rivers, streams, and ponds and form windbreaks along agricultural fields are hardwoods. Farther west, the perpetually green softwoods that cover the Rocky Mountains are frequently interrupted by patches of white-stemmed aspen and other hardwoods. In the far West, hardwoods grow in valleys below softwood-covered mountains. Softwoods dominate forests of the deep South, the far North, the mountainous West, and the extreme Northwest (Figure 1.1). Softwoods also are predominate in the mountains and coastal regions of Alaska.

Hardwood growing stock globally has been estimated to exist in volumes almost double that of softwoods. The area covered by hardwood forests is greater as well. Sedjo and Lyon (2017) noted that hardwoods predominate in 57% of the area covered by closed-canopy forests and that softwoods predominate in the remaining 43%. North America, Russia, and Europe account for over 92% of the world's softwood forests. Hardwoods, on the other hand, are more widely distributed than softwoods. Natural forests of the tropics are almost totally hardwood. South America alone contains about one-third of the world's hardwood forests and South America and Asia account for over one-half of hardwood forests worldwide. In addition, Africa's forests are 99% hardwood. The hardwood forests of North America, Russia, and Europe together comprise about one-third of the world's hardwood forests. Overall, forests cover some 3869 million hectares (9556 million acres) worldwide, of which just over 95% are natural forests (Table 1.2).

In the United States about one hundred wood-producing and commercially important species reach tree size; only about thirty five of these are softwoods. Roughly the same is true of Europe. However, throughout the world, and particularly the tropical regions, the number of wood-producing species of tree size exceeds ten thousand. Of these, the number of softwoods is small – only about five hundred. The wet tropics are particularly rich in species; a few hectares may contain several hundred species. The large number of species complicates efforts to fully utilize the tropical rain forest. Despite considerable research to determine properties, use potential, and processing technology, work has been completed

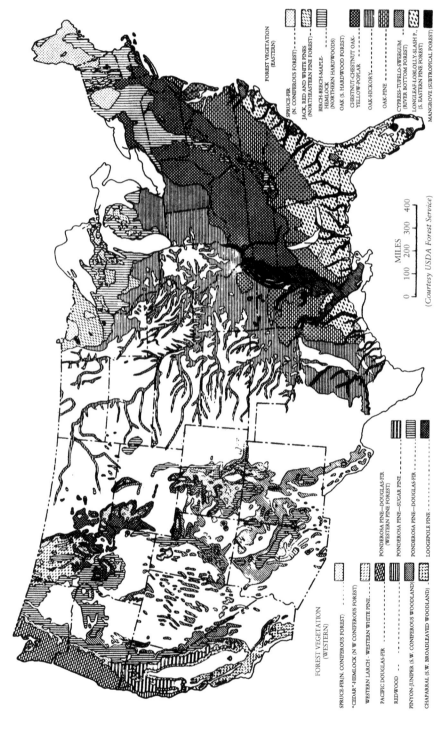

FIGURE 1.1. Forest vegetation of the continental United States.

FOREST VEGETATION
(WESTERN)

SPRUCE-FIR (N. CONIFEROUS FOREST) ----------

"CEDAR"-HEMLOCK (N.W CONIFEROUS FOREST) -

WESTERN LARCH - WESTERN WHITE PINE ----

PACIFIC DOUGLAS-FIR ------------------

REDWOOD -- ------------------

PINYON-JUNIPER (S.W. CONIFEROUS WOODLAND) -

CHAPARRAL (S.W. BROADLEAVED WOODLAND) --

PONDEROSA PINE—DOUGLAS-FIR
(WESTERN PINE FOREST) ------------

PONDEROSA PINE—SUGAR PINE ----------

PONDEROSA PINE—DOUGLAS-FIR ----------

LODGEPOLE PINE ----------

FOREST VEGETATION
(EASTERN)

SPRUCE-FIR (N. CONIFEROUS FOREST) --------

JACK, RED AND WHITE PINES
(NORTHEASTERN PINE FOREST) --------

BIRCH-BEECH-MAPLE-
HEMLOCK
(NORTHERN HARDWOODS) -------

OAK (S. HARDWOOD FOREST) ------------

CHESTNUT-CHESTNUT OAK-
YELLOW-POPLAR -----------

OAK-HICKORY- ----------

OAK-PINE- ----------

CYPRESS-TUPELO-SWEEGUM
(RIVER BOTTOM FOREST) ------------

LONGLEAF-LOBLOLLY-SLASH P.
(S. EASTERN PINE FOREST) ----------

MANGROVE (SUBTROPICAL FOREST) --------

MILES
0   100 200 300 400

(Courtesy USDA Forest Service)

TABLE 1.2.   Forest area by region and type.

	Total forest (natural forests and forest plantations)					
	Land area (million ha)	Area (million ha)	Percentage of land area	Percentage of world's forest	Natural forest (million ha)	Plantation (million ha)
Africa	2978	650	22	17	642	8
Asia	3085	548	18	14	432	116
Europe	2260	1039	46	27	1007	32
North and Central America	2137	549	26	14	532	18
Oceania	849	198	23	5	194	3
South America	1755	886	51	23	875	10
World total	13064	3869	30	100	3682	187

Source: FAO (2005).

on only a relatively small number of species. Today, approximately twenty-five hundred tree species have commercial importance.

Wood – a Collection of Small Cells

A close look at wood shows it to be made up of tiny cells or fibers that are so small that they generally cannot be seen without a magnifying glass or microscope. Illustrated in Figure 1.2 is a type of cell that makes up most of the volume of a softwood such as white fir; the cell has a hollow center (*lumen*), is closed at the ends, and is perforated with openings known as pits in the sidewall.

Figure 1.3 shows how a tiny block of white fir would look if magnified. Unmagnified, this block would occupy only about $1/50,000 \, cm^3$. Rays, which are composed of a number of individual ray cells and provide for horizontal movement of substances in a standing tree, can be seen cutting across the near-left (radial) surface; rays in an end view can be seen on the right vertical (tangential) surface.

In Figure 1.3, note that the three different surfaces of the block – labeled transverse (cross-section), radial, and tangential – look quite different (see also Figure 2.1). A cross-sectional or *transverse* surface is formed by cutting a log or piece of lumber to length; radial and tangential surfaces result from cutting along the grain. A *radial* surface is made by cutting longitudinally along the radius of a round cross-section. Tangential surfaces result from cutting perpendicular to a radius or a tangent to the growth rings. The names *transverse*, *radial*, and *tangential* are frequently encountered in the study of wood science.

Basic Processes in Tree Growth

The mechanisms that result in tree growth are remarkably complex, a reality that is becoming more and more evident with advances in science. In the following section, tree growth is discussed in rather general terms to provide a basic understanding of the processes involved.

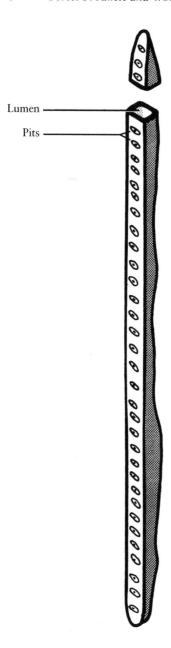

Lumen

Pits

FIGURE 1.2. Typical longitudinal tracheid of softwood.

Production of Wood and Bark

Wood (*xylem*) is found inside a covering of bark, which is composed of an inner layer (*phloem*) and an outer protective layer (*outer bark*). As a tree grows, it adds new wood, increasing the diameter of its main stem and branches. Bark is also added in the process of growth to replace that which cracks and flakes off as the stem grows larger.

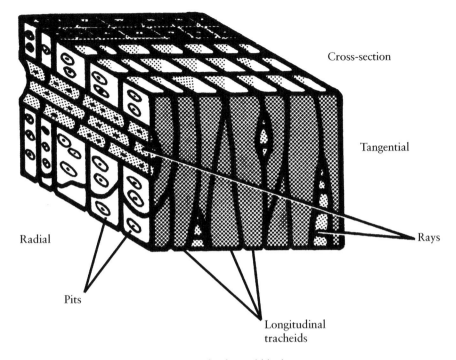

Cross-section

Tangential

Radial

Rays

Pits

Longitudinal
tracheids

FIGURE 1.3.   Three-dimensional drawing of softwood block.

Like all green plants, a tree can manufacture its own food through the process of photosynthesis, which takes place in the leaves. It needs only water (from the soil), carbon dioxide (from the atmosphere), and light (from the sun) to do this. Water, along with nutrients, is taken up by the roots and moved through the outer part of the xylem up to the leaves. The wood cells provide pathways for unbroken fluid columns that link the roots to the leaves. Note that, according to physics, capillary action is influenced by pore diameter and surface tension; in trees the combination of these suggests that the maximum column height is about 116 m (380 ft), the height of the tallest trees. Carbon dioxide is taken in through tiny openings in leaf surfaces called stomata. With the help of the sun, water and air are combined in the presence of chlorophyll to make sugars that provide energy to the growing tree. Some sugars are used in making new leaves, some in making new shoots, and some in making new wood. Some of the sugar moves to special locations in the wood or to the roots, where it is stored for later use; some consumed through respiration. Sugars used in making new wood move down a tree through the phloem.

Sugar is transported throughout the tree in the form of sap, a solution containing various sugars and water as well as growth regulators (hormones) and other substances. The term sap is also used to refer to the mineral-rich water that is taken up by roots and moved upward through the outer portion of the xylem.

A thin layer between the xylem and phloem produces new xylem and phloem tissue. This layer, called the *cambium*, completely sheaths the twigs, branches, trunk, and roots, meaning that a season of growth results in a new continuous layer of wood throughout the tree (Figure 1.4).

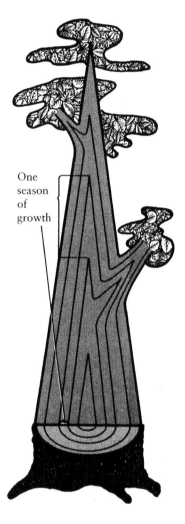

One
season
of
growth

FIGURE 1.4.   New growth occurs as a sheath covering the main stem, branches, and twigs.

Because sap moves down the tree through the phloem but is necessary for food in the cambium, a way is needed for it to travel horizontally toward the center of the tree. *Wood rays* provide for this horizontal movement. Rays also function in storing carbohydrates and may serve as avenues of horizontal transport for stored materials from near the center of the tree outward following periods of dormancy.

Figure 1.5 illustrates the relative position of various portions of a tree stem. Careful examination should help in gaining an understanding of the relationship between various layers of tissue.

Development of a Young Stem

To begin a study of the development process, the growth of a young pine seedling will be considered (Figure 1.6). The seedling shown has a well-developed root system and crown typical of a one- to two-year-old tree. With the beginning of growth in early spring, buds

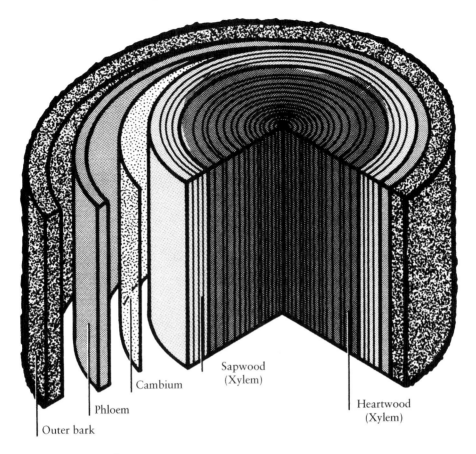

FIGURE 1.5. Parts of a mature tree stem.

at the tip of each branch swell as tissue expands through the formation and growth of cells. These regions in which cells divide repeatedly to form new cells are called *meristematic regions*. Highlighted in Figure 1.6 is an expanding bud at the apex of a young pine. Buds of similar appearance occur at the tip of each branch. The meristematic zone at the apex of the main stem is of special significance because it controls, to some extent, the development of branches and shoots; it is called the *apical meristem*.

Cell division at the apical meristem serves to lengthen the main stem. New cell production at this location is followed by cell elongation, resulting in height growth. As the stem is built through production of new cells during growth periods, the terminal bud moves upward, leaving new and expanding cells behind. Because trees grow in height from the apex rather than from the base, nails driven into a tree at, for example, 2 m above ground level will remain 2 m off the ground regardless of the height to which the tree grows.

Cell production at the stem tip and subsequent cell lengthening are followed by a sequence of changes as newly formed cells mature. We explain this entire process using a representation of a section of a growing stem tip that shows various tissue layers (Figure 1.7). The student should be cautioned that Figure 1.7 and the accompanying discussion present a greatly simplified picture of the actual growth process.

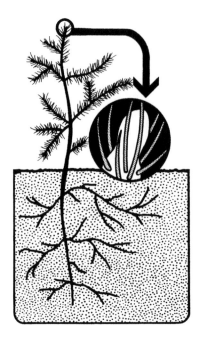

FIGURE 1.6.   Pine seedling.

New cells are produced in the several layers of cells in the area designated as section I. Soon after formation, these newly formed cells begin to differentiate with changes in size, shape, and function. Tissue at the outer edge of the young stem forms an *epidermis* composed of one layer of cells, which have thick, wax (cutin)-covered outer walls serving as protection from moisture loss. Nearer the center of the stem, cells undergo a developmental process, changing size and shape to eventually form an unbroken ring about the stem center. This region is called the *procambium* and is the precursor of a new meristematic region that develops a little later. At the very center of the stem, cells develop differently still, forming a layer dissimilar to the wood that will later surround it. This is the *pith*. The pith, procambium, and epidermis can be seen in section II of Figure 1.7.

The process of change continues. The procambium reaches a maximum size (lower edge of section II); then cells that make it up undergo further differentiation. As depicted in section III, inner cells of the procambium continue to undergo change to become similar to xylem, which will form later. Cells of the outer portion of the procambium assume characteristics similar to those of phloem, formation of which will also follow. These two new tissue layers are called *primary xylem* and *primary phloem*. The transformation to primary xylem or primary phloem continues until eventually a ring of procambium tissue only one to several cells in width remains (sections IV–V).

As a final step in the developmental process that began at the stem apex, the remaining ring of procambial tissue becomes active, its cells dividing repeatedly to form xylem and phloem. Vascular cambium (or simply cambium) is the name given to this meristematic layer.

The new lateral meristem, the vascular cambium, is considered of secondary origin because it forms after the terminal meristem. Xylem and phloem cells produced in this new meristem are thus correctly called *secondary xylem* and *secondary phloem*, respectively. It is interesting to note that in monocotyledons (such as palms) all procambial cells

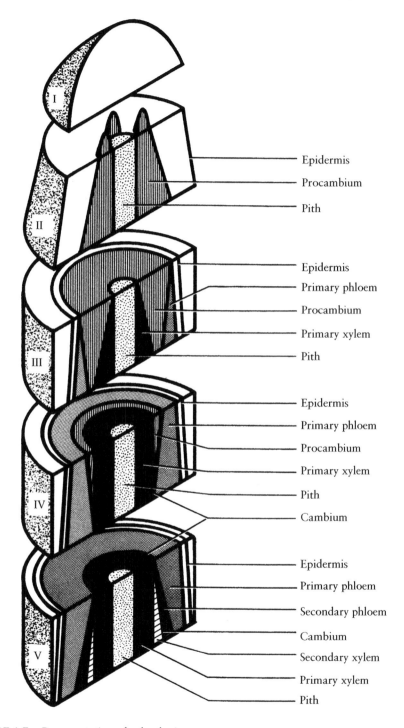

FIGURE 1.7.   Representation of a developing stem.

typically differentiate into primary xylem or phloem, leaving no vascular cambium. These plants, therefore, do not produce secondary xylem and phloem.

After being formed at a given location in the tree, the vascular cambium remains active throughout the life of the tree (or tree part). Because new cell formation in the cambium serves to increase stem diameter, the very large stature attained by mature trees is traceable to cell division in the cambium.

The stem depicted in section V of Figure 1.7 would be about one year old. Note that the tissue layers that sheathed the woody stem soon after its formation have become thinner; this is because of compression forces resulting from diameter expansion. The same is true of the primary phloem. At this stage of stem development, no cell division occurs in any of the layers outside the secondary phloem. Because of this, the circumference of these layers cannot keep pace with the expanding stem diameter. As explained in more detail in Chapter 7, the epidermal layer fractures with stem expansion and flakes off, giving way to a new outer bark layer. A new cell-producing layer forms outside the secondary phloem, and eventually (normally within the second growing season) all tissue originally formed outside the secondary phloem is shed.

Vascular Cambium

Because the cells that compose wood are formed in the vascular cambium, growth processes in this part of the tree are examined more closely. As before, what are in reality very complex processes are summarized in simplified terms.

Composition

In the previous section, the vascular cambium was described as consisting of a one- to several-cell-width ring of meristematic cells. An artist's conception of a cambium layer that has been isolated from surrounding wood and bark tissue is shown in Figure 1.8. Two kinds of cells can be seen to make up the cambium layer. The long, slender cells are called *fusiform initials*; these divide repeatedly to form either new *cambial initials* or new xylem and phloem cells (Figure 1.9). The short, rounded cells shown in Figure 1.8 are ray initials; division of these creates either new xylem or phloem rays or new ray initials. Division parallel to the stem surface in a tangential plane that results in formation of either xylem or phloem cells is called *periclinal division*. Production of new initials by radial partitioning is termed *anticlinal division*.

Development and Growth of Xylem and Phloem

Periclinal division of a fusiform initial results in formation of two cells, one of which remains meristematic and a part of the cambium. The other cell becomes either a xylem or phloem mother cell. The *mother cell* immediately begins to expand radially and may itself divide one or more times before developing into a mature xylem or phloem element. Maturation of new xylem cells involves growth in diameter and length, with growth accompanied by thickening of the cell walls and finally lignification.

Note that not all types of cells grow in both diameter and length. For example, longitudinal cells formed in late summer by pines, spruces, and other softwoods grow

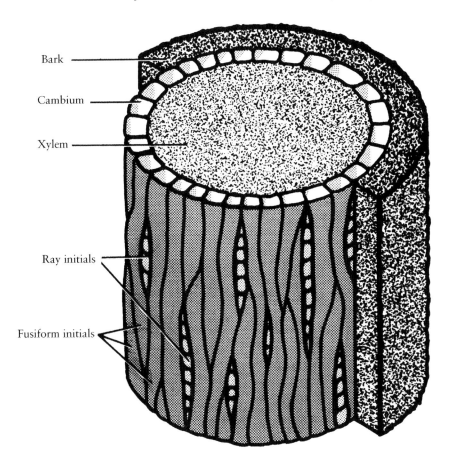

FIGURE 1.8.   Three-dimensional representation of the vascular cambium.

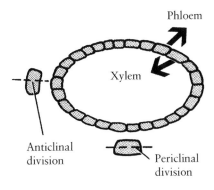

FIGURE 1.9.   Cambial cell division.

considerably in length but little in diameter. Vessel elements that characterize hardwood (*broadleaf*) species grow little or may even shrink slightly in length but may expand up to 50 times in diameter.

Two consecutive periclinal divisions of a fusiform initial are illustrated in Figure 1.10. Beginning at (a), a fusiform initial prepares to divide, as chromosomes split, and then separate. In (b), a cell plate begins to form and becomes a new cell wall at (c). Both cells begin to grow in diameter (d) and length (e). The innermost cell becomes part of the xylem, pushing outward the other portion that remains part of the cambium. In (f), the cycle begins again.

Early in the growing season, new cell production in primary meristems at stem tips occurs rapidly. Intervals of only 8–18 hours between successive divisions in the primary meristems of white cedar (*Thuja*) have been reported (Zimmermann and Brown 1971, p. 78). This rate diminishes as the season progresses. Cambial initial division typically commences later than division at the apical meristems, and once initiated, division occurs relatively slowly. Bannan (1955) found in experiments with northern white cedar (*Thuja occidentalis*) that successive cambial divisions occurred about once weekly (each seven days) in the early spring. Zimmermann and Brown (1971), after working with the same species, agreed with the finding of a seven-day interval. They found, however, that the rate of division increased several weeks into the growing season. Only three to four cambial initial divisions were noted in the first few weeks of activity, after which the rate increased to one division per four- to six-day interval during the period of most-rapid earlywood formation. Wilson (1964), on the other hand, found a 10-day interval between successive divisions of cambial initials during earlywood formation in white pine. These rates of cell division translate to the production of about 30–50 divisions of each cambial initial during each growing season. This total meshes with the findings of Deslauriers and Morin (2005), who tracked longitudinal tracheid production in balsam fir; they found the average number of cells produced each year to be rather uniform, varying from 37 to 41 over a three-year period.

The process of periclinal division of cambial initials is again illustrated in Figure 1.11; subsequent development of new cells is also depicted. In Figure 1.11a, the one-cell-wide cambium (C), with mother cells (M) adjacent on either side, can be seen sandwiched between the xylem (X) and phloem (P). Line 1 of Figure 1.11a shows the appearance of cells in and near the cambial zone during a period of cambial activity. To the left of the cambial initial are two xylem mother cells and one cell of mature secondary xylem. To the right of the cambial initial is one phloem mother cell and a secondary phloem cell. Line 2, which represents the cambial zone a short time later, shows that the xylem mother cell nearest the cambium has divided to form two mother cells. The xylem mother cell farthest from the cambium has begun to enlarge (E). No activity is noted on the phloem side of the cambium or in the cambium itself. By line 3, one of the xylem mother cells formed in the previous period has divided again. The other has begun to enlarge. Although no cambial activity is noted in this period, outward movement of the cambium and phloem cells has occurred as the result of new cell formation through division of a xylem mother cell. At the stage of activity illustrated by line 4, the cambial initial has divided; a portion of the initial remains in the cambium, and the other half becomes a xylem mother cell. Enlargement of yet another xylem mother cell has begun. Growth of the innermost mother cell has ceased and cell wall thickening has begun. This cell (X_1) is mature and cannot divide further. To the outside of the cambium, the phloem mother cell seen earlier

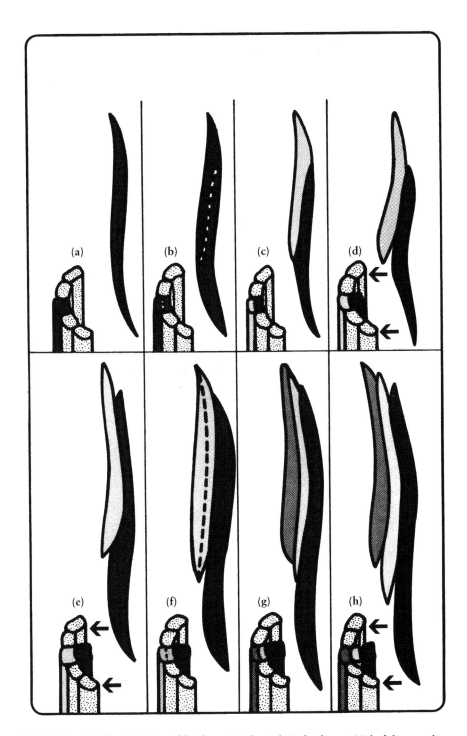

FIGURE 1.10. Periclinal division of fusiform initials. (a, b) A fusiform initial of the vascular cambium prepares to divide. (c) The initial divides, forming two new cells, one of which remains in the cambium (shaded), the other becoming a xylem mother cell (black). (d, e) Both cells begin to increase in diameter and length, increasing the diameter of the stem and pushing the vascular cambium outward. (f, g) After a period of rest, the initial once again divides, resulting as before in a fusiform initial and a new xylem mother cell. (h) As both the new initial and mother cell increase in diameter and length, the vascular cambium is again pushed outward.

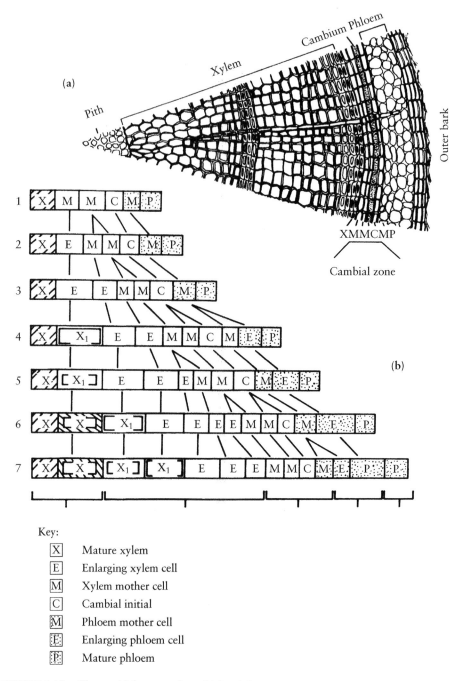

Key:

X	Mature xylem
E	Enlarging xylem cell
M	Xylem mother cell
C	Cambial initial
M	Phloem mother cell
E	Enlarging phloem cell
P	Mature phloem

FIGURE 1.11.   The cambial zone and cambial activity.

has divided into two cells, one of which has begun to enlarge. This sequence of events is repeated again and again as growth continues.

Figure 1.11 and the previous discussion describe the cambium as being one cell in width. However, the research of Catesson (1984) suggests that any of the cells in the cambial zone can, given the proper stimulus, either divide to form new cells or differentiate into vascular tissue. This raises the possibility that specific initials may be frequently replaced as the initials themselves differentiate to become cells of the xylem or phloem. Kozlowski and Pallardy (1997) recognize the likelihood of a single layer of cambial initials, but acknowledge that it is difficult to conclusively identify such a layer. They suggest use of the term *cambial zone* to refer to the entire zone of cambial initials and xylem and phloem mother cells.

Expansion of the Cambial Layer

As a tree expands in diameter, the cambium is pushed progressively outward. Thus, the cambium must expand in circumference to remain an unbroken layer around the stem. Such growth of the cambium is achieved in several ways, the most important of which is anticlinal division of fusiform initials.

Anticlinal division of a fusiform initial (see Figure 1.9) results in two cells, both of which remain in the cambium. Assuming that the new cells survive, they begin to grow in length almost immediately. After a short rest, the new meristematic cells may divide again, either periclinally or anticlinally.

As with new initials that result from periclinal division, survival of new fusiform initials formed anticlinally is dependent upon the availability of adequate nutrition. Availability of nutrition is, in turn, dependent upon proximity to rays. A fusiform initial without sufficient ray contact may die or further divide to form one or more ray initials (Zimmermann and Brown 1971, p. 74). Initials in the process of failing in this manner have been called *declining initials* (Philipson et al. 1971, p. 26). Kozlowski and Pallardy (1997) report that survival of new cells produced through anticlinal division is high in rapidly growing trees, and that new fusiform initials and cambial deriviatives are short. In older trees the rate of anticlinal division is said to be slower and the survival rate lower, giving rise to longer fusiform initials and derivatives. This contrasts with the findings of Bannan (1960), who reported that only 20% of new cambial initials formed in rapidly growing northern white cedar remained as part of the cambium. The figure was 50% for slower-growing trees. The fact that ray contact is needed for survival and normal development means that long cells, which nearly always have adequate contact, survive while very short cells nearly always decline. There is evidence that very small fusiform initials usually decline even where ray contact is extensive (Philipson et al. 1971). This mechanism ensures that fusiform initials will maintain a long average length, even in periods of rapid growth (Panshin and de Zeeuw 1980).

In addition to expansion through an increase in the number of fusiform initials, the cambium also expands through an increase in the length of these cells. The length of initials progressively increases over time so that the length of functioning initials in a mature tree is much greater than those present at the seedling to sapling stage. In one study of white pine, initials at the age of 60 years were found to be four to five times longer than those present in the first year of growth. Kozlowski and Pallardy (1997) note that this occurs in part because the rate of survival of newly formed cambial initials declines in older trees. This, in turn, means that surviving initials have more room in which to expand, translating to longer fusiform initials and cambial derivatives.

Other factors leading to circumferential expansion of a cambium are diameter growth of fusiform and ray initials and increases in the number of ray initials (Bailey 1923).

References and Supplemental Reading

Avery, G.A. Jr., Burkholder, P.R., and Creighton, H.B. (1937). Production and distribution of growth hormone in shoots of *Aesculus* and *Malus* and its probable role in stimulating cambial activity. *Am. J. Bot.* 24: 51–58.

Bailey, I.W. (1923). The cambium and its derivative tissues. IV. The increase in girth of the cambium. *Am. J. Bot.* 10: 499–509.

Bannan, M.W. (1955). The vascular cambium and radial growth in *Thuja occidentalis* L. *Can. J. Bot.* 33: 113–184.

Bannan, M.W. (1960). Ontogenic trends in conifer cambium with respect to frequency of anticlinal divisions and cell length. *Can. J. Bot.* 38: 795–802.

Bannan, M.W. (1962). The vascular cambium and tree-ring development. In: *Tree Growth* (ed. T.T. Kozlowski), 6–9. New York: Ronald Press.

Bausch, J. and Dünish, O. (2000). Comparison of growth dynamics and wood characteristics of plantation-grown and primary forest *Carapa guianensis* in Central Amazonia. *IAWA J.* 21 (3): 321–333.

Borchert, R. (1991). Growth periodicity and dormancy. In: *Physiology and Trees* (ed. A.S. Raghavendra), 221–245. New York: Wiley.

Catesson, A.M. (1984). La dynamique cambiale. *Ann. Sci. Nat. Bot. Biol. Veg.* 6: 23–43.

Creelman, R.A., Mason, S., Bensen, R.J. et al. (1990). Water deficit and abscisic acid cause differential inhibition of shoot versus root growth in soybean seedlings: analysis of growth, sugar accumulation, and gene expression. *Plant Physiol.* 92: 205–214.

Deslauriers, A. and Morin, H. (2005). Intro-annual tracheid production in balsam fir stems and the effect of meteorological variables. *Trees* 19 (4): 402–408.

Digby, J. and Wareing, P.F. (1966). The relationship between endogenous hormone levels in the plant and seasonal aspects of cambial activity. *Ann. Bot.* 30: 607–622.

Dougherty, P.M., Whitehead, D., and Voge, J.M. (1994). Environmental influences on the phenology of pine. In: *Environmental Constraints on the Structure and Productivity of Pine Forest Ecosystems: A Comparative Analysis*, vol. 43 (ed. H.L. Gholz, S. Linder and R.E. McMurtie), 64–75. Copenhagen: Ecological Bulletins.

Eames, A.J. and MacDaniels, L.H. (1947). *An Introduction to Plant Anatomy*, 2e. New York: McGraw-Hill.

Food and Agricultural Organization of the United Nations. 2005. State of the World's Forests 2005. Rome. (http://www.fao.org/docrep/007/y5574e/y5574e00.htm)

Jacoby, G.C. (1989). Overview of tree-ring analysis in tropical regions. *IAWA Bull.* 10 (2): 99–108.

Kienholz, R. (1934). Leader, needle, cambial, and root growth of certain conifers and their relationships. *Bot. Gazette* 96: 73–92.

Kozlowski, T.T. (1971). *Growth and Development of Trees*, vol. 1. New York: Academic Press.

Kozlowski, T.T., Kramer, P.J., and Pallardy, S.G. (1991). *The Physiological Ecology of Woody Plants*. San Diego: Academic Press.

Kozlowski, T.T. and Pallardy, S.G. (1997). *Growth Control in Woody Plants*, 34–67. San Diego: Academic Press.

Kramer, P.J. and Kozlowski, T.T. (1979). *Physiology of Woody Plants*. New York: Academic Press.

Lanner, R.M. (1964). Temperature and the diurnal rhythm of height growth in pines. *J. For.* 62 (7): 493–495.

Larson, P.R. (1964). Some indirect effects of environment on wood formation. In: *Formation of Wood in Forest Trees* (ed. M.H. Zimmermann), 345–365. New York: Academic Press.

Leitch, M.A. and Savidge, R.A. (1995). Evidence for auxin regulation of bordered-pit positioning during tracheid differentiation in Larix laricina. *IAWA J.* 16 (3): 289–297.

Little, C.H.A. and Savidge, R.A. (1987). The role of plant growth regulators in forest tree cambial growth. *Plant Growth Reg.* 6: 137–169.

Nitsch, J.P. (1957). Growth response of woody plants to photoperiodic stimuli. *Proc. Am. Soc. Hortic. Sci.* 70: 512–525.

Ogden, J. (1981). Dendrochronological studies and the determination of tree ages in the Australian tropics. *J. Biogeogr.* 8: 405–420.

Oppenheimer, H.R. (1945). Cambial wood production in stems of Pinus halapensis. *Palest. Bot. Rehovot. Ser.* 5: 22–51.

Pallardy, S.G. (2007). *Physiology of Woody Plants*, 3e. San Diego: Academic Press.

Panshin, A.J. and de Zeeuw, C. (1980). *Textbook of Wood Technology*, 4e, 70. New York: McGraw-Hill.

Philipson, W.R., Ward, J.M., and Butterfield, B.G. (1971). *The Vascular Cambium*, 26. London: Chapman & Hall.

Phillips, I.D.J. and Wareing, P.F. (1958). Studies in the dormancy of sycamore. I. Seasonal changes in growth substance content of the shoot. *J. Exp. Bot.* 9: 350–364.

Priestley, J.H. (1930). Studies in the physiology of cambial activity. III. The seasonal activity of the cambium. *New Phytol.* 29: 316–354.

Priestley, J.H. and Scott, L.I. (1933). Phyllotaxis in the dicotyledon from the standpoint of developmental anatomy. *Biol. Rev. Camb. Phil. Soc* 8: 241–268.

Reimer, C.W. (1949). Growth correlations in five species of deciduous trees. *Butler Univ. Bot. Stud.* 9: 43–59.

Roberts, L.W., Gahan, P.B., and Aloni, R. (1988). *Vascular Differentiation and Plant Growth Regulators*, 24. New York: Springer-Verlag.

Romberger, J.A. (1963). Meristems, growth, and development of woody plants. *USDA For. Serv. Tech. Bull.* 1293.

Sedjo, R.A. and Lyon, K.S. (2017). *The Long Term Adequacy of World Timber Supply*. Taylor & Francis Group: Florence, KY.

Thimann, K.V. (1972). The natural plant hormones. In: *Plant Physiology*, vol. 6B (ed. F.C. Steward), 3–145. New York: Academic Press.

Tomlinson, P.B. and Longman, K.A. (1981). Growth phenology of tropical trees in relation to cambial activity. In: *Age and Growth Rate of Tropical Trees*, vol. 94 (ed. F.H. Bormann and G.P. Berlyn), 7–19. New Haven: Yale University of School Forestry Environmental Studies Bull.

Wareing, P.F. (1951). Growth studies in woody species. IV. The initiation of cambial activity in ring-porous species. *Physiol. Plant.* 4: 546–562.

Wareing, P.F. (1956). Photoperiodism in woody plants. *Ann. Rev. Plant Physiol.* 7: 191–214.

Wareing, P.F. (1958). The physiology of cambial activity. *J. Inst. Wood Sci.* 1: 34–42.

Wareing, P.F., Haney, C.E.A., and Digby, J. (1964). The role of endogenous hormones in cambial activity and xylem differentiation. In: *The Formation of Wood in Forest Trees* (ed. M.H. Zimmermann), 323–344. New York: Academic Press.

Wareing, P.F. and Phillips, I.D.J. (1970). *The Control of Growth and Differentiation in Plants*. New York: Pergamon Press.

Wilson, B.F. (1964). A model for cell production by the cambium of conifers. In: *The Formation of Wood in Forest Trees* (ed. M.H. Zimmermann), 19–36. New York: Academic Press.

Worbes, M. (1989). Growth rings, increment and age of trees in inundation forests, savannas and a mountain forest of the neotropics. *IAWA Bull.* 10 (2): 109–122.

Worbes, M. (1995). How to measure growth dynamics in tropical trees—a review. *IAWA J.* 16 (4): 337–351.

Zimmermann, M.H. and Brown, C.L. (1971). *Trees: Structure and Function*. New York: Springer-Verlag.

2

Macroscopic Character of Wood

A number of features of wood can be detected by casual observation and are termed *macroscopic* because a microscope is not needed for detection. The macroscopic characteristics of wood are of interest because they often give clues to the conditions under which the wood was grown, provide an indication of physical properties, and serve as an aid in wood identification.

Wood identification is a hobby, a business, and an issue of major international importance. Wood workers delight in knowing exactly what species they are working with. Similarly, the public takes pride in knowing that their floors are red oak, their table is cherry, their paneling is beaded spruce, etc. Different species have different performance characteristics, thus knowing that rafters or roof trusses are made from yellow pine versus grand fir versus something else has structural implications. The Convention on International Trade in Endangered Species (CITES) lists species that are endangered and illegal to trade worldwide. Paired with the Lacey Act in the USA, it is mandated that importers declare with certainty which species, by botanical name, are coming into the country, and in a global trade environment this issue is of major importance. Contemporary tools for wood identification range from razor blades and hand lenses to high-power cameras with sophisticated image analysis.

Three Distinct Surfaces of Wood

Illustrated in Figure 2.1 is a wedge-shaped piece of wood as it would appear if cut from a round cross-section. Notice that the macro features of the cross-sectional, radial, and tangential surfaces appear quite different (see also Figure 1.3).

Wood differs not only in appearance depending upon the direction from which it is viewed, but (as will be explained later) in physical properties as well. Thus, in a solid wood product such as lumber, boards are classified by the surface of the wood that corresponds with the widest face (Figure 2.2).

Forest Products and Wood Science: An Introduction, Seventh Edition. Rubin Shmulsky and P. David Jones.
© 2019 John Wiley & Sons Ltd. Published 2019 by John Wiley & Sons Ltd.
Companion website: www.wiley.com/go/shmulsky

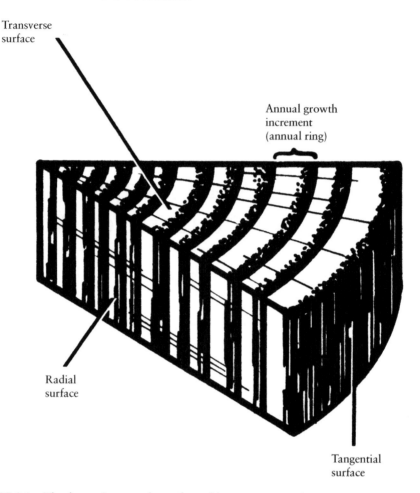

Transverse
surface

Annual growth
increment
(annual ring)

Radial
surface

Tangential
surface

FIGURE 2.1.   The three primary surfaces of wood (gross appearance).

Growth Rings

The seasonal nature of growth was indicated in Chapter 1. Growth in temperate zones is characterized as proceeding rapidly in early spring and slowing in late summer before ceasing in the fall. For reasons explained in the succeeding paragraphs, this kind of growth pattern usually results in the formation of one growth ring annually and more or less distinct layers of early- and late-formed wood within each ring.

Appearance

Figure 2.3 is a photograph of a magnified thin cross-section of redwood. At this magnification it is easy to see why wood formed in the latter part of a growing season appears different to the unaided eye than that formed early in the year. The latewood tissue is of

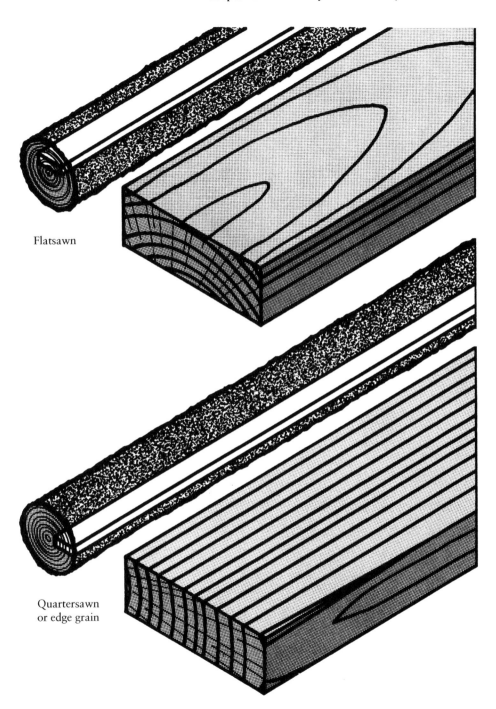

Flatsawn

Quartersawn
or edge grain

FIGURE 2.2.   Classification of lumber by the manner of cutting.

Bark →

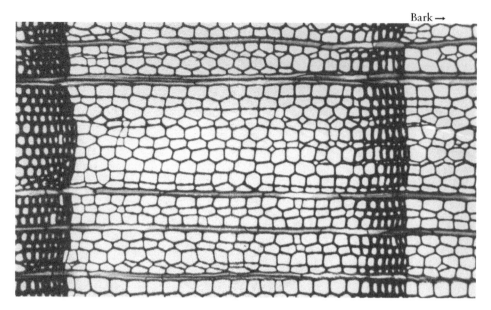

FIGURE 2.3.   Latewood cells appear as distinct bands. Transverse view of redwood (*Sequoia sempervirens*); ×85. Source: Courtesy of Ripon Microslides Laboratory.

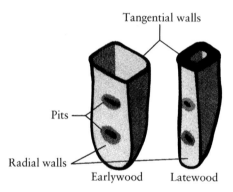

FIGURE 2.4.   Tapered ends of earlywood and latewood longitudinal tracheids of a softwood.

greater density, being composed of cells of relatively small radial diameter, with thick walls and small lumens. Latewood forms the darker-colored portion of the growth ring. Illustrations of portions of earlywood and latewood cells are presented in Figure 2.4.

   Growth rings do not always appear as distinct alternating bands of earlywood and latewood. Some hardwoods, for example, form large-diameter pores early in a growing season and much smaller pores later in the year (Figure 2.5); such woods are called *ring-porous*. Other hardwoods exhibit little variation in cell structure across a growth increment, thus forming rings that are difficult to detect. Because the pores are about the same size throughout the growth ring, these woods are termed *diffuse-porous*. A diffuse-porous hardwood having indistinct rings is pictured in Figure 2.6.

Bark →

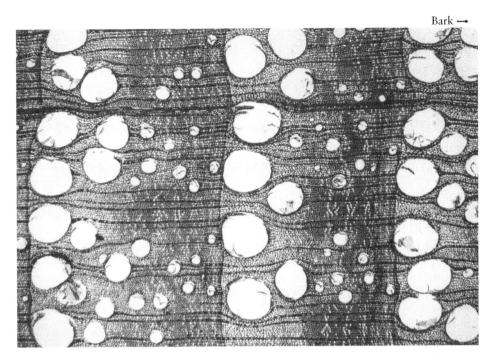

FIGURE 2.5.   Ring-porous hardwood. Transverse view of southern red oak (*Quercus falcata*). ×30. Source: Reproduced with permission from Wheeler et al. (1986).

Bark →

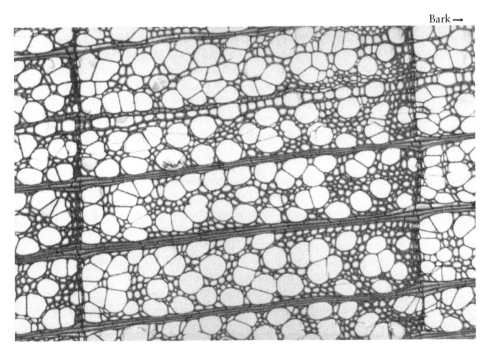

FIGURE 2.6.   Diffuse-porous hardwood. Transverse view of yellow poplar (*Liriodendron tulipifera*). ×80. Source: Courtesy of the Department of Wood and Paper Science, North Carolina State University.

Trees growing in the tropics sometimes produce one growth ring annually, but more commonly will form more than one ring each year. Often there is little or no difference in the structure and appearance of early- and late-formed wood within such rings or in wood formed within the span of one year; thus, it is not possible to determine the ages of tropical trees based on a ring count.

Formation

Scientists have sought to explain the causes of earlywood and latewood formation for more than 100 years and the riddle is still not completely solved. Strong evidence does exist, however, indicating that this is related to photosynthate availability and the presence of auxins.

A number of investigators have concluded that two major characteristics of *latewood* – cells of relatively small radial diameter and thick cell walls – develop independently of each other (Zimmermann and Brown 1971, p. 96). The formation of large-diameter cells, characteristic of earlywood, is apparently dependent on an abundance of soil moisture (Kozlowski et al. 1991). Conversely, water deficits can induce early formation of latewood, and continued moisture deficits are reported to shorten the period of latewood formation. Development of thick cell walls, on the other hand, is generally related to a plentiful supply of photosynthate (Pallardy 2007, p. 93).

Using the deciduous softwood larch as an example, the spring season finds the tree with swollen buds but no leaves. Sugars used in the growth process come from storage sites in the roots and elsewhere. As new needles emerge and begin to produce photosynthate, they also begin to grow in size. At the same time, a flush of shoot growth at the apical meristems is typically under way and the rate of new cell formation in the cambium is rapid. Thus most available sugars at this time of the growing season are consumed in building new shoots; photosynthate reaching cambial regions supplies large numbers of new and developing cells. Later in the growing season, factors such as drought, low temperature, and shortened daylight hours provide less favorable growing conditions, resulting in a reduced rate of new cell production and the development of cells of lesser radial diameter. By this time, the growth of shoots and the development of new needles have largely ceased. Thus much of the photosynthate produced by the now full-grown needles is available for cell wall synthesis. Small-diameter and thick-walled cells are the result.

The effect of late-season abundance of photosynthate is less obvious in hardwoods than in softwoods. In ring-porous hardwoods, the formation of large-diameter vessels early in the season is commonly followed by the production of a more compact latewood, smaller- diameter vessels, and a greater proportion of fibers. Diffuse-porous hardwoods likewise often produce a higher proportion of fibers late in a growing season, and these are sometimes radially flattened as well (Zimmermann and Brown 1971, pp. 91–93).

The term *late-season* appears to more or less correspond to midsummer. Grotta et al. (2005), working with Douglas fir in western Oregon, noted that cambial growth in most trees began in mid-May and ceased in late August to early September, with 6 July the mean date of transition to latewood. Thus, earlywood was noted to form over a period of roughly three to three-and-a-half months and latewood over a period of about two months.

Effect on Wood Properties

The significance of varying proportions of earlywood and latewood in standing trees is not well documented. However, it is known that the strength of latewood is significantly greater than that of earlywood in distinct-ring softwoods (Mott et al. 2002), a reality that has major implications for properties of lumber and other wood products. Kretschmann et al. (2006) found that earlywood and latewood have different mechanical properties and different relationships of mechanical properties to other wood characteristics, even when in close proximity within the same growth ring and same tree. Latewood in general was found to have two to three times the strength and stiffness of earlywood.

Discontinuous Rings

Growth rings sometimes fail to form around the complete cross-section. This is a result of the cambium remaining dormant in one or more places around the stem. *Discontinuous rings* are occasionally found in trees having one-sided crowns (such as at forest-field borders) and in heavily defoliated, suppressed, and overmature trees (Kramer and Kozlowski 1979). The fact that discontinuous rings do occur indicates that caution should be used when using increment borings to determine tree age; borings from several locations around a stem should be used where the presence of discontinuous rings is suspected.

False Rings

Occasionally, normal seasonal growth is interrupted by events such as drought, late frost, or defoliation by insects or hail. If such an event results in slowing or cessation of terminal growth, auxin production will be reduced and may cause latewood-type cells to be produced. If events that cause slow growth are followed in the same growing season by conditions favorable to growth, normal growth patterns may be resumed, accompanied by the production of large and thin-walled earlywood cells (Pallardy 2007, p. 48). Casual observation of an annual ring formed under such circumstances will show two rings to have formed in a single year. The ring thus created is called a *false ring* (Figure 2.7). It is possible for several of these to form in a given year (Esau 1965). Such rings may form throughout the length of a stem, but more commonly they are restricted to upper regions of the crown.

    False rings of conifers can usually be distinguished from normal growth rings by examining latewood to earlywood transition (Panshin and de Zeeuw 1980). Normal growth rings are characterized by an abrupt change in cell size and wall thickness from the last-formed latewood of one seasonal ring to the earlywood of the next (Figure 2.7a). False rings exhibit a gradual change in cell character on both sides of the false latewood, resulting in a double gradation (Figure 2.7b).

(a)

Bark →

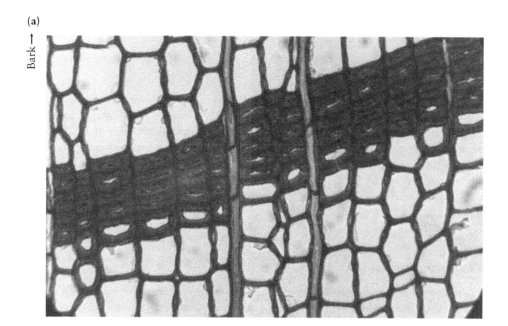

(b)

Bark →

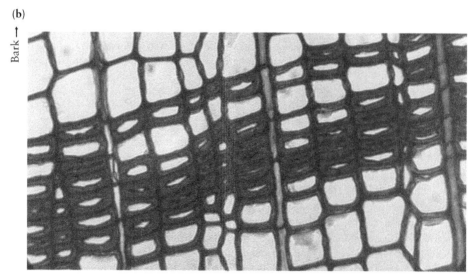

FIGURE 2.7.   (a) Normal versus (b) false growth rings. Transverse view of cypress (*Taxodium distichum*).

Heartwood and Sapwood

Trees were defined in Chapter 1 as one of nature's largest living organisms. Some are described as nature's oldest living creations as well. Indeed, one bristlecone pine growing in the White Mountains of California is estimated to be Earth's oldest living resident at a ripe old age of about 4600 years.

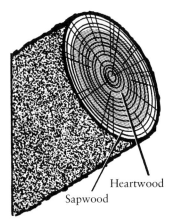

FIGURE 2.8.   Heartwood in a round cross-section.

Heartwood

Sapwood

An estimate that a living organism is almost 5000 years old is impressive, but it can also be misleading. Plant tissue seldom remains alive longer than several years, even when part of an old bristlecone pine. This apparent contradiction is explained by the fact that new cells are continually being produced while others cease to function. Even cambial initials are periodically replaced. It has been estimated that living cells of a tree may comprise as little as 1% of its total bulk (Mirov and Hasbrouck 1976). The longevity of trees such as the bristlecone pine provides evidence that wood will last indefinitely if conditions are favorable.

Examination of a stem cross-section often reveals a dark-colored center portion surrounded by a lighter-colored outer zone (Figure 2.8). The dark center area is known as heartwood and the lighter tissue as sapwood. It is important to note that the *heartwood* is occasionally the same color as *sapwood*. It is in the sapwood that the living cells are found. The inner region of the sapwood, in which most cells are dead, also serves to conduct water upward in a living tree. Heartwood no longer functions physiologically because all the cells in this region of the tree are dead.

Formation of Heartwood

Perhaps no term relating to wood has more mystery associated with it than the word *heartwood*. A typical belief is that because heartwood is older than sapwood, having aged and seasoned more slowly, it is better. Heartwood is reputed to be heavier, stronger, more highly figured, and more resistant to decay than sapwood. Some of these notions are true, but others are not. Before an understanding can be gained of what properties heartwood does and does not have, and why, it is first necessary to know what heartwood is.

When a tree or tree part is young and growing vigorously, it often contains no heartwood. After a number of years, however, heartwood typically begins to form near the center of a stem. The most common age at which the transformation from heartwood to sapwood occurs is reported to be 14–18 years (Hillis 1987). However, as reported by Dadswell and Hillis (1962), some species, such as eucalyptus, may begin to form heartwood at five years of age. Other studies confirm heartwood formation at relatively young tree ages. Gominho and Pereina (2000), for instance, reported that heartwood formation in nine-year-old plantation *Eucalyptus globulus* had been found in all trees; the heartwood was noted to extend to 60–75% of tree height, and to account for one-third of tree volume. Hoadley (1990) noted that black locust and catalpa maintain no more than three years of sapwood. In contrast, Dadswell and Hillis reported that some species, such as beech or European ash, may not begin to form heartwood until 60–100 years of age and that a few species apparently never form it. Once initiated, the transformation of sapwood into heartwood occurs continuously. In general, the number of rings of sapwood is directly related to the size of the crown of the tree (Hazenburg and Yang 1991; Yang and Hazenberg 1991). The boundary between heartwood and sapwood does not necessarily follow the growth rings.

Chapter 1 discusses the formation of new cells by the cambium. Recall that for a time these cells retain the ability to further subdivide, and then they lose this capacity as thickened and lignified cell walls form. Death of most cells follows cell wall thickening and is marked by the disappearance of nuclei and protoplasm (Esau 1965). However, some cells (from 5 to 40% of those that comprise sapwood) retain their protoplast; these are specialized storage cells known as *parenchyma* cells that occur as both longitudinal and ray cells (Kollmann and Côté 1968). The living cells of the sapwood carry on metabolic processes. It is the death of these cells that marks the formation of heartwood. An early explanation of heartwood development was that as the living parenchyma cells get older and progressively farther from the cambium, metabolic rates and enzymatic activity decline and the remaining living cells begin to fail. The cytoplasm begins to change chemically, with a reduction in starches, sugars, and nitrogenous materials. Cell nuclei become rounded, begin to deteriorate, and then disappear completely (Frey-Wyssling 1963), marking the death of these cells. More recently, however, it was noted that whereas a gradual decline of parenchyma occurs in some species, in others all parenchyma remain living up to the heartwood/sapwood boundary (Nobuchi et al. 1979). This suggests that factors other than, or in addition to, age and distance from the cambium are involved.

As noted by Taylor et al. (2002), "the process of heartwood formation remains poorly understood and requires further study." They observed that "Ultimately, the ability to manipulate the amount and quality of heartwood formed in trees would be of enormous practical value; however, many questions need to be resolved before control over heartwood formation is possible."

Factors suspected as playing a role in heartwood formation include the accumulation of carbon dioxide or ethylene gas; reduction of moisture content levels and/or cessation of water transport in inner regions of the stem; enzyme activity; gradual reduction of starch, sugars, or lipid reserves from the cambium inwards; and the formation of compounds known as *extractives* (Taylor et al. 2002).

Although the causes of heartwood formation continue to be debated, the death of cells is known to be accompanied by species-dependent biosynthesis of compounds termed *secondary metabolites*. Secondary metabolites are defined as substances that do not participate directly in tree growth and development. These metabolites accumulate in cell walls and lumens and are usually polyphenolic in nature. Polyphenols commonly associated with heartwood include oils, resins, gums, tannins, and aromatic and coloring materials (Koch 1972, p. 202). Other compounds include fats and waxes. Many heartwood secondary metabolites can be removed or extracted from heartwood by soaking or boiling in water or alcohol or by other extraction processes. Because of this capacity to be extracted, as well as the fact that the number of individual components is great, these various compounds are commonly and collectively referred to as *extractives*.

Whether the death of remaining storage cells is caused by the development of secondary metabolites and other substances within these cells, or whether death is caused by water deficits, with the production of polyphenols being a secondary result of cell failure, is not clear. Regardless of what causes the death of parenchyma cells at the stem center, it appears likely that photosynthates that are not consumed in growth processes in the cambial zone play a direct role in polyphenol development. Hillis (1968) found, for example, that rapid growth and efficient utilization of carbohydrate are associated with low amounts of heartwood polyphenols. More recently, it has been noted that "although starch has been detected in the [heartwood/sapwood] transition zone of some species, the translocation of a considerable amount of primary metabolites from the sapwood to the sapwood/heartwood transition zone is required to form the high levels of extractives found in some heartwoods" (Hillis 1987, p. 187). Ten years later Pallardy (2007, p. 51), after a review of sometimes conflicting scientific evidence, supported the findings of Hillis and others, concluding that the weight of evidence indicates that extractives are formed from *transported metabolites* (products of physical and chemical processes associated with the maintenance of life) by living parenchyma cells at the heartwood boundary.

It should be noted that the findings of Hillis (1968) referenced above do not necessarily imply that rapid growth decreases heartwood formation. In fact, a relatively recent study by Kärenlampi and Riekkinen (2002) of heartwood formation in pine found heartwood content to be independent of either growth rate or tree size.

As indicated previously, heartwood formation is often marked by a reduction in water content of cells. Moisture content of heartwood versus that of sapwood is listed in Table 2.1 for a number of species. Huber observed in 1956: "We are virtually certain that the water columns in the xylem are under tension and that these tensions play a role in the movement of water. It is equally certain that, sooner or later, water columns under tension must break and the vessels embolize, probably irreversibly." An embolism is an obstruction, caused in this case by formation of air bubbles as liquid pressure is reduced; embolism is more likely to occur as height from the ground increases. The reduction of water content of cells is thought to be the first step in heartwood formation (Zimmermann 1983). This is said to be the reason for the existence of higher proportions of heartwood with increasing height in a tree; it also may explain why the wood of roots contains very little heartwood. This explanation is slightly muddled by the recent findings of Zwieniecki et al. (2001) that resistance to transport through dead hardwood vessels can be substantially increased by increasing the ion concentration of the solute. This finding raises the possibility that the mechanism controlling the reduced rates of flow and water content of cells that precede the onset of heartwood formation is more complex than earlier thought.

From the preceding discussion, it is evident that the basic cell structure is unchanged in the transformation to heartwood and that the primary change is the presence of extractable chemicals. Recognition of this fact is important in understanding differences in properties of heartwood and sapwood. One structural change sometimes associated with heartwood formation in softwoods is aspiration of many of the bordered pits in tracheids. Pit structure and aspiration are covered in Chapter 3.

TABLE 2.1.   Moisture content of heartwood and sapwood when in green condition.

Species	Moisture content	
	Heartwood	Sapwood
	(%)	
Hardwoods		
Red alder	—	97
Apple	81	74
Ash		
Black	95	—
Green	—	58
White	46	44
Aspen	95	113
American basswood	81	133
American beech	55	72
Birch		
Paper	89	72
Sweet	75	70
Yellow	74	72
Black cherry	58	—
American chestnut	120	—
Black cottonwood	162	146
Elm		
American	95	92
Cedar	66	61
Rock	44	57
Hackberry	61	65
Pecan Hickory		
Bitternut	80	54
Water	97	62
Hickory (true)		
Mockernut	70	52
Pignut	71	49
Red	69	52
Sand	68	50
Magnolia	80	104
Maple		
Silver	58	97
Sugar	65	72
Oak		
California black	76	75
Northern red	80	69
Southern red	83	75
Water	81	81
White	64	78
Willow	82	74
Sweetgum	79	137
American Sycamore	114	130
Tupelo		
Black	87	115
Swamp	101	108
Water	150	116
Black walnut	90	73
Yellow-poplar	83	106

TABLE 2.1.   *Continued.*

Species	Moisture content	
	Heartwood	Sapwood
	(%)	
Softwoods		
Baldcypress	121	171
Cedar		
Alaska	32	166
Eastern red	33	—
Incense	40	213
Port Orford	50	98
Western red	58	249
Douglas fir, coast type	37	115
Fir		
Grand	91	136
Noble	34	115
Pacific silver	55	164
White	98	160
Hemlock		
Eastern	97	119
Western	85	170
Western larch	54	119
Pine		
Loblolly	33	110
Lodgepole	41	120
Longleaf	31	106
Ponderosa	40	148
Red	32	134
Shortleaf	32	122
Sugar	98	219
Western white	62	148
Redwood (old-growth)	86	210
Spruce		
Eastern	34	128
Engelmann	51	173
Sitka	41	142
Tamarack	49	—

Source: USDA Forest Service, Forest Products Laboratory (2010).

Properties of Heartwood

As the differences between heartwood and sapwood are almost totally chemical, the presence of these chemicals is primarily responsible for giving heartwood its unique properties, a number of which are as follows:

1. *Heartwood may be darker in color than sapwood.* Most polyphenols found in temperate zone trees are colorless in regular light. However, they are also unstable and degrade with time. The death of parenchyma cells allows secondary metabolites

to diffuse in the wood, and these subsequently darken as a result of random oxidation reactions. In some woods, heartwood and sapwood show no color difference; this lack does not necessarily mean an absence of heartwood but may simply indicate that no dark-colored extractives have formed. Hardwoods exhibit a wider range of heartwood coloration than softwoods.

2. *Heartwood may be highly decay or insect resistant, or both.* When woods are naturally resistant to decay and insects, such as the heartwood of cypress, redwood, and most cedars, it is because some of the extractives provide protection from decay fungi and insects. Recent research (Goodell et al. 1999; Illman 2003; Schultz and Nicholas 2000) suggests that the initial stages of brown-rot decay, the most serious form of decay affecting softwood lumber used in residential construction, involve reactions of hydrogen peroxide, which is produced by fungi, with ferrous ions. It is believed that to impart decay resistance, extractives must have both antioxidant and antifungal properties. The heartwood of many woods does not contain extractives exhibiting these dual properties, or has lower levels than do the naturally decay-resistant woods; such heartwood has either no or only slight decay resistance. Because decay resistance is imparted by chemicals that occur only in heartwood, sapwood of all species is readily susceptible to decay.

3. *Heartwood may be difficult to penetrate with liquids (such as preservative chemicals).* Resistance to penetration is the result of: (a) the presence of extractable oils, waxes, and gums that may serve to plug tiny passages in cell walls; (b) the closure of cell-to-cell passageways in softwoods through slight rearrangement of tiny membranes in the passageways (called pit aspiration); (c) the blocking of bordered pit membranes of hardwoods and softwoods through progressive deposition of incrusting materials (such as phenolic compounds) over a period of time (Sano and Nakada 1998); or (d) blocking of pores in hardwoods by the ballooning of parenchyma cell sacs into vessel lumens (called tyloses; see Chapter 5). When a wood is both susceptible to decay and difficult to penetrate, its usefulness for certain applications is limited. Douglas fir is an example of a wood not highly resistant to decay; furthermore, its heartwood is difficult to treat for improved durability.

4. *Heartwood may be difficult to dry.* Drying difficulties are generally traceable to the same factors that inhibit penetration and permeability.

5. *Heartwood may have a distinct odor.* When it does, this is usually due to the presence of aromatic extractive compounds. Most cedars contain pungent-smelling compounds.

6. *Heartwood may have a slightly higher weight-per-unit volume than sapwood.* When this occurs, it is from the presence of significant quantities of extractives.

Several other properties such as low hygroscopicity and a reduced fiber saturation point (see Chapter 8) might be listed, but note that strength does not appear on the preceding list. There is no difference in the strength of heartwood and sapwood, with one relatively rare exception. In some instances, as in redwood, western red cedar, and black locust, considerable amounts of infiltrated material may somewhat increase the weight of wood and its resistance to crushing.

Rays

Rays provide an avenue by which sap can travel horizontally either to or from the phloem layer. Virtually all woods contain rays. In some hardwoods such as oak, rays are quite large and readily visible in a cross-section (Figure 2.9). In softwoods and a number of hardwoods, rays are very narrow and in some cases difficult to see even with a magnifying glass (Figure 2.10). Many highly valued hardwoods used for paneling and furniture and in other decorative ways are characterized by distinct ray patterns on radial and tangential surfaces (Figure 2.11). These are often helpful when identifying wood species.

Rays also have an effect on wood properties. For example, rays restrain dimensional change in the radial direction, and their presence is partially responsible for the fact that upon drying, wood shrinks less radially than it does tangentially. Rays also influence strength properties because they constitute radially oriented planes of weakness. Because of this effect upon strength, splitting may occur along rays in veneer-slicing operations if the veneer knife is improperly oriented. Splitting can also develop along rays when wood is dried.

A close look at a ray layout on a cross-section (Figure 2.9) reveals that rays extend from the cambium and bark inward. Few rays can be traced to the stem center. Further observation shows that the distance between rays remains relatively constant in growth rings near the pith and outward.

Review of the way in which rays form helps explain why rays are arranged as they are. Ray cells of the xylem and phloem are produced by division of ray initials in the cambium. Ray initials, in turn, arise either from division of other ray initials or from an unequal division of a fusiform initial. An undernourished or unusually small product of cell division will invariably fail to redivide; such a cell is likely to fail or may convert to a ray initial (Larson 1994). Because the cause of inadequate nourishment is lack of proximity to rays, uniform spacing of rays is methodically assured. As soon as the distance between rays begins to increase, the chances decrease that a cell midway between them will obtain sufficient nutrients and thus develop normally, and the chances increase that a new ray initial will be formed.

Yet other ways exist for additional rays to appear. At any given point along the length of a stem cross-section, a "new" ray may appear as a result of an increase in the height of an existing ray through cell division that results in new cells located above or below the parent cells. In some hardwood species, rays can increase in width through division of cells tangentially; such division can lead to rays that are many cells in width when viewed tangentially.

Evidence suggests that conversion of fusiform initials to ray initials and expansion of existing rays are regulated by hormones and influenced specifically by the presence of ethylene that originates in the xylem (Kozlowski and Pallardy 1997, p. 145). The finding that availability of specific chemical compounds, and not simply inadequate nourishment, controls or influences conversion of fusiform intals to ray initials suggests greater complexity in mechanisms involved in ray initiation and development than previously recognized.

The fact that most rays can be traced from the xylem to the cambium and into the phloem is due to the permanence of the cambium. Initials, once formed, continue to divide while the cambium of which they are a part moves outward. A ray will be terminated (and thus not connect with the cambium) only if the ray initials that formed it fail and are not replaced by new ones.

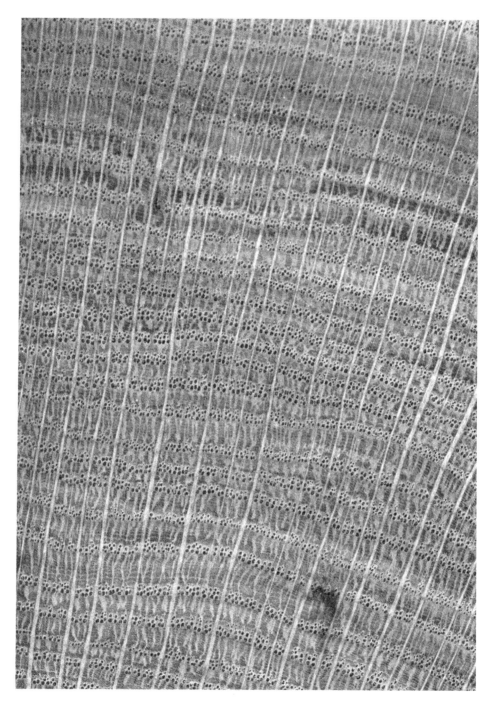

FIGURE 2.9.   Large rays of oak are readily visible even without magnification.
Source: Reproduced with permission from Wheeler et al. (1986).

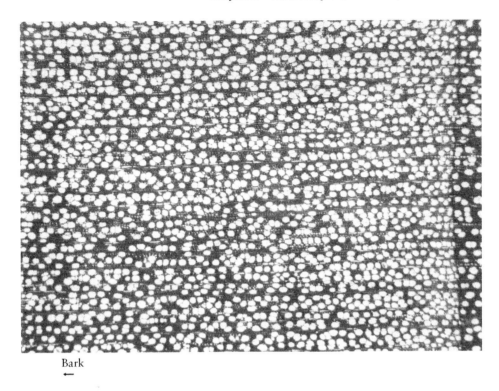

Bark
←

FIGURE 2.10.   Rays in some hardwoods are hard to see even under magnification. Transverse view of red gum (*Liquidambar styraciflua*); ×30.

(a)   (b)

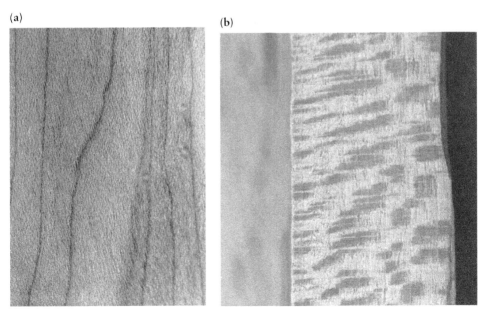

FIGURE 2.11.   Distinct ray patterns in hardwoods. (a) Tangential view of sugar maple (*Acer saccharum*); ×3. (b) Radial view of sycamore (*Platanus occidentalis*); ×3.

Grain Orientation

The direction parallel to the long axis of most of the long tapered fibers of wood is called the grain direction. Fibers are normally oriented as illustrated in Figure 2.12, with their length essentially parallel to the long axis of the stem. Not uncommon, however, is fiber arrangement at a slight angle to the stem axis rather than precisely parallel to it. In fact, angled grain orientation may be the rule rather than the exception (Beals and Davis 1977). Occasionally, the deviation from parallel is large, resulting in an obvious spiraling grain pattern. This kind of grain orientation can significantly affect wood properties.

Spiral Grain

Trees in which fibers are spirally arranged about the stem axis (Figure 2.13) are said to have *spiral grain*. This condition is apparently caused by anticlinal division in which new cambial cell formation occurs in one direction only; that is, walls formed during fusi form initial division consistently slant the same way (Bannan 1966). In utility poles, spiral grain can be readily identified by surface checks that occur in helical formations.

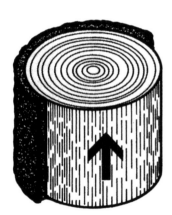

FIGURE 2.12.   Straight-grain orientation.

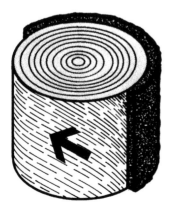

FIGURE 2.13.   Spiral-grain orientation.

FIGURE 2.14.   Reversing spiral grain (interlocked grain).

In at least one instance, spiral grain was shown to be the result of the entwinement of stemwood with a woody vine or liana (Lim 1996).

When logs exhibiting spiral grain are sawn, the lumber formed has a grain direction that is not parallel to the board length. Such lumber is said to have slope of grain (see Chapter 9, Figure 9.17); it is typically low in strength and stiffness and may tend to twist as it dries. Planing of such lumber to a high-quality surface may also be difficult.

Interlocked Grain

In some trees, grain may spiral in one direction for several years and then reverse direction to spiral in the opposite direction (Figure 2.14). Wood produced in this way is said to have *interlocked grain.*

Reversing spiral grain is evidently genetically controlled, occurring very frequently in some species and seldom if at all in others. Woods with interlocked grain, such as elm, are difficult to split and thus are recognized as ones for the do-it-yourself firewood splitter to avoid. Wood with this characteristic may also shrink longitudinally upon drying, warp unpredictably, or both. Occurrence of interlocked grain is occasionally considered desirable from an appearance standpoint. Alternating grain directions cause light to reflect in varying patterns across radially cut wood, giving what is known as a *ribbon stripe* figure. When well developed, this feature can add considerably to the value of appearance-grade veneer.

Knots

The seasonal addition of new wood results in progressive layering over previously produced wood. As new growth increases the diameter of the main stem, branch bases become more and more deeply embedded in the trunk.

Examination of Figure 2.15 shows the living branch extending to the pith, the point at which most branches originate. The base of the branch is cone shaped, appearing as a tapered wedge when sectioned. The cone-shaped appearance arises from the fact that the cambium, which sheaths branches and the main stem, moves ever farther from the embedded branch base as the main stem grows larger, preventing further diameter increase at this location. Because main stem and branch growth are simultaneous, incorporation of living branches into the main stem results in knots that are an integral part of the surrounding wood. Such knots do not become loose or fall out upon drying and are called *intergrown* or *tight knots*.

When a branch dies, its cambial layer also dies, stopping diameter growth throughout its length. The cambial layer of the main stem or bole continues to grow, however, slowly encasing the dead branch in the process. Knots formed in this way are not an integral part of the surrounding wood and, if included in lumber, may fall out as drying takes place. These are called *loose* or *encased knots*.

Embedded branch stubs are usually free of surrounding bark. This was explained by Eames and MacDaniels (1947), who wrote:

> As the base of a [living] branch is buried by the formation of new xylem on the main axis, the phloem tissues about its insertion are forced outward ... so that the base of the branch is stripped of its phloem. In small branches in which the increase in diameter is relatively small as compared with that of the main axis ... this stripping is most marked. In this process the phloem tissues are thrown up into folds which often appear as concentric rings about the base of the partly buried branch.

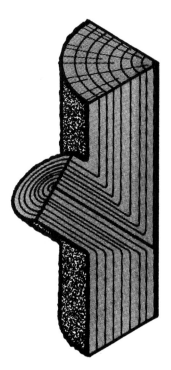

FIGURE 2.15.   Branch configuration in main stem.

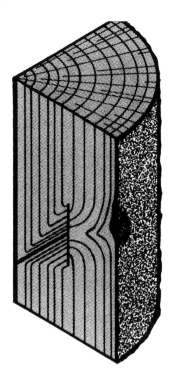

FIGURE 2.16. Progressive covering of branch stub by seasonal growth following pruning.

A similar process takes place in larger branches.

When branches are lost from a stem, such as through pruning, the surrounding cambium layer gradually overgrows the area (similar to the growth depicted in Figure 2.16) and within a few years begins to produce clear, knot-free wood. In some species, dead branches are persistent, so clear wood is not produced for many years after the lower branches on the tree die. Artificial pruning is sometimes used on valuable trees. The results of pruning are illustrated in Figure 2.16.

References and Supplemental Reading

Bannan, M.W. (1966). Spiral grain and anticlinal division in the cambium of conifers. *Can. J. Bot.* 44: 1515–1538.

Beals, H.O. and Davis, T.C. (1977). *Figure in Wood: An Illustrated Review*, 486. Auburn: Auburn University. Agricultural Experiment Station Bulletin.

Dadswell, H.E. and Hillis, W.E. (1962). Wood. In: *Wood Extractives and Their Significance to Pulp and Paper Industries* (ed. W.E. Hillis), 3–55. New York: Academic Press.

Eames, A.J. and MacDaniels, L.H. (1947). *An Introduction to Plant Anatomy*, 2e, 156–165. New York: McGraw-Hill.

Esau, K. (1965). *Plant Anatomy*, 2e, 249–250. New York: Wiley.

Frey-Wyssling, A. (1963). Cytology of aging ray cells. In: *The Formation of Wood in Forest Trees* (ed. M.H. Zimmermann), 457. New York: Academic Press.

Gominho, J. and Pereina, H. (2000). Variability of heartwood content in plantation-grown *Eucalyptus globulus* (Labill). *Wood Fiber Sci.* 32 (2): 189–195.

Goodell, B., Jellison, J., Yuhui, Q., and Connelly, J. 1999. Understanding how structural timbers decay; mechanisms involved in the brown rot decay process. Proc: 1st Int. Conf. on Adv. Eng. Wood Composites, Bar Harbor, ME, 5–8 July.

Grotta, A., Gartner, B., Radosevich, S., and Huso, M. (2005). Influence of red alder competition on cambial phenology and latewood formation in Douglas-fir. *IAWA J.* 26 (3): 309–324.

Hazenburg, G. and Yang, H.C. (1991). Sapwood/heartwood width relationships with tree age in balsam fir. *IAWA Bull.* 12 (1): 95–99.

Hillis, W.E. (1968). Chemical aspects of heartwood formation. *Wood Sci. Technol.* 2 (4): 241–259.

Hillis, W.E. (1987). *Heartwood and Tree Exudates*. New York: Springer.

Hoadley, R.B. (1990). *Identifying Wood*, 107. Newtown, CT: The Taunton Press.

Huber, B. (1956). Die Gefässleitung. In: *Encylopedia of Plant Physiology*, vol. 3 (ed. W. Ruhland), 541–581. Berlin: Springer.

Illman, B. (2003). Synchrotron applications in wood preservation and deterioration. In: *Wood Deterioration and Preservation: Advances in Our Changing World*, vol. 845 (ed. B. Goodell, D.D. Nicholas and T. Schultz), 337–345. Washington, DC: American Chemical Society Symposium Series.

Kärenlampi, P. and Riekkinen, M. (2002). Pine heartwood formation as a maturation phenomenon. *J. Wood Sci.* 48 (6): 467–472.

Koch, P. (1972). *Utilization of the Southern Pines*, vol. 1, 420. Asheville, NC: USDA Forest Service Agriculture Handbook.

Kollmann, F.F.P. and Côté, W.A. Jr. (1968). *Principles of Wood Science and Technology*, vol. 1, 55. New York: Springer.

Kozlowski, T.T. and Pallardy, S.G. (1997). *Growth Control in Woody Plants*, 34–67. San Diego: Academic Press.

Kozlowski, T.T., Kramer, P.J., and Pallardy, S.G. (1991). *The Physiological Ecology of Woody Plants*, 258–268. San Diego: Academic Press, Inc. –269.

Kramer, P.J. and Kozlowski, T.T. (1979). *Physiology of Woody Plants*. New York: Academic Press.

Kretschmann, D., Cramer, S., Lakes, R., and Schmidt, T. (2006). Selected mesostructure properties in loblolly pine from Arkansas plantations. In: *Characterization of the Cellulosic Cell Wall* (ed. D. Stokke and L. Groom), 149–179. Ames, IA: Blackwell.

Larson, P.R. (1962). Auxin gradients and the regulation of cambial activity. In: *Tree Growth* (ed. T.T. Kozlowski), 97–117. New York: Ronald Press.

Larson, P.R. (1994). *The Vascular Cambium: Development and Structure*, 199–203. New York: Springer.

Lim, D.O. (1996). Spiral growth in *Cudrania tricuspidata* caused by liana entwinement. *IAWA J.* 17 (2): 133–139.

Mirov, N.T. and Hasbrouck, J. (1976). *The Story of Pines*, 16. Bloomington: Indiana University Press.

Mott, L., Groom, L., and Shaler, S. (2002). Mechanical properties of individual southern pine fibers. Part II. Compression of earlywood and latewood fibers with respect to tree height and juvenility. *Wood Fiber Sci.* 34 (2): 221–237.

Nobuchi, T., Takahara, S., and Harada, H. (1979). Studies on the survival rate of ray parenchyma cells with aging process in coniferous secondary xylem. *Bull. Kyoto Univ. For.* 51: 239–246.

Pallardy, S.G. (2007). *Physiology of Woody Plants*, 3e. San Diego: Academic Press.

Panshin, A.J. and de Zeeuw, C. (1980). *Textbook of Wood Technology*, 4e, 20–22. New York: McGraw-Hill.

Sano, Y. and Nakada, R. (1998). Time course of the secondary deposition of incrusting materials on bordered pit membranes in *Cryptomeria japonica*. *IAWA J.* 19 (3): 285–299.

Schultz, T.P. and Nicholas, D.D. (2000). Naturally durable heartwood: evidence for a proposed dual defense function of the extractives. *Phytochemistry* 54 (2000): 47–52.

Taylor, A., Gartner, B., and Morrell, J. (2002). Heartwood formation and natural durability: a review. *Wood Fiber Sci.* 34 (4): 587–611.

USDA Forest Service, Forest Products Laboratory. 2010. Wood Handbook: Wood as an Engineering Material. USDA For. Serv., For. Prod. Lab. Rep. FPL-GTR-113.

Wheeler, E.A., Pearson, R.G., LaPasha, C.A. et al. (1986). *Computer-aided Wood Identification*, 474. Raleigh, NC: North Carolina Agricultural Research Service Bulletin.

Yang, K.C. and Hazenberg, G. (1991). Relationship between tree age and sapwood/heartwood width in *Populus tremuoloides* Michx. *Wood Fiber Sci.* 23 (2): 247–252.

Zimmermann, M.H. (1983). *Xylem Structure and the Ascent of Sap*. New York: Springer.

Zimmermann, M.H. and Brown, C.L. (1971). *Trees: Structure and Function*. New York: Springer.

Zwieniecki, M.A., Melcher, P.J., and Holbrook, N.M. (2001). Hydrogel control of xylem hydraulic resistance in plants. *Science* 291: 1059–1062.

3

Composition and Structure of Wood Cells

As a building material, wood is one of the simplest, most easily used products; it can be cut and shaped with ease and fastened readily. At the same time, wood is one of our most complex materials. It is made up of tiny cells, each of which has a precise structure of tiny openings, membranes, and intricately layered walls. The ease with which wood is converted to a product and maintained depends upon practical knowledge of its structure. The cellular nature of wood was introduced in Chapters 1 and 2. In this chapter the molecular composition of wood cells is discussed.

Chemical Components

Wood is composed principally of carbon, hydrogen, and oxygen. Table 3.1 details the chemical composition of a typical North American wood and shows carbon to be the dominant element on a weight basis. In addition, wood contains inorganic compounds that remain after high-temperature combustion in the presence of abundant oxygen; such residues are known as *ash*. Ash is traceable to the occurrence of incombustible compounds containing elements such as calcium, potassium, magnesium, manganese, and silicon. The fact that domestic woods have a very low ash content, particularly a low silica content, is important from the standpoint of utilization; woods having a silica content of greater than about 0.3% (on a dry weight basis) dull cutting tools excessively. Silica contents exceeding 0.5% are relatively common in tropical hardwoods and in some species may exceed 2% by weight.

    The elemental constituents of wood are combined into a number of organic *polymers* (from Greek, *poly* plus *metros*, meaning "many parts") (Brown 1997): cellulose, hemicellulose, and lignin. Table 3.2 shows the approximate percentage of dry weight of each in

Forest Products and Wood Science: An Introduction, Seventh Edition. Rubin Shmulsky and P. David Jones.
© 2019 John Wiley & Sons Ltd. Published 2019 by John Wiley & Sons Ltd.
Companion website: www.wiley.com/go/shmulsky

TABLE 3.1.   Elemental composition of wood.

Element	Dry weight (%)
Carbon	49
Hydrogen	6
Oxygen	44
Nitrogen	0.1
Ash	0.2–0.5[a]

Source: Fengel and Wegner (1984).
[a] As high as 3.0–3.5 in some tropical species.

TABLE 3.2.   Organic constituents of wood.

Type	Cellulose (dry weight %)	Hemicellulose (dry weight %)	Lignin (dry weight %)
Hardwood	40–44	15–35	18–25
Softwood	40–44	20–32	25–35

Source: Kollmann and Côté (1968, pp. 57, 65).
Note: Pectins and starch commonly compose ~6% of the dry weight.

hardwood and softwood. Cellulose, perhaps the most important component of wood, constitutes slightly less than one-half the weight of both hardwoods and softwoods. The proportions of lignin and hemicellulose vary widely among species and between the hardwood and softwood groups.

Cellulose

Photosynthesis is the process by which water and carbon dioxide are combined in the leaves of green plants, employing the energy from sunlight to form glucose and other simple sugars, with oxygen as a by-product (Figure 3.1). Following its formation, glucose may be converted to starch, or to other sugars such as glucose 6-phosphate or fructose 6-phosphate and then to sucrose ($C_{12}H_{22}O_{11}$; Kozlowski and Pallardy 1997). Sucrose and other sugars are then transported in the form of sap to processing centers located at branch tips (apical meristems) and through the inner bark to meristems at the root tips and to the cambial region that sheaths the main bole, branches, and roots. Upon reaching the cytoplasm of individual cells in these various regions, sucrose is hydrolyzed (combined with water) to form glucose and fructose (both $C_6H_{12}O_6$). As indicated in Chapter 1, trees use these sugars to make leaves, wood, and bark.

   Cellulose is synthesized within living cells from a glucose-based sugar nucleotide. A nucleotide is a compound derived from combining a sugar with a phosphate group and a base that is a constituent of RNA or DNA. Complex and separate mechanisms are thought to control initiation of cellulose-chain formation, chain elongation, and termination of the synthesis process (Peng et al. 2002). The net effect of these processes is that glucose molecules are joined together end to end, with elimination of a water molecule

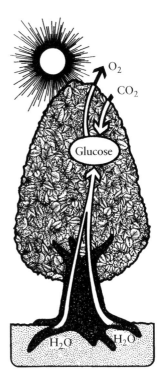

FIGURE 3.1. The process of photosynthesis.

for each chemical linkage formed between neighboring units. The ensuing linear long-chain polymer, cellulose $(C_6H_{10}O_5)n$, has a degree of polymerization, n, which may be as large as 10 000. The structural relationship between glucose and cellulose is formally depicted in Figure 3.2. Hemicelluloses are synthesized in a sequence similar to that outlined above, but they start from a different nucleotide. It is important to note that fructose may convert to glucose for use in synthesis of cellulose, or to mannose, or to other sugars used in making hemicellulose or other compounds.

Cellulose is a material with which people are somewhat familiar. Cotton, for example, is 99% pure cellulose. Fine writing papers are also manufactured largely from the cellulosic fraction of wood. Although it is a carbohydrate, cellulose is not a source of food for humans or most animals. In cellulose, the glucose units are inter-connected through β linkages, where the β designation refers to a specific spatial configuration of the glucosidic bond connecting successive sugar units. As a point of interest, the glucose residues of the *polysaccharide* (saccharide meaning *sugar unit*) starch are identical in every respect except one: they incorporate a glucosidic bond (actually, the mirror image of β) between glucose units. Although cellulose in the form of wood or cotton has as much food value as sucrose, cellulose cannot be digested by humans because the enzymes in body fluids can hydrolyze α but not β linkages. However, certain animals (ruminants) are able to utilize cellulose as food because they maintain intestinal colonies of microorganisms that produce enzymes known as cellulases, which convert cellulose to metabolically useful glucose. Termites have similar internal microflora.

(a)

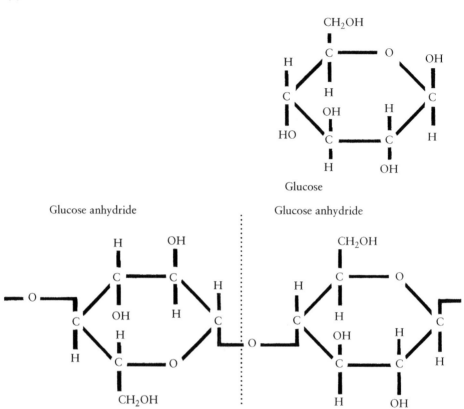

Glucose

Glucose anhydride   Glucose anhydride

Cellulose (*n* ≈ 10,000)

(b)

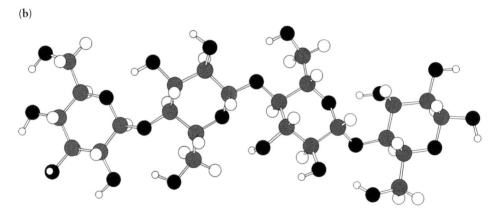

FIGURE 3.2.   (a) Glucose to cellulose. (b) Stereo view of a portion of cellulose molecule.
Source: drawing by Charles Frazier, using ChambridgeSoft, Chem3D, v. 4.0.

At this point it is important that the reader have an idea of the size of the material being discussed. After reading about a cellulose molecule made of up to 10 000 glucose units, a very large structure might be envisioned. Although large from a molecular viewpoint, the longest cellulose molecules are about 10 μm (1/1000 cm) in length and about 8 Å in diameter (1 Å = 1/10,000,000 cm), too small to be seen even with the use of an electron microscope.

Hemicellulose

Although glucose is the primary sugar produced in the process of photosynthesis, it is not the only one. Other six-carbon sugars, such as galactose and mannose, and five-carbon sugars, such as xylose and arabinose, are also manufactured in the leaves. These and other sugar derivatives such as glucuronic acid, along with glucose, are used within developing cells in synthesizing lower-molecular-weight polysaccharides called *hemicelluloses*. Most of the hemicelluloses are branched-chain polymers, in contrast to the straight-chain polymer cellulose, and generally are made up of sugar units numbering only in the hundreds (that is, the degrees of polymerization are in the hundreds rather than thousands or tens of thousands).

Lignin

Lignin is a complex and high-molecular-weight polymer built upon phenylpropane units (Figure 3.3). Although composed of carbon, hydrogen, and oxygen, lignin is not a carbohydrate nor even related to this class of compound. It is, instead, essentially phenolic in nature. Lignin is quite stable and difficult to isolate and occurs, moreover, in a variety of forms; because of this, the exact configuration of lignin within wood remains uncertain. One view is that lignin consists of a group of aromatic polymers, predominantly glycolignin – an ordered polymer made up of multiples of a repeating unit consisting of 18 phenylpropane units (Forss and Fremer 2003).

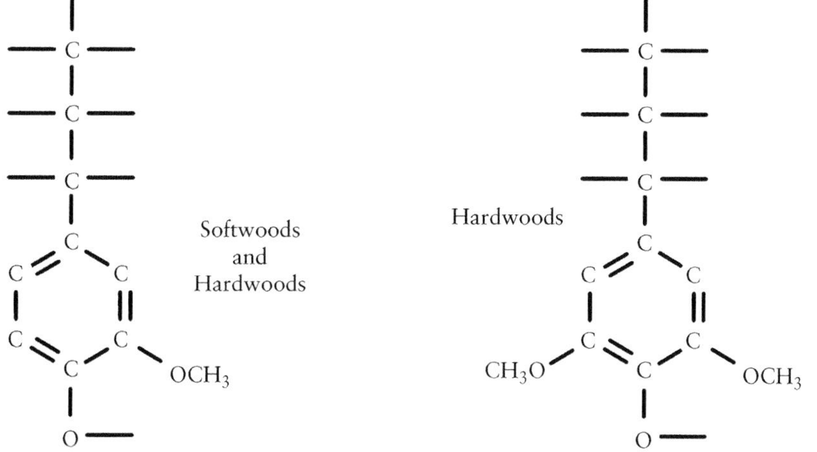

FIGURE 3.3.   Building blocks of lignin.

Lignin occurs between individual cells and within the cell walls. Between cells, it serves as a binding agent to hold the cells together. Within cell walls, lignin is very intimately associated with cellulose and the hemicelluloses, and it gives rigidity to the cell. Lignin is also credited with reducing dimensional change with moisture content fluctuation and has been said to add to wood's toxicity, thus making it resistant to decay and insect attack. The rigidity provided by lignin is an important determinant of wood properties. Recollection of the very soft nature of cotton (almost pure cellulose) and the compliant nature of seaweed (which has very little lignin) are indications of how nonrigid wood would be without a stiffening ingredient.

In its native form, lignin is only very lightly colored. However, even the mildest treatments available for removing lignin from wood cause appreciable degradation of its structure, resulting in a deepening of its color. Thus, chemical pulps (see Chapter 15) that contain residual lignin require considerable bleaching to make them white in color. Because the pulping process used in making newsprint involves mechanical separation of fibers and not lignin removal, only a light brown color develops, which is readily removed by chemical bleaching. However, when the lignin present in newsprint is exposed to air, particularly in the presence of sunlight, the resulting lignin derivatives tend to become yellow or brown with age; a small part of yellowing with age is also traceable to the hemicelluloses. Because of the lignin in mechanical pulp, newsprint has a notoriously short longevity owing to its high lignin content; it is also coarse, bulky, and of low strength because the fibers bond to one another with difficulty because of their inherent stiffness.

Cell Wall

Recall that a tree is sheathed by a thin cambial layer, which is composed of cells capable of repeated division. New cells produced to the inside of this sheath become new wood, and those moved to the outside become part of the bark. In this section, the chemical configuration of woody cell walls is examined, as are steps in the development of new cells.

Chemical Structure

A newly formed wood cell is encased in a thin, membrane-like and pectin-rich wall called a *primary wall*, and the cell is filled with fluid. *Pectins* are complex colloidal substances of high molecular weight that, upon hydrolysis, usually yield galacturonic acid and small amounts of arabinose and galactose. The precise structure of pectin is not completely understood. In a process that may take several days to several weeks to complete, the cell enlarges and the cell wall gradually thickens as biopolymers produced within the cells are progressively added to the inside (lumen side) of the wall (Figure 3.4). Eventually, the protoplasm that fills the cell is lost and the cell has a thickened wall, consisting of primary and secondary wall layers, and a hollow center (Figure 3.4d). Successive arrangements of biopolymer assemblies are responsible for the gradual thickening of a cell wall. But what are these *biopolymers?* They are the three distinct types of macromolecule described earlier: cellulose, hemicellulose, and lignin.

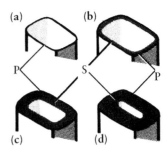

FIGURE 3.4.   Stages in development of a wood cell. Longitudinal cells in cross-section: (a) new cell has only ultrathin primary wall (P); (b, c) cell enlarges and then wall thickens as secondary wall (S) forms to inside of the primary wall; (d) wall continues to thicken with buildup of deposits.

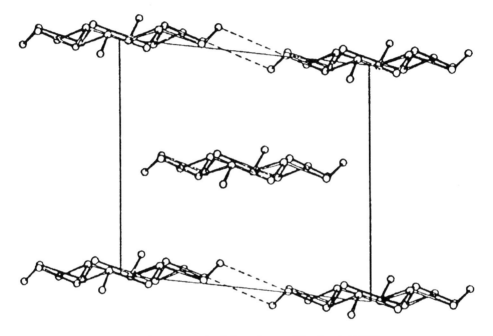

FIGURE 3.5.   Unit cell configuration of cellulose. Source: Woodcock (1979).

The buildup of biopolymers on the inner surfaces of the cell wall is not haphazard; it occurs in a very precise fashion. Cellulose, for example, is not incorporated into the cell wall as individual molecules but rather as intricately arranged clusters of molecules. The long-chain cellulose molecules are synthesized from anhydroglucose (actually glucose attached to a mononucleotide) in many specific locations at the inner surface of the cell wall itself. As these chains lengthen, they aggregate laterally in a well-defined way with their immediate neighbors, which are also growing, to form crystalline domains in a unit cell configuration (Figure 3.5). The cellulose crystal lattice is held together by intermolecular and dipolar interactions primarily in the form of hydrogen bonds; this arrangement is so stable that the individual chains cannot be dissolved in common solvents such as water or acetone. Cellulose can be dissolved in very exotic, highly polar solvents capable of disrupting hydrogen bonds (Heinze and Liebert 2001).

A number of studies of wood, with some dating back three decades, have determined that the highly crystalline networks of cellulose are united into larger structures within the wood cell wall. These structures are known as *microfibrils*.

As noted by Preston (1986), studies of cellulose in the late 1940s and early 1950s soon after introduction of the electron microscope showed the cell wall to be composed of long, thin strands. These strands, or microfibrils, were later shown to be cellulosic and composed of bundles of cellulose chains lying parallel to the microfibril length. Subsequent research in the mid-1950s through early 1970s, in which isolated microfibrils were subjected to chemical extraction, showed microfibrils to consist of an inner core consisting almost entirely of glucose constituent sugars (suggesting a cellulose core) and an outer core composed of 15% or more nonglucose sugars (Jane et al. 1970; Stamm 1964). These findings led to the conclusion that cellulose bundles are covered or sheathed in strongly adsorbed chains of sugars other than glucose (such as hemicellulose). Research also points to crystalline (highly ordered) and amorphous (less ordered) regions within microfibrils, with individual cellulose molecules running through several crystalline and amorphous regions (Kollmann and Côté 1968; Tsuchikawa and Siesler 2006).

Because of the difficulty of isolating cellulose that is part of a heavily lignified woody cell wall, researchers have focused on cells that can be more easily studied. One such kind of cell is that of the green algae, valonia, that has unlignified cell walls in which cellulose is organized into microfibrils. Other work has been done with the bacterium *Acetobacter xylinum* and the primary walls of tobacco leaf epidermal cells. Findings have been correlated with measured sizes of wood microfibrils to develop models of the cellulose structure in wood.

Fujita and Harada (1991) described cellulose microfibrils in a very straightforward manner, describing them as consisting of a "core crystalline region of cellulose surrounded by the paracrystalline [less highly ordered] cellulose and short chain hemicellulose." Ruben et al. (1989) proposed a more specific structure based on their studies of tobacco leaf cells and *A. xylinum*. Their work indicates that the extent of each crystalline domain is confined to just nine cellulose chains, which together may be viewed as a *subelementary fibril* 18Å in width. Three such subelementary fibrils are wound in a left-handed triple-helical fashion around one another to form an *elementary fibril* that is 37Å wide. These elementary fibrils then aggregate into microfibrillar bundles with some assistance from hydrogen bonding in which hemicellulose plays a role. A proposed model of the arrangement of elementary fibrils (Figure 3.6) shows crystalline and amorphous regions and variations in spacing between elementary fibrils in hardwoods and softwoods.

Later work led to the observation that electron micrograph analyses of microfibrils of valonia showed cellulose to be highly crystalline; furthermore, there did not appear to be any subunits in these microfibrils corresponding to elementary or subelementary fibrils (Fujita and Harada 1991). There is some speculation that the triple-stranded structures typify microfibrils in the primary cell wall, whereas microfibrils in the *secondary cell wall* are highly crystalline arrays of straight cellulose chains (Ruben et al. 1989). There is also now strong evidence that each microfibril in a woody cell wall is composed of 36 glucan chains (Dellmer and Armor 1995). In any case, it is clear that long-chain cellulose molecules are arranged within the cell wall in a rather precise fashion and are combined into larger structures, the microfibrils.

What, then, of the hemicellulose and lignin? The hemicelluloses, probably somewhat selectively, interact through hydrogen bonding with the cellulose and have been implicated

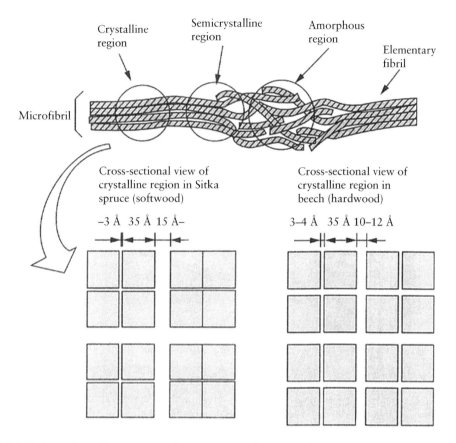

FIGURE 3.6.   Crystalline and amorphous regions within a microfibril. Source: Tsuchikawa and Siesler (2006).

in the aggregation of elementary fibrils into microfibrils. Hemicelluloses, as noted earlier, are known to sheath the microfibrillar bundles. Moreover, the hemicelluloses are chemically linked to lignin *macromolecules* and thus fulfill a particularly important function in maintaining cohesion between the architectural building materials of the wood cell wall. The way in which lignin is incorporated into the cell wall represents another area of disagreement among scientists. The longstanding view is that lignin is deposited between microfibrils during and after the wall-thickening process. Another view (Goring 1977) is that lignin is placed in a lamellar configuration between the microfibrils (Figure 3.7): it appears to occupy its allocated space in the form of undulating two-dimensional sheets with thicknesses of 16–20Å. Although the precise nature of lignin continues to elude researchers, it is known that the aromatic rings tend to lie parallel to one another within the wood cell wall. Yet neither lignin nor its chemical derivatives have ever been coaxed into a crystalline form from solution, whereas cellulose and many of the hemicelluloses crystallize quite readily.

In summary, the secondary layer of the wood cell wall can be viewed as a laminated filamentary composite. The cellulose molecules provide the structurally reinforcing

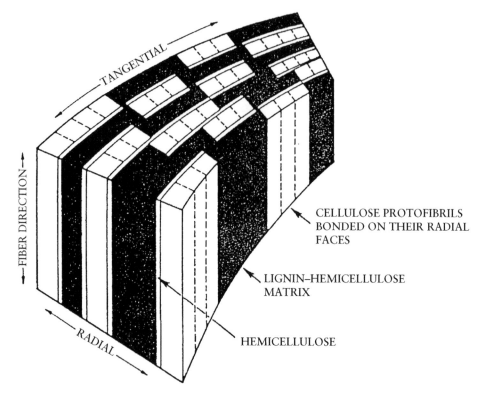

FIGURE 3.7.   Ultrastructural arrangement of lignin and polysaccharides in wood cell wall.

network. These are embedded in a matrix composed also of hemicelluloses and lamellar lignin "sheets" that are partly bound to one another through chemical bonds. There is, however, more to the story, and the part to come is no less important.

Layering

The primary wall, described previously as being pectin rich, later becomes heavily lignified. The primary wall is also reinforced with a more-or-less random network of microfibrils. This random arrangement contrasts with the very organized microfibril pattern in the secondary wall.

The first few microfibrils that are synthesized as the secondary wall starts to form are laid down in a particular way; they are spiraled around the cell interior, with the long axes of the microfibrils nearly perpendicular to the long axis of the cell. After a few layers form in this way, the orientation begins to change; microfibrils spiral about the cell at a much smaller angle to the cell axis. Just prior to final development of the cell, a change in orientation again occurs, with the last several layers arranged similarly to the first few layers. Thus the secondary part of a cell wall has three more-or-less distinct layers (Figure 3.8). For purposes of discussion, these layers are numbered according to the order in which they are formed: S_1, S_2, S_3. Study Figure 3.8 carefully. This intricate structure of the cell wall is the key to the behavior of wood. Note that the S_2 layer is much thicker

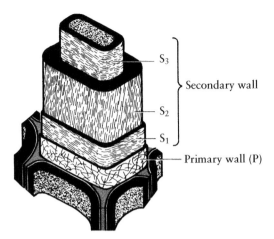

FIGURE 3.8.   Layering of a mature cell wall.

than the others. The S_1 and S_3 layers in a softwood are on the order of four to six layers of clustered microfibrils, or *lamellae*, thick; the number of lamellae comprising the S_2 may vary from 30 to 40 in thin-walled earlywood cells to 150 or more in latewood cells (Kollmann and Côté 1968, p. 26). In an earlywood cell, these proportions would translate to thicknesses of about 0.1 µm for S_1 and S_3 layers and 0.6 µm for the S_2. Because the S_2 layer is much thicker, this wall layer has the greatest effect on how the cell behaves.

As various layers of the secondary wall are built up as a series of uniformly thick sublayers or lamellae, microfibril angles in the cell wall change gradually from lamella to lamella, rather than abruptly as might be inferred from Figure 3.8. As described by Abe et al. (1995, 1997) and Abe and Funada (2005), the orientation of microfibrils in the S_1 layer of conifers shifts gradually in a clockwise direction from an S-helix (>90°) to a Z-helix (<90°) from the outer to the inner layer in the S_1 layer. In the S_2 layer the Z-helix is maintained, although the microfibril angle is much smaller than in the S_1 layer. In the S_3 layer the microfibril orientation again gradually shifts clockwise from a Z-helix to an S-helix at the innermost surface. An example of microfibril orientation is provided by radiata pine; in this species microfibril angles in the S_1, S_2, and S_3 layers have been measured at 79–113°, 1–59°, and 50–113°, respectively (Donaldson and Xu 2005).

It is known that a high microfibril angle significantly affects both the strength and the directional shrinkage properties of wood, thus creating concerns about the presence of juvenile wood in lumber and other structural and nonstructural wood products. Courchene et al. (2006) demonstrated that microfibril angle is equally important in raw material used in making paper, with test results showing paper tensile strength, stretch, modulus of elasticity, stiffness, and moisture-induced expansion in refined and unrefined pulps also significantly impacted by microfibril angle.

A look at the distribution of biopolymeric components in the various wall layers concludes this discussion of cell wall organization. It is important to recognize that cellulose, hemicellulose, and lignin all occur in each layer of the cell wall. This is illustrated in Figure 3.9, which was derived from chemical analysis of various cell wall layers of coniferous woods. Note that cellulose is present in only small amounts in the compound middle lamella, increasing as a proportion of the dry weight of the cell wall through the

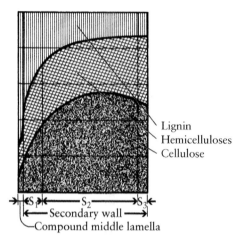

FIGURE 3.9.   Distribution of organic compounds within various cell wall layers of a softwood. Source: from Panshin and de Zeeuw (1980).

center portion of the S_2 layer. Lignin, on the other hand, is the dominant component between cells, with the concentration as a proportion of the cell wall decreasing as the lumen is approached. Considerable disagreement exists as to the proportion of lignin in the compound middle lamella. Figure 3.9 shows that although the largest concentration of lignin is in the middle lamella, the extreme thinness of this layer means that most of the overall quantity of lignin is found in the secondary wall.

The intricate structure of the woody cell wall that defines many of the physical and mechanical properties of wood and the fact that this intricacy develops within living cells have not escaped the attention of wood scientists. Relatively recent development of scientific tools that allow the measurement and imaging of extremely small structures laid the foundation for a new field of scientific inquiry and development known as *nanotechnology*. Most early work in this area has focused on carbon and the use of carbon *nanotubes* (perfectly straight tubules with diameters of nanometer size, and properties close to those of an ideal graphite fiber) for applications ranging from microelectrodes in electrochemical reactions to mechanical reinforcement of a range of products (Ajayan and Zhou 2001). For purposes of reference, a nanometer is one billionth of a meter, and a sheet of paper is about 100,000 nm thick.

Now, scientists are beginning to look at wood and the possibility of working at the nanoscale to create new, lighter, stronger, more absorptive products. A commentary about cellulose nanofibrils (cellulose molecules and microfibrils) as compared with carbon nanotubes included the observation that "cellulose nanofibrils exhibit a modulus roughly one-quarter to one-fifth that of carbon nanotubes, yet they are produced naturally without the need for energy consuming, high-temperature processing" (Atalla et al. 2006).

The potential gains from application of nanotechnology to wood science are large. If scientists were, for example, able to understand the intricate mechanisms that control formation of cellulose, hemicellulose, microfibrils, and lignin within the cytoplasm of developing cells under the guidance of genetic encoding, that might enable them to develop new materials that self-assemble from molecular building blocks. Or, it might be possible to insert nonwood materials with some precision into the cellular structure of wood for the purpose of developing or enhancing specific properties, or to create molecular scale holes or punctures in cell walls to enhance wood densification, wood chemical modification, and surface adhesion (Atalla et al. 2006; Moon et al. 2006). Considerable research into nanotechnology applications in wood science is likely in coming years.

Cell Wall Sculpturing

Wood cells that function primarily in the storage and conduction of food materials are known as *parenchyma*. These cells typically form thin secondary walls and are the last to remain functional prior to heartwood formation. Other kinds of cells, in contrast, serve principally as avenues of fluid conduction in the living tree; these often form thick secondary walls and thus are important in providing mechanical support to stems in which they occur.

Pitting.   All types of cells are characterized by secondary wall layers that are not continuous. Instead, walls are interrupted by regions in which the secondary portion of the wall is lacking. Known as *pits*, these regions generally appear quite different in parenchyma as compared with other kinds of cells.

Normally, pit placement in one cell is exactly matched by the position of pits in adjoining cells. Pits thus tend to occur as matched pairs.

Because pit regions are areas of the cell wall that lack secondary thickening, they are, in effect, thin spots in the cell wall. As such, these areas are much more readily penetrated by fluids and gases than are unpitted zones; thus *pit pairs* are the primary avenues of cell lumen-to-lumen transport (Figure 3.10).

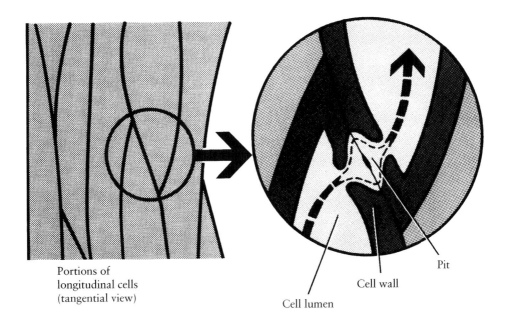

Portions of longitudinal cells (tangential view)

Cell lumen

Cell wall

Pit

FIGURE 3.10.   Pits provide tiny passageways for flow.

Pits that mark the walls of parenchyma cells are called *simple pits*. Pitting between two ray parenchyma cells is illustrated in Figure 3.11. Because both cells in this figure are of the parenchyma type, the pits shown form a simple pit pair. Note that whereas secondary wall material is lacking in the pit zone, the primary walls of the two adjacent cells remain. The primary walls and the thin layer of intercellular material that separates them form the *pit membrane*.

The type of pit typifying nonparenchyma cells is shown in Figures 3.12 and 3.13. This is called a *bordered pit*, so named because the pit aperture appears to be surrounded by a border when viewed frontally. Rather than consisting of a simple gap in the secondary part of the wall, a bordered pit is a conical depression in the secondary wall that is concave toward the middle lamella and has an opening leading to the cell lumen at the depth of the depression. A pair of this kind of pit, typical of those connecting two conductive cells, is shown in highly magnified profile view in Figure 3.14b. In this view, secondary walls are seen to overarch the primary wall, forming a pit cavity. As in simple pits, the primary walls of adjacent cells form a pit membrane. When storage (parenchyma) and conductive cells are in contact, each cell usually forms simple and bordered pits, respectively. The resulting pit pair is termed *half-bordered* (Figure 3.14c).

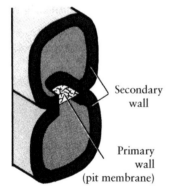

Secondary wall

Primary wall (pit membrane)

FIGURE 3.11.   Simple pitting in adjoining ray parenchyma cells (tangential view).

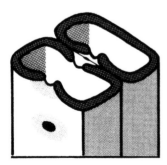

FIGURE 3.12.   Bordered pitting in softwood longitudinal tracheids.

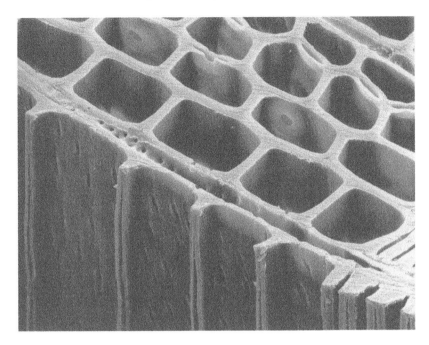

FIGURE 3.13.   Bordered pitting in longitudinal tracheids. Radial/transverse view of Sitka spruce (*Picea sitchensis*). ×400. Source: scanning electron micrograph by Josefina Gonzales, Forintek Canada Corp., Vancouver, BC.

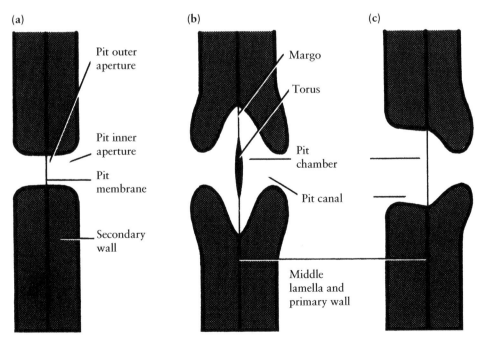

FIGURE 3.14.   Profile of various types of pit pairs: (a) simple; (b) bordered; and (c) half-bordered. Source: from Mark (1967).

In softwoods, the pit membrane between two bordered pits differs from that separating a simple or half-bordered pit pair. In the latter two kinds, the pectin-rich and microfibril-reinforced primary wall remains unmodified within the pit zone. The common primary walls separating two softwood bordered pits are, however, changed considerably as pits are formed.

Bordered pit formation apparently begins prior to the start of S_1 layer formation with the development of a ring of cellulose on the primary wall. This ring defines the outer boundary of the pit. Then, as secondary wall formation commences, the pit membrane undergoes modification. The membrane center becomes thickened through accumulation of densely packed and sometimes circularly arranged microfibrils. This thickening is called the *torus*. The area surrounding the torus is named the *margo*, and it too becomes different from the normal primary wall. A net of radially arranged microfibrils may form over the existing primary wall, connecting the torus to the pit exterior. At about the same time, the pectin-rich matrix of the compound middle lamella enzymatically decomposes, leaving a more or less open network (Figure 3.15). In at least one species, *Ginko biloba*, torus thickening of the pit membrane is the first step in bordered pit formation, occurring during the process of radial expansion of the longitudinal tracheid (Dute 1994). Finally, secondary wall thickening is completed through successive development of microfibrillar layers, thus forming the arch or conically shaped wall structure (Wardrop 1964).

Bordered pits are structurally similar in hardwood and softwood species except that the membranes are quite different. Membranes of all pit combinations in hardwoods are similar to those characterizing simple and half-bordered pit pairs in softwoods. Hence, no torus develops and there is no dissolution of portions of the primary wall. In such an

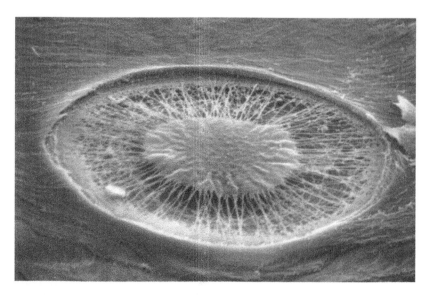

FIGURE 3.15.   Bordered pit membrane. Red pine (*Pinus resinosa*). ×3400 Source: scanning electron micrograph by John Crist and Ron Teclaw.

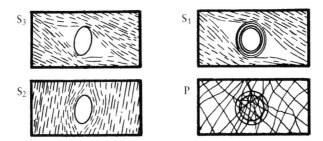

FIGURE 3.16.   Pits disrupt regular microfibril angles. P = primary wall; S_1, S_2, and S_3 refer to the secondary wall layers. Source: from Mark (1967).

unmodified wall, it has been reported that no openings are visible, even at magnifications of 3,100,000 (Kollmann and Côté 1968, p. 31). In comparison, filtration experiments with softwood bordered pit membranes have shown openings approximating 0.2 μm in size in the reinforced microfibril network (Liese 1954).

It has long been conjectured that pit regions substantially reduce fiber strength, a notion that is not generally substantiated by scientific observation (Mark 1967, p. 47). Bailey (1958) noted that bordered pits, which characterize stem-strengthening and fluid-conducting cells, appear to be configured to provide the maximum exposure of thin, readily penetrable wall area, with only minimal reduction in secondary wall reinforcement. Microfibril buildup around pit areas may also help to reduce the effect of pits upon wall strength.

The pits dotting the cell wall cause local variation in the normal S_1, S_2, and S_3 microfibril orientation discussed earlier. Sedighi-Gilani et al. (2006) used confocal laser scanning microscopy to study microfibril angles in spruce, showing clear evidence of microfibril angle disruption around large bordered pits on radial cell walls of earlywood, but little disruption on tangential cell walls and in cell walls of latewood. Figure 3.16 illustrates how microfibrils curve around the pit regions (Anagnost et al. 2002; Mark 1967, p. 13).

Spiral Thickening.   In some woods, formation of the S_3 layer is followed by development of spirally arranged ridges of microfibril bundles on the lumen side of the secondary wall. Such ridges are distinctly separate from the S_3 layer, as evidenced by the fact that they are relatively easily detached from it (Wardrop 1964), and they only rarely parallel the S_3 microfibril orientation. These ridges are termed spiral thickenings (Figure 3.17).

Spiral thickenings occur in cells of relatively few softwoods and thus, when present, are a valuable clue to a wood's identity. In hardwoods, spiral thickening is relatively common. Although helically arranged thickenings may form throughout the length of a cell, this feature in other cells may be restricted to only the tips or center portions. Because location of thickenings is often consistent within a given species of wood, this factor can also be of diagnostic significance.

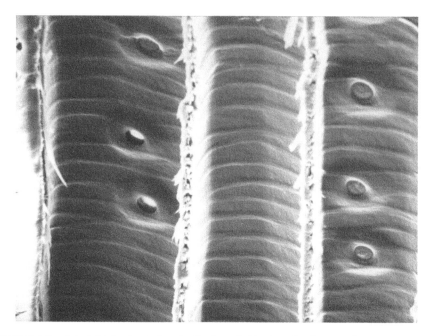

FIGURE 3.17.   Spiral thickening in longitudinal tracheids. Douglas fir (*Pseudotsuga menziesii*). ×830. Source: scanning electron micrograph by John Crist and Ron Teclaw.

References and Supplemental Reading

Abe, H. and Funada, R. (2005). Review – the orientation of cellulose microfibrils in the cell walls of tracheids in conifers. *IAWA J.* 26 (2): 161–174.

Abe, H., Funada, R., Ohtanti, J., and Fukuzawa, K. (1995). Changes in the arrangement of microtubules and microfibrils in differentiating conifer trachieds during the expansion of cells. *Ann. Bot.* 75: 305–310.

Abe, H., Funada, R., Ohtanti, J., and Fukuzawa, K. (1997). Changes in the arrangement of cellulose microfibrils associated with the cessation of cell expansion in tracheids. *Trees* 11 (3): 328–332.

Ajayan, P. and Zhou, O. (2001). Applications of carbon nanotubes. In: *Carbon Nanotubes, Topics, Applied in Physics* (ed. M. Dresselhaus, G. Dresselhaus and P. Avouris), 391–425. Heidelberg: Springer.

Anagnost, S., Mark, R., and Hanna, R. (2002). Variation of microfibril angle within individual tracheids. *Wood Fiber Sci.* 34 (2): 337–349.

Atalla, R. 1990. The structures of cellulose. In Materials Interactions Relevant to the Pulp, Paper, and Wood Industries: Proc.: Materials Res. Soc. Symp. D.F. Caulfield, J.D. Passaretti, and S.F. Sebcznski, eds. Vol. 197, pp. 89–98.

Atalla, R.; Beecher, J.; Caron, R.; Catchmark, J.; Deng, Y.; Glasser, W.; Gray, D.; Haigler, C.; Jones, P.; Joyce, M.; Kohlman, J.; Koukoulas, A.; Lancaster, P.; Perine, L.; Rodriguez, A.; Ragauskas, A.; Wegner, T.; and Zhu, J. 2006. Nanotechnology for the Forest Products Industry: Vision and Technology Roadmap. US Department of Agriculture, Forest Products Laboratory; US Department of Energy, Pacific Northwest National Laboratory; Technical Association of the Pulp and Paper Industry; and American Forest and Paper Association.

Bailey, I.W. (1958). The structure of tracheids in relation to the movement of liquids, suspensions, and undissolved gases. In: *The Physiology of Forest Trees* (ed. I.V. Thimann), 71–82. New York: Ronald Press.

Brown, W.H. (1997). *Introduction to Organic Chemistry*, 113. Philadelphia: Saunders College Publishing.

Côté, W.A. Jr. (1967). *Wood Ultrastructure – An Atlas of Electron Micrographs*. Seattle: University of Washington Press.

Courchene, C., Peter, G., and Litvay, J. (2006). Cellulose microfibril angle as a determinant of paper strength and hygroexpansivity in *Pinus taeda* L. *Wood Fiber Sci.* 38 (1): 112–120.

Dellmer, D.P. and Armor, Y. (1995). Cellulose biosynthesis. *Plant Cell* 7: 987–1000.

Donaldson, L. and Xu, P. (2005). Microfibril orientation across the secondary wall of radiata pine tracheids. *Trees* 19 (6): 644–653.

Dute, R. (1994). Pit membrane structure and development in *Ginko biloba*. *IAWA J.* 15 (1): 75–90.

Fengel, D. and Wegner, G. (1984). *Wood: Chemistry, Ultrastructure, Reactions*, 26–65. Berlin: deGruyter.

Forss, K. and Fremer, K. (2003). *The Nature and Reactions of Lignin – A New Paradigm*. Atlanta: TAPPI Press.

Fujita, M. and Harada, H. (1991). Ultrastructure and formation of wood cell wall. In: *Wood and Cellulosic Chemistry* (ed. D.N.-S. Hong and N. Shiraishi), 3–57. New York: Marcel Dekker.

Goring, D.A.I. (1977). A speculative picture of the delignification process. *Cellul. Chem. Technol.* ACS Symposium Series 48: 273–277.

Heinze, T. and Liebert, T. (2001). Unconventional methods in cellulose functionalization. *Prog. Polym. Sci.* 26 (9): 1689–1762.

Jane, F.W., Wilson, K., and White, D.J.B. (1970). *The Structure of Wood*, 170. London: Adam and Charles Black.

Kollmann, F.F.P. and Côté, W.A. Jr. (1968). *Principles of Wood Science and Technology*, vol. 1. New York: Springer.

Kozlowski, T.T. and Pallardy, S.G. (1997). *Physiology of Woody Plants*, 2e, 163–164. San Diego: Academic Press.

Liese, W. 1954. Der Feinbau der Hoftopfel im Holz der Koniferen. Proc. Int. Conf. Electron Microscopy, London, pp. 550–554.

Mark, R.E. (1967). *Cell Wall Mechanics of Tracheids*. New Haven, CT: Yale University Press.

Meier, H. (1964). General chemistry of cell walls and distribution of the chemical constituents across the walls. In: *The Formation of Wood in Forest Trees* (ed. M.H. Zimmermann), 137–151. New York: Academic Press.

Moon, R., Frihart, C., and Wegner, T. (2006). Nanotechnology applications in the forest products industry. *For. Prod. J.* 56 (5): 4–10.

Okamura, K. (1991). Structure of cellulose. In: *Wood and Cellulosic Chemistry* (ed. D.N.S. Hong and N. Shinaishi), 89–112. New York: Marcel Dekker.

Panshin, A.J. and de Zeeuw, C. (1980). *Textbook of Wood Technology*, 4e, 93–96. New York: McGraw-Hill.

Peng, L., Kawagoe, Y., Hogan, P., and Delmer, D. (2002). Sitosterol-B-glucoside as primer for cellulose synthesis in plants. *Science* 295 (5552): 147–150.

Preston, R.D. (1986). Natural celluloses. In: *Cellulose: Structure, Modification, and Hydrolysis. R.A. Young, and* (ed. R.M. Rowell), 278–298. New York: John Wiley & Sons.

Ruben, G.C., Bokelman, G.H., and Krakow, W. (1989). Triple-stranded left-hand helical cellulose microfibril in *Acetobacter xylinum* and in tobacco primary cell wall. In: *Plant Cell Wall Polymers – Biogenesis and Biodegradation*, ACS Symposium Series, vol. 399, 278–298. Washington, DC: American Chemical Society.

Sedighi-Gilani, M., Sunderland, H., and Navi, P. (2006). Within-fiber non-uniformities of microfibril angle. *Wood Fiber Sci.* 38 (1): 132–138.

Stamm, A.J. (1964). *Wood and Cellulose Science*, 31. New York: Ronald Press.

Tsuchikawa, S. and Siesler, H. (2006). Near infrared spectroscopic monitoring of the diffusion process of deuterium-labeled molecules in wood. In: *Characterization of the Cellulosic Cell Wall* (ed. D. Stokke and H. Groom), 123–137. Ames, IA: Blackwell.

Wardrop, A.B. (1964). The structure and formation of the cell wall in xylem. In: *The Formation of Wood in Forest Trees* (ed. M.H. Zimmermann), 87–134. New York: Academic Press.

Woodcock, C. 1979. The x-ray crystallographic analysis of the structure of native ramie cellulose. MS thesis. SUNY College of Environmental Science and Forestry.

4

Softwood Structure

Softwoods have traditionally been the mainstay of the wood products industry in North America, and these woods continue to be extremely important today. The homogeneous, straight-grained, and lightweight softwood is preferred for construction lumber and plywood for the housing market. Tall, straight-boled softwoods are used for poles and pilings. Because they are typically composed of long fibers, softwoods are also a premium raw material in the manufacture of strong papers. A knowledge of the physical nature of softwood xylem is basic to an understanding of wood and wood products. The structural characteristics of this important group of woods are examined in this chapter.

The xylem of softwoods is quite simple. Most species have no more than four or five different kinds of wood cells, and only one or two of these occur in appreciable numbers. Because of this simplicity and uniformity of structure, softwoods tend to be similar in appearance.

Longitudinal Tracheids

The great majority of softwood volume, 90–95%, is composed of long, slender cells called *longitudinal tracheids*. Such cells are oriented parallel to the stem axis (Figure 4.1).

Configuration

Longitudinal tracheids are about 100 times greater in length than in diameter and are rectangular in cross-section (Figure 4.2). Tracheids have hollow centers (lumens) but are closed at the ends, and their shape is blunt or rounded radially and pointed tangentially. The pits in tracheids are normally bordered. A longitudinal tracheid can be visualized by thinking of a soda straw, pinched shut at both ends. In this way the straw is similar in both appearance and relative proportions to a longitudinal tracheid. The tracheid is much smaller, however, averaging only 25–45 μm in diameter and 3–4 mm in length.

Forest Products and Wood Science: An Introduction, Seventh Edition. Rubin Shmulsky and P. David Jones.
© 2019 John Wiley & Sons Ltd. Published 2019 by John Wiley & Sons Ltd.
Companion website: www.wiley.com/go/shmulsky

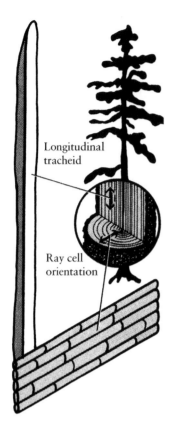

Longitudinal
tracheid

Ray cell
orientation

FIGURE 4.1.   Cell orientation in a standing softwood.

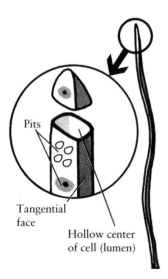

Pits

Tangential
face

Hollow center
of cell (lumen)

FIGURE 4.2.   Longitudinal tracheid.

FIGURE 4.3.   Radial file of tracheids representing one year
of growth (earlywood – upper right; latewood – lower left).

As explained in Chapter 2, the softwood cells formed early in a growing season differ from those formed later in the year. A review of earlywood and latewood differences
is perhaps best accomplished by looking at one radial file of tracheids representing one
year of growth. Thin-walled earlywood cells with relatively large radial diameters are
seen to the right in Figure 4.3; thicker-walled and smaller-diameter cells are seen to the
left. The abrupt change from thin- to thick-walled cells depicted in the figure is characteristic of only some softwood species, such as the hard pines, larch, and Douglas-fir. In
these woods, latewood is sharply delineated from the earlywood part of the ring. In other
species such as true fir and hemlock, the transition in wall thickness and radial diameter
progresses gradually from early- to late-formed wood. Rings in these woods are less
clearly defined than in those having abrupt transition. Abrupt and gradual transition in
growth rings is shown in Figure 4.4.

Pitting

Again referring to Figure 4.3, most of the numerous pits that mark the radial cell walls
are of the bordered type. Such pits typify tracheid-to-tracheid linkages, and thus their
location is usually matched with a pit in an adjacent longitudinal tracheid. The rows of
small, lemon-drop-shaped pits mark the points at which ray parenchyma cells contact the
longitudinal tracheids.

The characteristics of the softwood bordered pit membrane were outlined in
Chapter 3. Recall that the softwood pit membrane in most species has a thickened central
torus surrounded by a microfibrillar network known as the *margo*. At least one softwood
species – western red cedar – lacks tori in bordered pit membranes.

The typical softwood bordered pit membrane is the source of several significant
use-related problems: a shift in the membrane from its normal central position can result
in difficulties in both drying and preservative-treatment. Because the membrane is
flexible, it can shift to one side of the pit cavity, resulting in the blockage of the aperture
by the impenetrable torus (Figure 4.5). A pit in this condition is said to be aspirated.
Wood with aspirated pits is resistant to penetration by protective chemicals, such as creosote, or waterborne preservatives, such as CCA (chromated copper arsenate) or ACQ

(a)   Bark ⟶

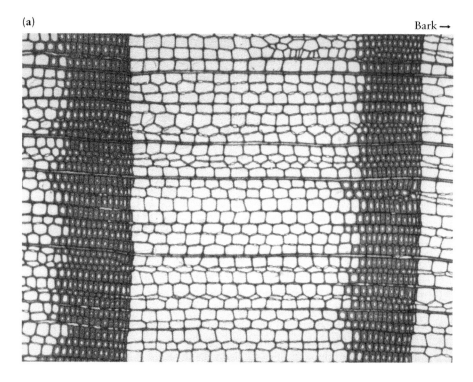

(b)   Bark ⟶

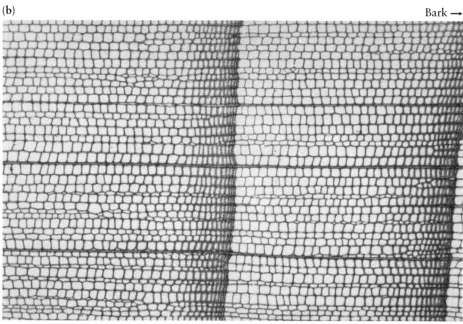

FIGURE 4.4.   Earlywood to latewood transition in softwoods. (a) Abrupt transition in transverse view of western larch (*Larix occidentalis*). ×85. (b) Gradual transition in transverse view of balsam fir (*Abies balsamea*). ×85. Source: Courtesy of Ripon Microslides Laboratory.

(a)

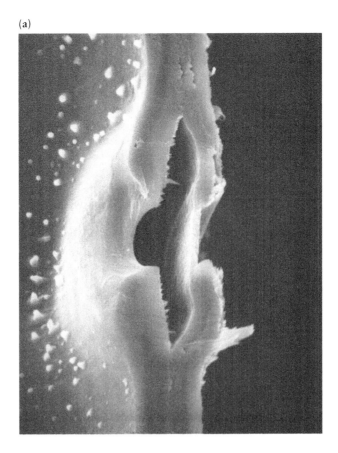

(b)

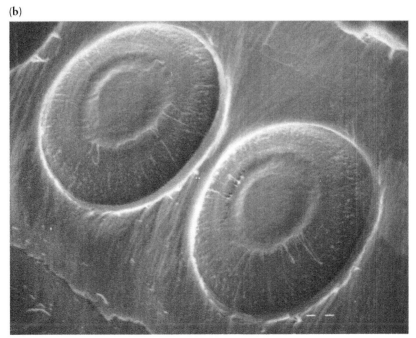

FIGURE 4.5. Aspirated pits in pine. (a) Sugar pine (*Pinus lambertiana*). Cross-section of pit. ×4000. (b) Red pine (*Pinus resinosa*). ×2800. Tori are blocking inner apertures of pits; rims of outer apertures are in foreground. Source: Scanning electron micrograph by John Crist and Ron Teclaw.

(ammoniacal copper quat). Such wood also dries slowly. Drying can, in fact, be the cause of pit aspiration from differences in pressure that may develop on different sides of pit membranes (Hart and Thomas 1967; Tsoumis 1964). Once aspiration occurs, it is apparently a permanent condition (Jane et al. 1970, p. 282); hydrogen bonding between the torus and the overarching secondary wall can fix the position of the membrane.

Pit aspiration happens as a result of liquid tensions that can occur in standing trees or in processed wood that is being dried. A situation that results in aspiration, for example, is one in which there is a closed system of water under tension in the lumen on one side of a bordered pit and an air/water-filled lumen, with resulting meniscus at the air/water interface, on the other (Hart and Thomas 1967). Evaporation in any part of the system or transpirational pull on the closed-water system can produce sufficient liquid tension to cause pit aspiration. Hart and Thomas reported that aspiration is always toward the closed-water system side of the membrane. Aspiration can also occur in zones characterized by air/water-filled lumens on both sides of pit pairs. In this case, evaporation from one or both lumens can create a pressure differential, causing deflection of pit membranes toward the lower air pressure side. It is interesting to note that it is possible to avoid pit aspiration during drying by impregnating wood with a low-surface-tension liquid such as alcohol.

Although the kinds of situations described above can occur in functioning sapwood, pit aspiration is much more common during the transition from sapwood to heartwood. This at least partially explains why the sapwood of some species, such as Douglas-fir, is readily penetrated by treating chemicals, whereas little or no penetration can be achieved in the heartwood. The impenetrability of heartwood may also result from a buildup of encrusting materials in the margo, resulting in pit membranes that are progressively less porous over time.

Pit aspiration develops more frequently in earlywood (characterized by large pits) than in latewood (characterized by small bordered pits). This explains why the end-grain penetration of treating chemicals often extends for some distance along the grain in latewood zones, whereas adjacent earlywood zones are free from chemicals.

Other Longitudinal Cells

In some softwood species, such as fir and hemlock, the longitudinal tracheid is commonly the only kind of longitudinal cell present. In other species, including redwood, the cedars (genera *Juniperus*, *Libocedrus*, *Chamaecyparis*, and *Thuja*), and the pines, several other kinds of cells make up minor portions of the volume.

Longitudinal Parenchyma

A small portion of the volume of some softwoods is composed of longitudinally oriented parenchyma cells. When mature, these cells have the same general shape as longitudinal tracheids, although they often subdivide a number of times along their length prior to forming secondary walls. The result is that mature parenchyma usually occur as longitudinal strands of short cells butted end-to-end in series. The thin-walled and simple pitted parenchyma account for as much as 1 or 2% of the volume of some softwoods.

Epithelium

Structures known as resin canals are found in certain softwood species. They are consistently found in the genera *Pinus* (pines), *Picea* (spruce), *Larix* (larch), and *Pseudotsuga menziesii* (Douglas-fir); this is another feature that assists in the identification of softwoods. Normal longitudinal resin canals are always accompanied by horizontal canals, which occur in some of the rays.

A *resin* canal is an intercellular space surrounded by specialized parenchyma cells that secrete resin into the canal. This resin is believed to play an important role in the healing of damaged tissue and in repelling attack by insects or other would-be invaders. A cut through the inner bark of pine, for example, begins a flow of resin to the wound area and may even be accompanied by the production of new resin-producing cells near the wound (Mirov and Hasbrouck 1976, p. 36). Kozlowski and Pallardy (1997) note that the capacity to secrete resin decreases over time, but that wounding, insect infestation, or bacterial infection can increase resin production in existing resin-producing cells and stimulate the development of new resin-producing cells in the vicinity of the wound or infection.

Longitudinal resin canals arise following cell formation in the cambium and are formed through separation at the adjoining corners of several undifferentiated longitudinal cells. The cells that surround the space then fail to develop like the surrounding tracheids, form a series of crosswalls, and remain thereafter thin walled. These units are the resin-secreting cells and are known as *epithelial cells* (Esau 1965). Radially oriented resin canals form similarly. Several longitudinal resin canals with surrounding epithelium are shown in the transverse view of ponderosa pine in Figure 4.6. Epithelial cells may be thin walled (*Pinus*) or thick walled (*Larix, Picea, Pseudotsuga*).

Production of resin canals in response to injury or other traumatic events is not restricted to those genera that produce resin canals of the normal type. Although the tendency to produce traumatic canals is greater in some woods than in others, resin canals of this type may occasionally occur in almost any of the softwoods. Hemlock (*Tsuga*), redwood (*Sequoia*), and true fir (*Abies*) are examples of woods that do not exhibit canals of the normal type but commonly form resin canals in response to injury.

Because the presence or absence of resin canals is used in wood identification, occasional development in some species could hamper use of this feature for this purpose if it were not possible to distinguish between normal and traumatic canals. Fortunately, it is relatively easy to distinguish the difference. Although traumatic canals form in exactly the same way as normal ones (that is, through postcambial separation of the middle lamella of adjacent cells), they are usually larger and often occur in tangential bands at the start of the growth ring. Traumatic canals, moreover, rarely occur in both longitudinal and radial orientation in the same piece of wood. Traumatic resin canals are pictured in Figure 4.7.

Rays

Uniformly narrow rays characterize softwoods except where horizontal resin canals are present. Viewed tangentially, softwood rays are from one to many cells in height but are usually only one cell wide (*uniseriate*). Rays of redwood are typically two cells in width

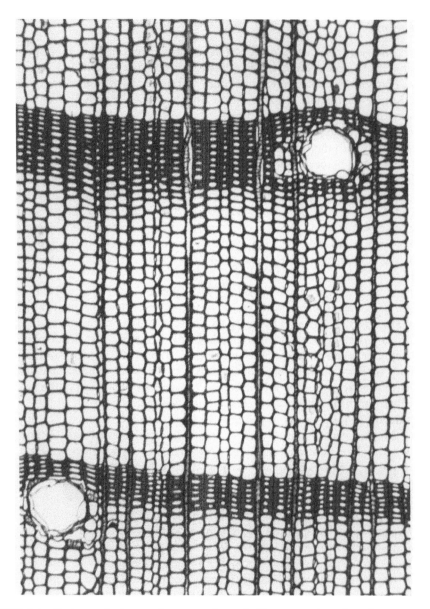

FIGURE 4.6.   Resin canals characterize some softwoods. Transverse view of ponderosa pine (*Pinus ponderosa*). ×85. Source: Courtesy of Ripon Microslides Laboratory.

(*biseriate*). In Figure 4.8 a typical softwood ray is shown in contact with the radial row of cells depicted earlier.

The cells composing softwood rays may be either ray parenchyma or ray tracheids. *Ray tracheids* are similar to longitudinal tracheids in that they have thick cell walls and bordered pits. In the hard pines (ponderosa, lodgepole, jack, red, and southern) ray tracheids form secondary walls that are locally thickened in the vicinity of pits. The ridge-like thickenings look much like teeth extending into the lumen. Such tracheids are called

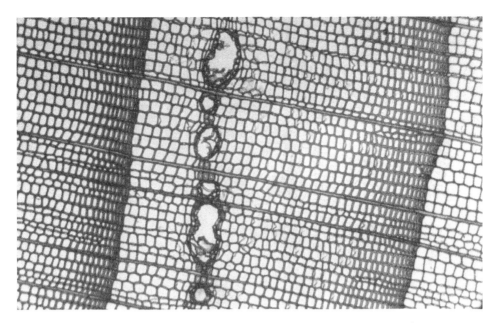

FIGURE 4.7.   Traumatic resin canals. Transverse view of incense cedar (*Calocedrus decurrens*). ×80.

FIGURE 4.8.   Ray orientation. Contacting radial faces of longitudinal tracheids.

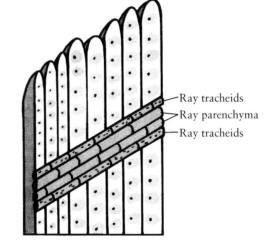

dentate ray tracheids (Figure 4.9). Ray cells of the parenchyma type, on the other hand, may be either thin or thick walled. Very thin-walled *ray parenchyma* commonly form no pitting, whereas simple pits typically perforate the thicker-walled variety of ray parenchyma cells.

An individual ray may be composed entirely of parenchyma, entirely of tracheids, or of both ray parenchyma and ray tracheids. A close-up of a uniseriate softwood ray (Figure 4.10a) shows it to be composed of both ray parenchyma and ray tracheids. Another uniseriate ray might be entirely ray parenchyma or ray tracheids. Uniseriate rays

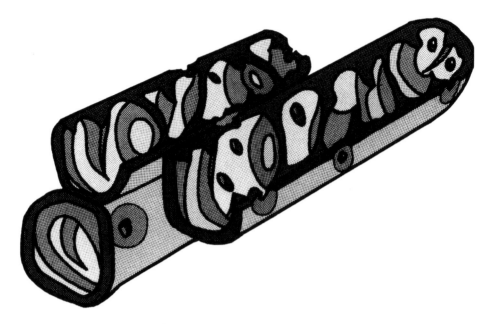

FIGURE 4.9. Dentate ray tracheids. Cutaway section showing ridge-like thickenings of secondary cell wall. Source: Adapted from Howard and Manwiller (1969).

that are constructed entirely of ray parenchyma or entirely of ray tracheids are termed homogeneous. A heterogeneous uniseriate ray contains both ray tracheids and ray parenchyma.

In Figure 4.10b, a special kind of ray is shown that occurs only in some woods. Thick-walled and dentate tracheids are seen on the upper and lower margins, bracketing rows of ray parenchyma. A resin canal and surrounding epithelium can be seen in the center. When a ray contains a resin canal, it is known as a *fusiform ray*. As shown in Figure 4.10b, a fusiform ray normally contains both ray tracheids and ray parenchyma in addition to epithelial cells. Depending upon the species, the ray tracheids may or may not be of the dentate type. Fusiform rays are not uniseriate, appearing swollen at the center where the resin canal and accompanying epithelial cells occur. In woods that contain them, only about one ray in 20 has the fusiform configuration.

Numerous pits interconnect ray cells; there are bordered pairs between ray tracheids, simple pairs between ray parenchyma (of the thickened variety), and half-bordered pairs between ray parenchyma and ray tracheids. Pitting on the ray sidewalls connects pits to longitudinal cells.

Crossfield pitting occurs at the intersections of longitudinal tracheids and ray parenchyma, and the half-bordered pits are of unique form. Crossfield pit apertures, when viewed radially, vary in shape as follows: lemon drops, cat's eyes, extended slits, and an expansive window-like form (Figure 4.11; see also Figures 4.2 and 4.3). The type, size, and number of crossfield pits are fairly consistent within a species; thus, this feature is quite useful in determining the identity of softwood timbers. A microscope is needed to view crossfield pitting.

(a)   (b)

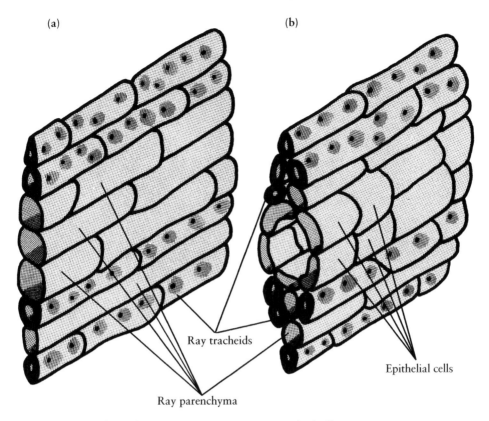

Ray tracheids

Epithelial cells

Ray parenchyma

FIGURE 4.10.   Softwood ray structure: (a) uniseriate ray; (b) fusiform ray.

Wood Identification

For those interested in learning how to identify the wood of various species, several texts are available that deal with this subject; see Kribs (1968), Edlin (1969), Jane et al. (1970), Core et al. (1979), Desch (1981), Hoadley (1990), and http://insidewood.lib.ncsu.edu/search.

Summary

A convenient way of reviewing the structure of softwoods is to study a three-dimensional block of a species that exhibits all of the features listed earlier. A carefully drawn representation of southern pine (Figure 4.12) serves this purpose.

Transverse Surface (I)

Examination of the transverse surface of Figure 4.12 reveals numerous longitudinal tracheids in cross-section. Portions of two growth rings are shown. Latewood tracheids of one annual growth layer lie to the left (1); these are followed to the immediate right by earlywood (2–2a) and latewood (3–3a) of the succeeding annual ring. Transition in zone 3–3a is relatively abrupt. A row of tracheids visible at 4–4a is traceable to an anticlinal

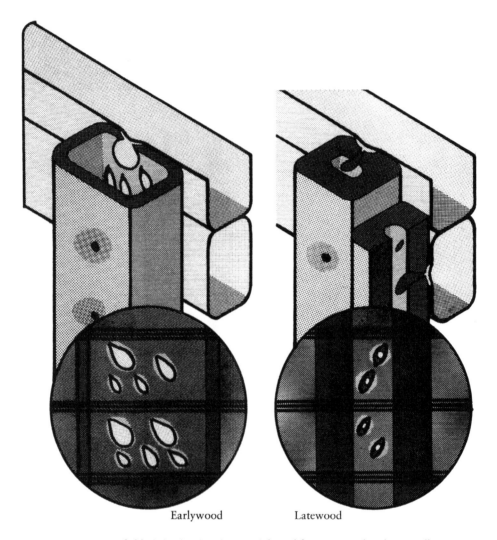

Earlywood   Latewood

FIGURE 4.11.   Crossfield pitting in pine. Source: Adapted from Howard and Manwiller (1969).

division of a fusiform initial that occurred when the position indicated by 4 marked the outer extremity of the xylem and the position of the tangentially oriented cambium. Bordered pits that have been sectioned transversely are pictured at A–D.

A resin canal, pictured at 5, is surrounded by short, thin-walled epithelial cells (E). Thin-walled longitudinal parenchyma (F) lie to the outside of the epithelial cells. A transversely sectioned row of ray tracheids can be seen at 6–6a.

Radial Surface (II)

Numerous conical-shaped bordered pits dot the radial surface of longitudinal tracheids, marking locations of matching pits in adjacent rows of tracheids. Earlywood tracheids are bluntly tapered in this view, whereas end walls of the narrower latewood are more angular.

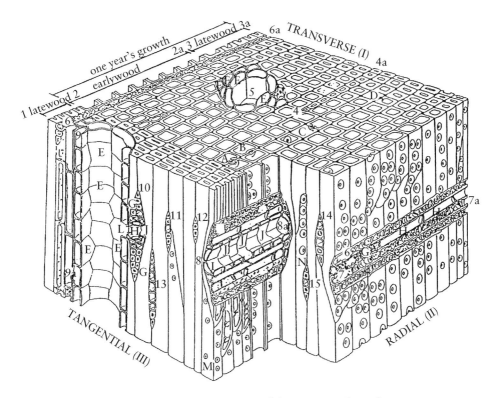

FIGURE 4.12.   Three-dimensional representation of distinct-ring softwood.

A longitudinally sectioned uniseriate and heterogeneous ray composed of both ray tracheids (G) and ray parenchyma (H) is shown at 7–7a. A sectioned fusiform ray is shown at 8–8a. This ray is built around a resin canal (I) and contains short, bricklike epithelial cells (J), ray parenchyma, and ray tracheids. An unsectioned ray tracheid is shown at (K).

Tangential Surface (III)

To the extreme left of the tangential surface a septated longitudinal parenchyma cell is visible (9–9a). This is adjacent to a large longitudinal resin canal, which is surrounded by epithelium (E). An opening can be seen connecting transverse and longitudinal resin canals (L).

A fusiform ray at 10 exhibits ray tracheids (G), ray parenchyma (H), ray epithelium (J), and a large-diameter transverse resin canal (I). Uniseriate heterogeneous rays are shown at 11, 13, 14, and 15. A homogeneous type ray, composed of ray tracheids, is seen at 12.

Longitudinal tracheids appear sharply tapered tangentially rather than rounded as in radial view. Pitting is sparse but can be seen in occasional tracheids at M (latewood) and N (earlywood).

References and Supplemental Reading

Core, H.A., Côté, W.A. Jr., and Day, A.C. (1979). *Wood Structure and Identification*, 2e. Syracuse, New York: Syracuse University Press.

Desch, H.E. (1981). *Timber, its Structure, Properties, and Utilization*, 6e. Forest Grove, OR: Timber Press.

Edlin, A.L. (1969). *What Wood Is That?* New York: Viking Press.

Esau, K. (1965). *Plant Anatomy*, 2e, 62–254. New York: Wiley.

Hart, C.A. and Thomas, R.J. (1967). Mechanism of bordered pit aspiration as caused by capillarity. *For. Prod. J.* 17 (11): 61–68.

Hoadley, R.B. (1990). *Identifying Wood*. New Town, CT: The Taunton Press.

Howard, E.T. and Manwiller, F.G. (1969). Anatomical characteristics of southern pine stemwood. *Wood Sci.* 2 (2): 77–86.

Jane, F.W., Wilson, K., and White, D.J.B. (1970). *The Structure of Wood*. London: Adam & Charles Black.

Kozlowski, T.T. and Pallardy, S.G. (1997). *Physiology of Woody Plants*, 2e, 181–183. San Diego: Academic Press.

Kribs, D.A. (1968). *Commercial Foreign Woods on the American Market*. New York: Dover.

Meylan, D. and Butterfield, B.G. (1972). *Three Dimensional Structure of Wood: A Scanning Electron Microscope Study*. Syracuse, NY: Syracuse University Press.

Mirov, N.T. and Hasbrouck, J. (1976). *The Story of Pines*. Bloomington: Indiana University Press.

Panshin, A.J. and de Zeeuw, C. (1980). *Textbook of Wood Technology*, 4e. New York: McGraw-Hill.

Tsoumis, G. (1964). Light and electron microscope evidence on the structure of the membrane of bordered pits in tracheids of conifers. In: *Cellular Ultrastructure of Woody Plants* (ed. W.A. Côté Jr.), 305–307. Syracuse, NY: Syracuse University Press.

5

Hardwood Structure

The wood formed by hardwoods is very different than that produced by softwoods. Softwoods are composed of only a few cell types, have a relatively simple structure, and frequently lack a distinctive appearance. Hardwoods, on the other hand, are composed of widely varying proportions of markedly different kinds of cells and are thus often uniquely and even spectacularly figured. Because of the unique figure possessed by many hardwood species, their woods are widely used for furniture, paneling, and other decorative purposes.

Differences Between Hardwood and Softwood Xylem

It was mentioned in the introduction that softwoods are uniform in structure whereas hardwood structure is complex. This and other differences are summarized in the following:

1. Softwoods are composed of only a few significant cell types – hardwoods of many (Figure 5.1). Long cells known as *longitudinal tracheids* comprise 90–95% of the volume of softwoods. *Ray cells* (either ray tracheids or ray parenchyma) constitute the remainder of softwood xylem. Although a few other types of cells may occur, they make up an insignificant part of the volume of softwoods. Hardwoods are composed of at least four major kinds of cells (Table 5.1); each of these may constitute a significant portion of hardwood volume.
2. Only hardwoods contain vessels, structures composed of *vessel elements*. The nature of vessel elements is discussed in the next section.
3. Hardwood ray widths vary within and between species. They are often wider than the (mostly) uniseriate rays found in softwoods (Figure 5.2). Except for fusiform rays, softwood rays are one cell (or occasionally two cells) in width when viewed tangentially. Collectively, ray cells comprise about 5–7% of total softwood volume. Hardwood rays range in width from 1 cell to 30 or more in some species. These rays can constitute more than 30% of the volume of hardwood xylem, the average being around 17%.

Forest Products and Wood Science: An Introduction, Seventh Edition. Rubin Shmulsky and P. David Jones.
© 2019 John Wiley & Sons Ltd. Published 2019 by John Wiley & Sons Ltd.
Companion website: www.wiley.com/go/shmulsky

(a)

FIGURE 5.1.   Principal cell types of hardwoods and softwoods. (a) Softwood: Sitka spruce (*Picea sitchensis*). ×75. Source: Scanning electron micrograph by Josefina Gonzales, Forintek Canada Corp., Vancouver, BC.

4. Straight radial rows of cells characterize softwoods, but generally not hardwoods (Figure 5.3). Softwood cells are aligned in straight radial rows in parallel form, with straight spoke-like rays; each row of cells is formed by a single fusiform initial in the cambium. Hardwood rays are seldom aligned in straight radial rows, nor are other hardwood elements. Distortion from a purely radial orientation occurs in the vicinity of large vessel elements.

5. Hardwood fibers are considerably shorter than softwood tracheids (Figure 5.4). The softwood tracheids average 3–4 mm in length; hardwood fibers, in contrast, have an average length of less than 1 mm. This fact explains why softwood tracheids are often preferred as raw material for paper manufacture. Fiber length is an important determinant of paper strength; thus, long fibers are a necessary ingredient of kraft paper used for unbleached paper products such as corrugated cartons and grocery bags.

(b)

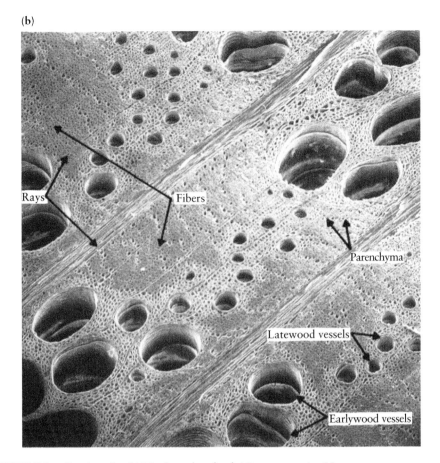

Rays

Fibers

Parenchyma

Latewood vessels

Earlywood vessels

FIGURE 5.1.   *Continued.*   (b) Hardwood: red oak (*Quercus* spp.). ×55.

TABLE 5.1.   Major hardwood cell types.

Cell type	Proportion of xylem volume accounted for by cell type[a] (%)
Fiber tracheid[b]	15–60
Vessel element	20–60
Longitudinal parenchyma	0–24
Ray parenchyma	5–30

[a] Within a species, the relative proportion of various cell type is quite consistent. Between species and species groups (genera), the proportions of various cells can vary widely.
[b] Included in this category are several kinds of cells: variations of true fiber tracheids and transition elements between fibers and vessel elements or between fibers and longitudinal parenchyma.

It is important to note that the terms *hardwood* and *softwood* have no relevance to the actual hardness or strength of the wood produced. Many softwoods produce wood that is harder and more dense than wood produced by some hardwoods.

(a)

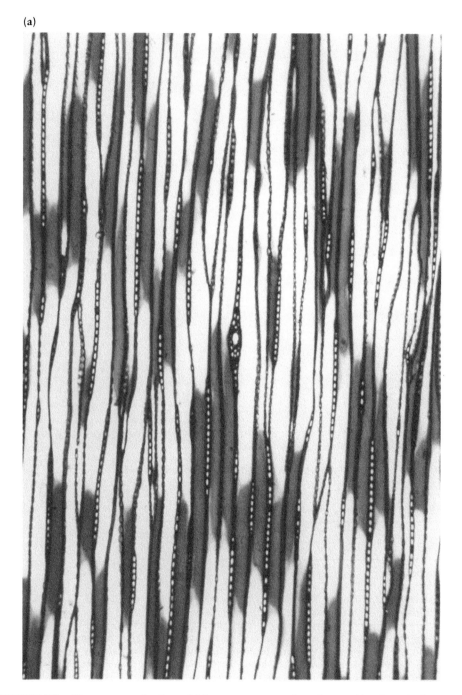

FIGURE 5.2.   Narrow rays of softwood (a) versus narrow to broad rays of hardwoods (b, c).
Tangential view. Source: Courtesy of Ripon Microslides Laboratory. (a) Western larch (*Larix occidentalis*). ×85. Source: Courtesy of Ripon Microslides.

(b)

FIGURE 5.2.   *Continued.*   (b) Quaking aspen (*Populus tremuloides*). ×80.

(c)

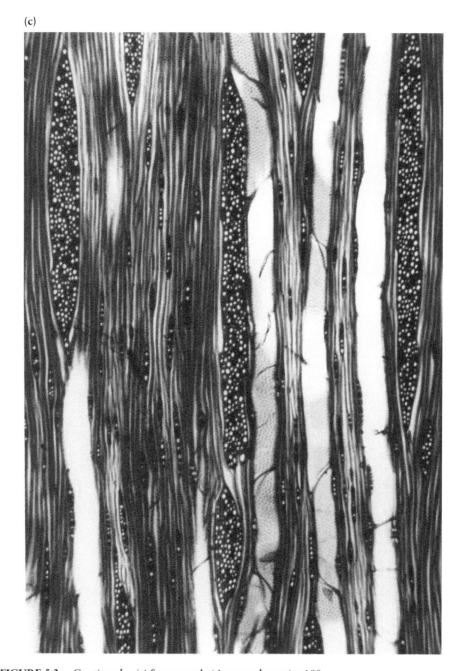

FIGURE 5.2.   *Continued.*   (c) Sugar maple (*Acer saccharum*). ×100.

(a)

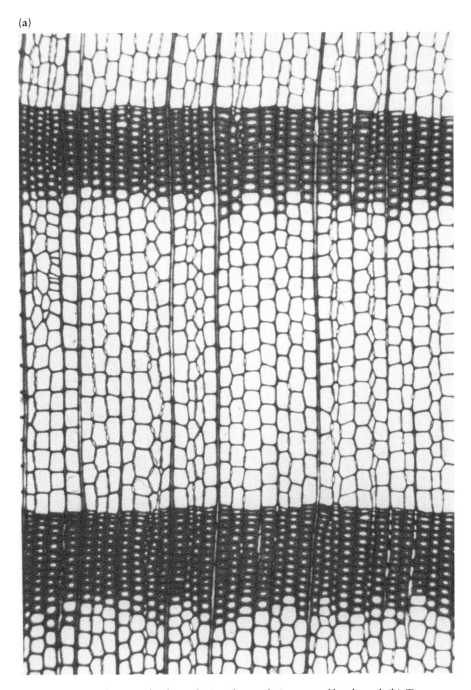

FIGURE 5.3.   Straight rays of softwood (a) and meandering rays of hardwood. (b). Transverse view. (a) Western larch (*Larix occidentalis*). ×85.

(b)

FIGURE 5.3.   *Continued.*   (b) White oak (*Quercus alba*). ×55.

Longitudinal Cells

Although longitudinal cells of hardwoods vary considerably in size and general configuration, all of these different cell types can be produced by a single fusiform initial in the cambium. Newly produced cells appear quite similar. The differences between types develop during the process of cell maturation.

Vessel Elements – Unique Cells of Hardwoods

Several differences exist between hardwood and softwood xylem, but the fundamental anatomical difference is that hardwoods contain specialized conducting cells called *vessel elements*. This cell type is found in virtually all hardwoods but rarely in softwoods. The wood of a few dicotyledons does not contain vessels; however, the number and economic importance of species exhibiting this feature are small. Vessel elements are generally much larger in diameter than other types of longitudinal cells. Figure 5.4 compares the size and shape of a softwood tracheid, a typical hardwood fiber, and a hardwood vessel element. Note that vessel elements are shorter than both hardwood and softwood fibers but larger in diameter. The short length of vessel elements is traceable to the fact that they often do not grow in length during the maturation process and may become even shorter than the cambial initials from which they were produced (Jane et al. 1970). Normally, a number of vessel elements link end to end along the grain to form long tube-like structures known as *vessels*. Such vessels are seldom arranged in a precise parallel and vertical alignment; instead, within a growth ring, vessels form a network with considerable tangential variation from a straight vertical orientation. This arrangement ensures that each branch of the crown receives water from many different roots, providing a safety feature against crown damage from the loss of one or more roots (Zimmermann 1983).

Vessel Arrangement.   Because of their large diameter, vessels often appear as holes when viewed in cross-section; in this view, they are often referred to as *pores*. Both the size and arrangement of pores are used to classify hardwoods for purposes of identification. Figure 5.5a is a drawing of a magnified cross-section of a hardwood. Only the vessels and rays are illustrated. Vessels of large diameter are concentrated in the earlywood, with vessels of much smaller diameter in the latewood. This type of wood is called *ring-porous* because the earlywood vessels form a visible ring in a tree cross-section. Figure 5.5b shows a hardwood that has pores of uniform size distributed fairly evenly across the growth ring. This wood is classified as *diffuse-porous*. The majority of hardwoods are diffuse-porous, but in northern temperate regions some of the most valuable woods, such as oak (*Quercus* spp.), ash (*Fraxinus* spp. [at the time of this edition, becoming increasingly rare because of the emerald ash borer]), and pecan (*Carya illinoensis*), are ring-porous. When hardwoods are sawn into lumber, the lengthwise sectioning of vessels results in a distinctive scratchlike pattern on radial and tangential surfaces. Sectioning of large earlywood vessels in ring-porous woods forms a very deep and sometimes spectacular pattern of vessel scratches (vessel lines) that is interrupted by latewood regions that have little texture. Photographs of ring- and diffuse-porous woods are shown in Chapter 2 (see Figures 2.5 and 2.6). See Figure 5.19 for an artist's conception of a three-dimensional diffuse-porous hardwood.

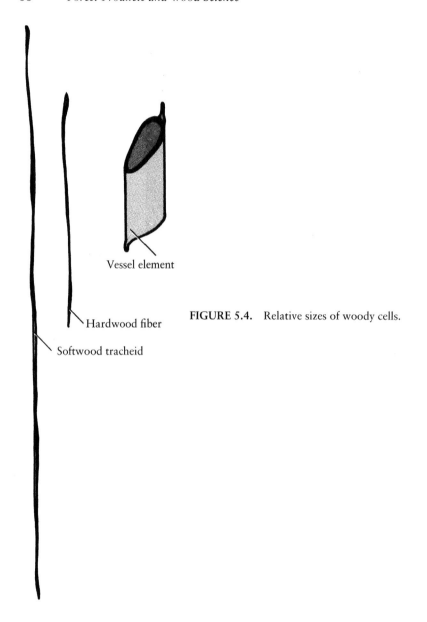

Vessel element

Hardwood fiber

Softwood tracheid

FIGURE 5.4.   Relative sizes of woody cells.

The lack of radial alignment of cells in hardwoods has been mentioned. Recall that all types of longitudinal cells arise from the same fusiform initial in the cambium and that all longitudinal cells are quite similar in size and shape immediately after formation. Because nothing occurs to disrupt alignment, newly formed hardwood xylem cells tend to be arranged in neat radial files corresponding to the initials that produced them. During the maturation process, however, cells begin to change, and eventually assume the characteristics of the mature units. In the case of vessel elements, one of these character-istics is a large diameter. Thus, cells that will mature to become vessels begin marked diameter growth, expanding from 2 to 50 times their original diameter. Other cells

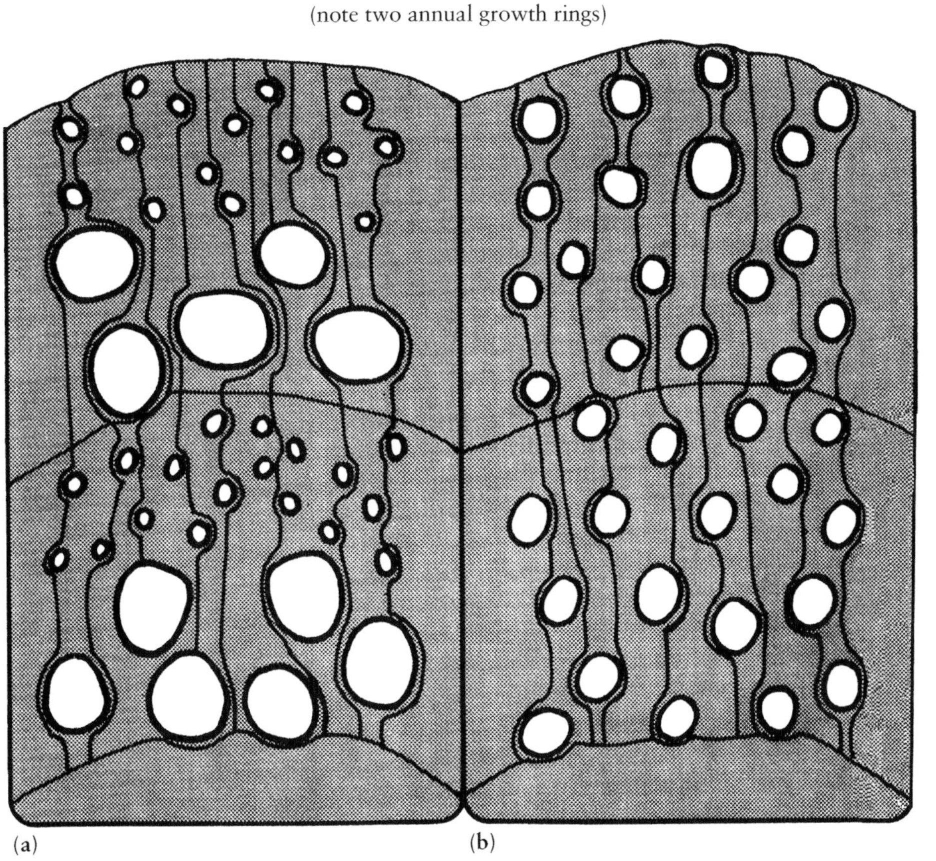

(note two annual growth rings)

(a)   (b)

FIGURE 5.5.   (a) Ring-porous hardwood. (b) Diffuse-porous hardwood.

expand little in cross-section. This growth in diameter of vessel elements pushes cells out of radial alignment. Follow the path of rays around the large earlywood vessels pictured in Figure 5.3; it is evident that the meandering ray pattern is caused by vessel growth.

End-to-end Connection of Vessel Elements.   Vessels are uniquely suited to serve as avenues of fluid conduction. Relatively small and membrane-divided pit pairs connect other cells, such as fiber tracheids, end to end. Common end walls of longitudinally linked vessel elements are, however, perforated by unrestricted holes. To facilitate discussion about this feature, names are given to the common vessel element end walls (perforation plates) and the holes in them (*perforations*).

Perforations develop near the end of the cell maturation process. Certain enzymes contained in the protoplast of developing vessel elements (such as cellulase) are apparently responsible for this dissolving of portions of the perforation plates (Roelofsen 1959). Some rearrangement of cell wall material may also be involved in formation of perforations (Frey-Wyssling 1959). It is interesting to note that perforations do not develop in a random fashion; instead, they form in one of several definite patterns, as depicted in Figure 5.6. A photograph of scalariform perforations is shown in Figure 5.7.

Simple   Scalariform   Foraminate

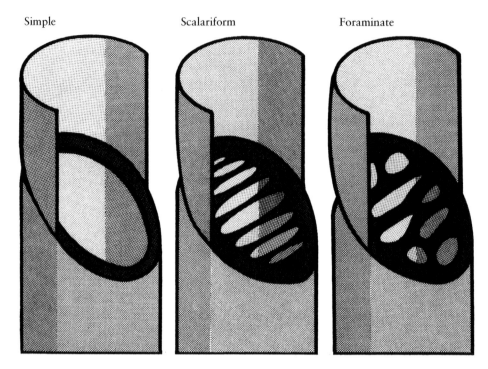

FIGURE 5.6. Types of vessel perforation plates.

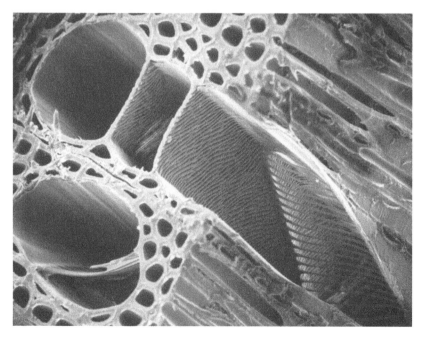

FIGURE 5.7. Scalariform perforation between vessel elements of white birch (*Betula papyrifera*). ×400. Source: Scanning electron micrograph by John Crist and Ron Teclaw.

Within a given species, the pattern of perforations is commonly the same in all perforation plates. Because of this, the nature of vessel perforations is often useful as an aid in the identification of hardwood timbers. Perforation plates invariably slope at an angle toward the radial plane. This surface should be examined microscopically to determine the type of perforation.

Side-to-side Connection of Vessels.   Lateral communication from vessel to vessel is provided by numerous pairs of bordered pits. Closely packed bordered pits are depicted in Figure 5.8. As is the case with perforation plates, the shape and arrangement of vessel-to-vessel pitting is often consistent within a given species and can be of assistance in wood identification. Photographs of adjacent vessels are presented in Figures 5.9 and 5.10, in which intervessel pitting is clearly visible.

Connections between Vessels and Other Cells.   Vessels often occur adjacent to fiber tracheids, longitudinal and ray parenchyma, or other kinds of cells. Although fiber tracheids and vessels are sometimes not linked by pitting, other kinds of cells typically form pits where they contact vessel elements.

Tyloses – Their Significance.   *Tyloses* are outgrowths of parenchyma cells into the hollow lumens of vessels. They commonly form in many hardwoods as a result of wounding and effectively act to prevent water loss from the area around damaged tissue (Zimmermann 1983). Tyloses also form in a number of species during the transition from sapwood to heartwood and may also develop as a result of infection from fungi or bacteria (Kozlowski and Pallardy 1997).

Just prior to tylosis formation, enzymatic action partially destroys membranes in vessel-to-parenchyma pit pairs. At about the same time, the cytoplasm of the parenchyma cells begins to expand with protrusion of the parenchyma cell membrane through pit pairs into the vessel lumen; this protrusion is called a tylosis. Several studies have

(a)   (b)   (c)

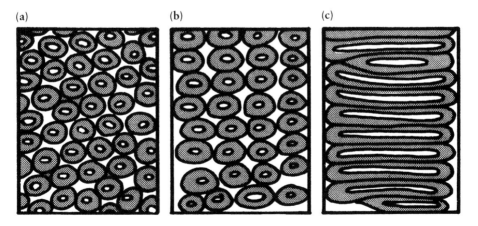

FIGURE 5.8.   Vessel-to-vessel pitting arrangements. (a) Alternate pitting. (b) Opposite pitting. (c) Sclariform pitting.

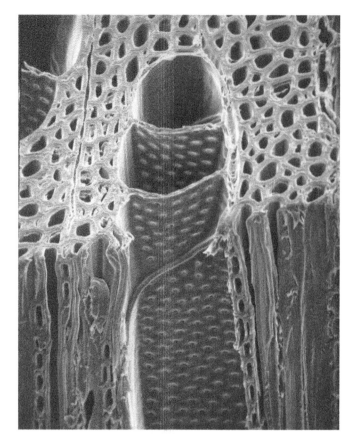

FIGURE 5.9.   Abundant vessel-to-vessel pits in alternate arrangement. Tangential view of *Populus* spp. ×450. Source: Scanning electron micrograph by John Crist and Ron Teclaw.

indicated that a special membrane-like meristematic layer forms in parenchyma cells, completely encasing the cytoplasm, prior to tylosis formation. This layer, known as the *protective layer*, is believed to actually form tyloses (Meyer and Côté 1968; Schmidt 1965). The membrane forming the tylosis may remain quite thin, or the walls may thicken in much the same way that they do in developing cells. Pits may even form where one tylosis contacts another (Foster 1967).

Tyloses are significant in that they partially or often completely block the vessels in which they occur, a situation that can be either detrimental or beneficial depending upon the use to which the wood is put. The existence of tyloses in the heartwood vessels of white oak, and the relative lack of them in red oak (Figure 5.11), is the reason white oak is preferred in the manufacture of barrels, casks, and tanks for the storage of liquids. White oak heartwood, with its tightly plugged vessels, is almost universally used in the manufacture of whiskey, scotch, bourbon, and wine barrels, for example, whereas the open-vesseled red oak is avoided for this use. In contrast to this beneficial feature of tyloses, wood in which they are well developed may be difficult to dry or to impregnate with decay-preventive or stabilizing chemicals. A radial view of thick-walled tyloses in hickory is shown in Figure 5.12.

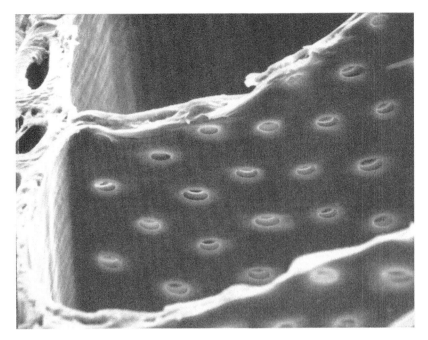

FIGURE 5.10. Highly magnified bordered pits in adjacent vessels in alternate arrangement. *Populus* spp. ×1750. Source: Scanning electron micrograph by John Crist and Ron Teclaw.

(a)

FIGURE 5.11. Tyloses fill earlywood vessels of (a) white oak (×16) not (b) red oak (×16). Source: Courtesy of the Department of Wood and Paper Science, North Carolina State University.

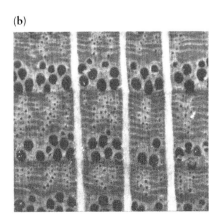

(b)

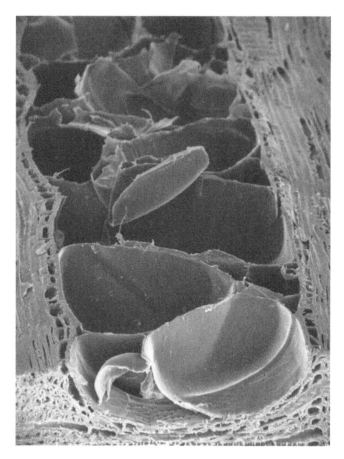

FIGURE 5.12.   Tyloses in vessel. Hickory (*Carya* spp.). ×170. Source: Scanning electron micrograph by John Crist and Ron Teclaw.

An interesting discovery (Dute et al. 1999) is that tyloses form in softwood tracheids in the vicinity of abscission scars following separation of needles from loblolly pine stems. As in hardwoods, tyloses arise from parenchyma cells adjacent to longitudinal tracheids.

Fibers

The term *fiber* is often used in a general way to refer to all wood cells isolated in pulping processes. However, in the context of wood morphology, the term *fiber* refers to a specific cell type. Thus fibers, or *fiber tracheids* as they are more properly called, are long, tapered, and usually thick-walled cells of hardwood xylem. A casual look suggests a great similarity to the longitudinal tracheids of softwoods, but closer examination reveals several significant differences.

As noted earlier, hardwood fibers are considerably shorter than softwood tracheids. Hardwood fibers also tend to be rounded in cross-section as compared with the nearly rectangular shape of softwood tracheids (Figure 5.13). However, fibers are sometimes

(a)

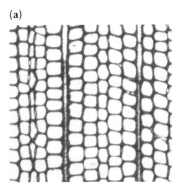

FIGURE 5.13. Softwood tracheids (a) and hardwood fibers (b) in transverse view.

(b)

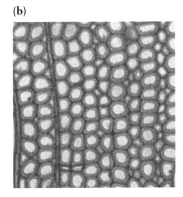

flattened radially in last-formed latewood in much the same way that latewood tracheids are in softwoods. Fibers are also characteristically very thick walled and have bordered pits with less-developed borders than softwood tracheids (Esau 1965, p. 239).

Although hardwood fibers and softwood tracheids are similar, the function of the fiber is more specialized. Longitudinal tracheids of softwoods serve as primary avenues of conduction while also being almost totally responsible for the strength of the wood of which they are a part. The presence of a high proportion of thin-walled earlywood tracheids is invariably related to low wood strength. The situation is somewhat different in hardwoods, where two kinds of longitudinal cells – fibers and vessel elements – are common. Most conduction occurs through the specialized vessels, leaving the thick-walled fibers the primary function of mechanical support. Fibers are most highly specialized as supporting elements in those woods that have the most specialized vessel members (Esau 1965, p. 239). The density, and thus strength, of hardwoods is therefore generally related to the portion of wood volume occupied by fibers relative to that accounted for by vessels. As a general rule, the higher the proportion of thick-walled fibers is, the higher the strength.

The walls of fiber tracheids are marked by pits of the bordered type. Fiber-to-fiber pit pairs are normally bordered, whereas fiber-to-parenchyma pitting is typically half-bordered. A variation of the fiber, known as a *libriform fiber,* is marked by simple, rather than bordered, pits. Libriform fibers occur in considerable numbers in some species. Fibers and vessels are seldom connected by pit pairs.

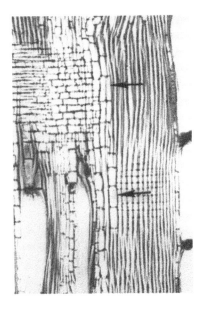

FIGURE 5.14.   Longitudinal strand parenchyma. Radial view of Honduras mahogany (*Swietenia macrophylla*). ×85; Strand parenchyma indicated by arrows. Source: Courtesy of Ripon Microslides Laboratory.

Longitudinal Parenchyma

Parenchyma cells are thin-walled storage units. In hardwoods, such cells occur in the form of long, tapered longitudinal cells, short, brick-shaped epithelium around gum canals (in only a few species), and ray cells. The longitudinal form of parenchyma is often divided into a number of smaller cells through the formation of crosswalls during the process of cell maturation (Figure 5.14).

Parenchyma cells on occasion are thin walled to the point that no secondary wall forms. Because a *pit* is defined as a gap in the secondary wall, a cell with an unthickened wall is therefore unpitted. Pits do form in parenchyma cells that form thickened walls, and in accordance with rules set forth in Chapter 3, simple pit pairs connect cells of the parenchyma type. Pitting "rules" are often broken where thickened parenchyma contacts vessels or fibers; in this case, the pit pairs formed are usually half-bordered but may be of the simple or bordered type (Esau 1965, p. 239).

Whereas *longitudinal parenchyma* is relatively rare in softwood species (no more than 1–2% of the volume of those woods in which it does occur), the longitudinal form of parenchyma is often quite significant in hardwoods. Certain species of hardwoods contain no longitudinal parenchyma. Some domestic hardwoods may, however, have up to 24% of their volume made up of longitudinal parenchyma cells; this figure may even exceed 50% for a few tropical hardwoods (Panshin and de Zeeuw 1980). In these woods, the longitudinal parenchyma is commonly arranged into definite and unique patterns that are readily visible in a transverse section (Figure 5.15). Because both the proportion and arrangement of longitudinal parenchyma are genetically reproduced, this kind of cell is often of value in the identification of hardwood timbers.

Gum canals occur in a few hardwoods and are similar to resin canals of softwoods. The hardwood canals are sometimes lined with parenchyma-type epithelial cells.

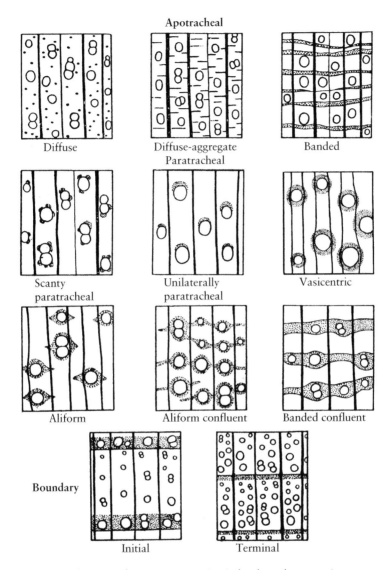

FIGURE 5.15. Parenchyma configurations occurring in hardwoods as seen in transverse view. Source: From Jane et al. (1970).

Other Kinds of Longitudinal Cells

In addition to vessels, fibers, and longitudinal parenchyma, other kinds of longitudinal cells occur in a few hardwoods, contributing to the variable nature of this group of woods. These other cells are mostly transition elements between major cell types and as such have features typical of each kind of cell to which they are related.

An example of a transition element is a *vascular tracheid*; this cell has a shape like a vessel element, but it lacks perforations in the end walls, having instead bordered pits in this location similar to those found in fibers. Another kind of cell known as a vasicentric tracheid looks much like parenchyma in cross-section, yet it is covered with numerous bordered pits.

Rays

As listed in the summary of hardwood–softwood differences, hardwood rays range in width tangentially from 1 to 30 or more cells. Softwood rays in comparison are generally one or, rarely, two cells in width. Also, unlike softwoods, the cells of hardwood rays are all of the parenchyma type (although two distinct types of ray parenchyma are formed).

Ray Size

Hardwoods characterized by very large rays, such as oak, exhibit distinctive ray patterns on both tangential and radial faces (Figure 5.16); such rays often add to a wood's esthetic appeal. Note that a highly magnified tangential view of this wood (Figure 5.16d) reveals

(a)

FIGURE 5.16.   Ray patterns in white oak (*Quercus alba*). (a) Radial surface (unmagnified).
Source: Courtesy of Ripon Microslides Laboratory.

(b)

FIGURE 5.16. *Continued.*   (b) Radial surface (×85).

(c)

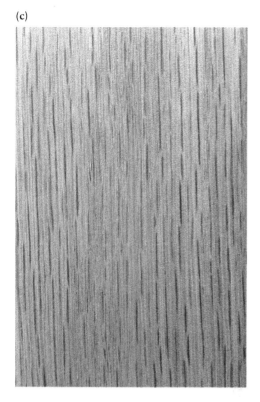

FIGURE 5.16. *Continued.* (c) Tangential surface (unmagnified).

numerous narrow rays in addition to the wide ones; the rays seen without magnification in Figure 5.16a and c represent only the largest of these.

Not all hardwoods exhibit wide rays. Woods such as aspen (*Populus tremuloides*) and cottonwood (*Populus deltoides*) have rays that are of the uniseriate type only (see Figure 5.2b). These woods lack a visible ray pattern unless viewed under high magnification.

Types of Ray Cells

Although all ray cells are of the parenchyma type, there are, nonetheless, different types of hardwood ray cells. The difference is in cell shape or configuration.

The ray parenchyma cells of hardwoods are sometimes almost square when viewed radially, but more commonly such cells have a rectangular shape. In most woods, these rectangular ray cells are arranged such that the long dimension is perpendicular to the axes of longitudinal cells (Figure 5.16b). Because ray cells arranged in this way appear to be lying down, they are said to be *procumbent*. In some hardwood species, part of the rectangularly shaped ray cells appear to stand on end with their long axes parallel to the grain direction (Figure 5.17); these cells are logically called *upright ray cells*. Upright or *square ray cells* usually occur along the upper and lower margins of rays.

The significance of ray cell configuration is that this feature can be used in wood identification because upright and square ray cells occur as a constant feature in only some species. An example is provided by cottonwood and willow – two easily confused

(d)

FIGURE 5.16.   *Continued.*   (d) Tangential surface (×85).

species. Positive identification is based upon the fact that rays of willow consistently have upright cells along the margins, whereas cottonwood rays do not.

In some hardwood species, the rays tend to be arranged into definite tiers as viewed on a tangential surface. In these woods, rays in each layer are roughly the same height, and all begin and end at about the same level along the grain (Figure 5.18a). Such woods are said to have *storied rays*, and they often exhibit a readily visible

FIGURE 5.17.   Upright ray cells on ray margins. Radial view of Andiroba (*Carapa guianensis*). ×200. Source: Reproduced with permission from Wheeler et al. (1986).

banded pattern on tangential surfaces (Figure 5.18b). A storied cell arrangement is not restricted to ray cells. Almost any type of hardwood cell can occur in storied arrangement, and the resulting pattern is often similar to that produced by storied rays. This pattern will show on both tangential and radial surfaces, whereas that from storied rays will be seen only on the tangential. Storying of elements is primarily of interest for wood identification.

Wood Identification

As with softwoods, no discussion of hardwood identification is included in this text. For a listing of books that deal with this subject, the reader is referred to Chapter 4.

Summary

A three-dimensional drawing of a diffuse-porous hardwood is presented in Figure 5.19 and is used as a means of reviewing structural features of hardwoods. The figure is drawn to the same scale as the softwood block depicted in Figure 4.12.

(a)

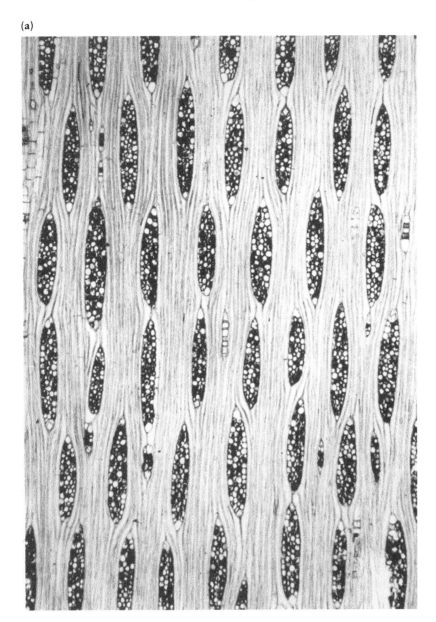

FIGURE 5.18.   Storied rays on tangential surface of sapele (*Entandrophragma cylindricum*). (a) Tangential view. ×50. Rays occur in a definite tier arrangement. Source: reproduced with permission from Wheeler et al. (1986).

Transverse Surface (I)

Portions of two annual growth rings appear on the transverse surface (Figure 5.19). The latewood of one growth ring can be seen on the left (1–1a) followed on the right by a portion of the earlywood of the succeeding growth layer (1a–1b). Growth rings in this

(b)

FIGURE 5.18.   *Continued.*   (b) Rays show up as ripple marks on unmagnified tangential surface.

case are delineated by a difference in wall thickness and radial diameter between early-wood fibers (c) and latewood fibers (d); latewood fibers have thicker walls, and in the outermost part of the growth ring these cells are flattened radially.

Thin-walled longitudinal parenchyma can be seen in cross-section (e). Ray parenchyma is much in evidence, with three rays visible in cross-section (2a–2b, 2c–2d, 2e–2f). Vessel-to-vessel pitting can be seen (f), as can vessel-to-ray parenchyma pits (g) and pitting between vessels and fibers (h).

Radial Surface (II)

Several radially sectioned vessels are visible on this surface, revealing the scalariform perforation plates between vessel elements (i). Also visible are earlywood and latewood fibers (c, d).

A ray (j) is seen to be made entirely of procumbent ray parenchyma. Pitting of ray cells can be seen as small dots on sidewalls and as gaps in common end walls.

Tangential Surface (III)

Rays appear in the end view on the tangential surface (k), providing an opportunity to judge ray size. The rays in this wood are not in a storied arrangement and thus begin and end at different levels along the grain. Rays vary from two to five cells in width.

A long, hollow vessel appears on the tangential surface, interrupted by a perforated remnant of the plates marking the ends of individual vessel elements (i); the perforation

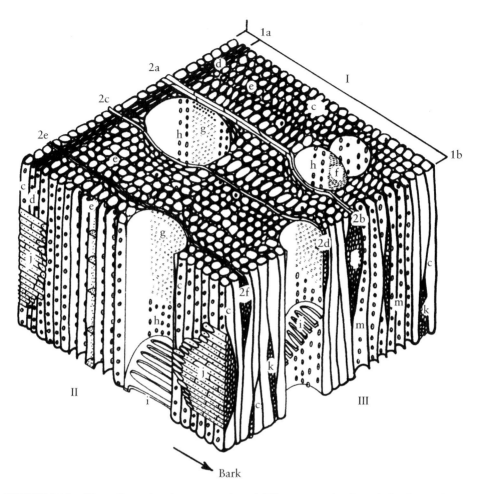

FIGURE 5.19.   Three-dimensional representation of diffuse-porous hardwood. (birch; *Betula* spp.). (I) Transverse view: 1–1a = latewood; 1a–1b = earlywood of succeeding growth ring; c = earlywood fiber; d = latewood fiber; e = longitudinal parenchyma; f = vessel-to-vessel pitting; g = vessel-to-ray parenchyma pitting; h = vessel-to-fiber pitting; 2a–2b, 2c–2d, and 2e–2f = rays. (II) Radial view: c = earlywood fiber; d = latewood fiber; e = longitudinal parenchyma; g = vessel-to-ray parenchyma pitting; h = vessel-to-fiber pitting; i = perforation plate between vessel elements; j = ray composed of procumbent ray parenchyma. (III) Tangential view: c = fibers; i = perforation plate between vessel elements; k = rays in end view; m = fiber-to-fiber pitting (bordered).

is of the scalariform type. Between vessels and rays are thick-walled fibers (c). Note the small bordered pits connecting adjacent fibers (m). Because the cut forming the tangential face was made through a transition area between earlywood and latewood, no fibers of the type found in 1–1a are seen on this surface.

References and Supplemental Reading

Dute, R.R., Duncan, K.M., and Duke, B. (1999). Tyloses in abscission scars of loblolly pine. *IAWA J.* 20 (1): 67–74.

Esau, K. (1965). *Plant Anatomy*, 2e. New York: Wiley.

Foster, R.C. (1967). Fine structure of tyloses in three species of the Myrtaceae. *Aust. J. Bot.* 15 (1): 25–34.

Frey-Wyssling, A. (1959). *Die pflanzliche Zellwand*. Berlin: Springer.

Jane, F.W., Wilson, K., and White, D.J.B. (1970). *The Structure of Wood*, 108. London: Adam & Charles Black.

Kozlowski, T.T. and Pallardy, S.G. (1997). *Physiology of Woody Plants*, 2e, 51. San Diego: Academic Press.

Meyer, R.W. and Côté, W.A. Jr. (1968). Formation of the protective layer and its role in tyloses development. *Wood Sci. Technol.* 2 (2): 84–94.

Pallardy, S.G. (2007). *Physiology of Woody Plants*, 3e. San Diego: Academic Press.

Panshin, A.J. and de Zeeuw, C. (1980). *Textbook of Wood Technology*, 4e, 186. New York: McGraw-Hill.

Roelofsen, P.A. (1959). The plant cell wall. In: *Handbuch der Planzenanatomie*, Band 3, Part 4. Berlin: Gerbruder Borntraeger.

Schmidt, R. (1965). The fine structure of pits in hardwoods. In: *Cellular Ultrastructure of Woody Plants* (ed. W.A. Côté Jr.). Syracuse, NY: Syracuse University Press.

Wheeler, E.A., Pearson, R.G., La Pasha, C.A. et al. (1986). *Computer-aided Wood Identification*. Raleigh, NC: North Carolina Agricultural Research Service, North Carolina State University Bull. 474.

Zimmermann, M.H. (1983). *Xylem Structure and the Ascent of Sap*. New York: Springer.

6

Juvenile Wood, Reaction Wood, and Wood of Branches

The first five chapters dealt only with the formation, composition, structure, and gross features of wood formed in the main stem of mature, upright trees. The character of wood in the centers of young trees, in trees that are leaning rather than vertical, and in branches and roots is considerably different from the normal wood of the mature *bole*. Such wood commonly has properties that affect the ways in which it may be processed and utilized.

Wood formed in the early (or juvenile) stages of growth of a tree stem is called *juvenile wood*. Juvenile wood is formed as part of the developmental process of tree growth and is found in the center portions of stem cross-sections. It is present in every tree, and virtually every living tree, regardless of age, continues to form juvenile wood during each growing season.

A different kind of wood – *reaction wood* – is produced in response to the tipping of a tree. This kind of wood, although abnormal, is nonetheless rather common.

Branchwood and *rootwood* properties are increasingly important as more emphasis is being placed on maximizing the use of all material in a tree. Thus, a working knowledge of wood must include an awareness of these variations in wood form.

Juvenile Wood

In Chapter 1 it was noted that an undefined mass of tissue known as the pith marks the stem center, and this region is surrounded by a thin layer of primary xylem. Both the pith and primary xylem are wholly formed in the first year of the life of a stem, and both types

Forest Products and Wood Science: An Introduction, Seventh Edition. Rubin Shmulsky and P. David Jones.
© 2019 John Wiley & Sons Ltd. Published 2019 by John Wiley & Sons Ltd.
Companion website: www.wiley.com/go/shmulsky

of tissue differ from secondary xylem produced later by the cambium. An important point is that secondary xylem produced for the first 5–15 years is different from secondary xylem produced after this juvenile period.

Juvenile wood is the secondary xylem at the center of a tree formed throughout the life of the tree. The width of the juvenile wood zone decreases upward to the tree crown. The width is species specific, can be affected by environmental conditions, and is the result of an aging process of the cambial initials (Yang and Benson 1997). With respect to the latter factor, Yang et al. (1994) found that the "width of juvenile wood is highly correlated with the age of formation of cambial initials counted from the year of seed formation." They noted that older cambial initials produce juvenile wood for a fewer number of years than do younger initials. The result is a conically shaped juvenile wood zone that narrows from a broad base toward the top of the tree (Figure 6.1). A recent in-depth study of a 48-year-old loblolly pine suggests more variation in the juvenile zone than earlier reported. Groom et al. (2002) evaluated a number of wood properties growth ring by growth ring and along the stem length and found the juvenile zone to be biconical, tapering from the stump to just below the live crown and then again from the live crown to the bole tip. This is attributed to two regions that promote juvenility: the area at the base of the stem (stump height) and the live crown.

By most measures, juvenile wood is lower in quality than mature wood; this is particularly true of the softwoods. In both hardwoods and softwoods, for example, juvenile wood cells are shorter than those of mature wood. Mature cells of softwoods may be three to four times the length of juvenile wood cells, whereas the mature fibers of hardwoods are commonly double the length of those found near the pith (Bendtsen and Senft 1986; Dadswell 1958; DeBell et al. 2002; Koubaa et al. 1998). In addition to differences in cell length, cell structure also differs. There are relatively few latewood cells in the juvenile zone, and a high proportion of cells have thin wall layers. The result is low density and a corresponding low strength in comparison with mature wood. In conifers of the United States, density is typically 10–15% lower in the juvenile core, with the strength of such material reported to range from only slightly lower to commonly 15–30% and as much as 50% less than normal mature wood for some strength properties (Bendtsen 1978). These reductions appear mild when compared with the findings of Senft et al. (1986). In a study of 60-year-old Douglas-fir, they found an average specific gravity difference of 32% when wood formed in the first 15 years was compared with wood formed thereafter. Moreover, although they found the average strength of mature wood to be about 40–60% higher than that of juvenile wood, differences as high as several hundred percent were found when comparing the stiffness of the first several growth rings with rings formed much later. Similar numbers were documented for loblolly pine and cottonwood (Bendtsen and Senft 1986). Even larger differences have been documented in southern pine. In 36-year-old loblolly pine, McAlister and Clark (1991) found juvenile wood to have 76% of the density, only 39% of the stiffness (MOE), and 54% of the bending strength (MOR – see Chapter 9) of mature wood from the same trees. Spectacular differences were also found in a study of plantation-grown Caribbean pine (*Pinus caribaea*). In this material, density was found to be only about 50% of that of wood from forest-grown trees, with stiffness as little as 26% of published values for the species (Boone and Chudnoff 1972). As a reminder that rules in nature are seldom absolute, at least one species has been found to produce wood in the first 6–10 growth rings that is *more* dense than wood formed thereafter.

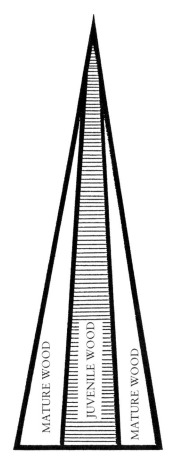

FIGURE 6.1. Juvenile wood zone in stem that has grown beyond the juvenile period. Source: Yang and Benson (1997).

Again, comparing juvenile and mature woods, there appears to be a greater tendency for spiral grain in juvenile wood (Noskowiak 1963; Zobel et al. 1972). Within the cell, the microfibril angle in the S-2 part of the secondary wall is characteristically greater in juvenile wood. Deresse et al. (2003), for example, recorded mean ring microfibril angles of 30° in two-year-old red pine compared with 15–18° microfibril angles at age 20. This kind of secondary wall microfibril orientation also occurs in compression wood that commonly develops in juvenile wood zones. As indicated in Chapter 3, the large S-2 microfibril angle causes a high degree of longitudinal shrinkage and a corresponding decrease in transverse shrinkage; along-the-grain shrinkage of juvenile wood has been reported to average from 3 times (McAlister and Clark 1992) to 9 or 10 times that of mature wood (Boone and Chudnoff 1972; Senft et al. 1986). Several investigators have noted, however, that not all juvenile wood shows excessive longitudinal shrinkage and that pieces may actually increase in length upon drying, possibly owing to growth stresses (Koch 1972, pp. 298–299; McAlister and Clark 1992). Large fibril angles are also associated with low tensile strength (Krahmer 1986; Page et al. 1972). In addition, veneer produced from juvenile wood has been found to be rougher and to contain more splits and deeper lathe checks, thereby producing greater thickness variation (Kellogg and Kennedy 1986).

The kinds of differences indicated in the previous paragraphs often translate to problems when using juvenile wood. For example, a comparison of lumber yield by grade from rapidly versus more slowly grown loblolly pine of the same diameter (20 versus 50 years old, respectively) showed rapidly grown logs to yield only one-fifth to one-half as much high-grade dimension lumber as more slowly grown materials (Fight et al. 1986). In another study of slash pine, the value of lumber obtained from 20-year-old, 0.36 m (14.3 in.) diameter trees was only 66% of the value obtained from 50-year-old, 0.38 m (15.1 in.) diameter trees (MacPeak et al. 1990). Biblis et al. (1995) reported that only 52–64% of lumber from 35-year-old loblolly pine met the stiffness requirements of existing lumber standards. These proportions are well below the percentage of lumber from 50-year-old loblolly found to meet such standards – 94% (Biblis and Carino 1999). Another study resulted in the observation that unless juvenile wood is separated from mature wood, there is a considerable loss in efficiency in utilizing lumber from fast-grown trees when attempting to use reliability-based design, particularly when stiffness is a critically important factor (Tang and Pearson 1992). Senft et al. (1985) reported, moreover, that juvenile wood is a matter of "obvious concern to construction in general and to the laminating industry in particular," owing both to low strength and to high longitudinal shrinkage. Considering all of these factors – reduced strength, occurrence of spiral grain, a high degree of longitudinal shrinkage, and problems in use – juvenile wood is generally undesirable when used in many wood products (Table 6.1).

As a raw material for high-grade and high-strength paper, juvenile wood has long been regarded as inferior, in part because of its low cellulose and high lignin content. It has been viewed with less disfavor by pulp and paper specialists as more has become known about it. Early research found juvenile wood to have significantly lower density and to yield less pulp per ton of material processed than mature wood. Higher chemical consumption in the pulping process and up to a 10% increase in manufacturing costs were also noted (Zobel and Kellison 1972). Yields of turpentine and possibly tall oil by-products of *kraft pulping* (see Chapter 15) have also been reported lower when processing juvenile wood (Foran 1984; McKee 1984), and pulp from juvenile wood has been found to be of low strength. However, juvenile wood appears better when the results of other investigations are reviewed. Several investigators have found, for example, that paper from juvenile wood has low tear strength (as much as 30% lower than paper made from mature wood) but unusually high burst and folding strength (Gooding and Smith 1972; Posey 1964; Semke 1984). *Burst strength* is measured by applying gradually increasing fluid pressure to a small area of the surface of a paper sheet and measuring the force required to rupture the sheet. Other researchers (Barefoot et al. 1964; Zobel et al. 1978) found that wood from the juvenile core produced paper with a higher tensile strength. Hatton (1993) and Hatton and Gee (1994), working with second-growth jack and lodgepole pine and bleachable-grade kraft pulps, confirmed earlier findings regarding strength; they noted that paper made from juvenile wood and topwood pulps (see the section "Branchwood," later in this chapter) exhibits better interfiber bonding than paper made from mature wood pulp. Both of these studies concluded that finer fibers from juvenile wood will provide new opportunities to tailor-make pulps with specific properties sought by papermakers. In a subsequent study, which involved Douglas-fir as well as jack pine (Hatton 1997), strength relationships were again confirmed. When kraft pulping juvenile wood, pulp yields at a given lignin content were found to be consistently lower (~5% lower) than when using mature wood as raw material. Kraft pulping of juvenile wood yielded pulps with shorter average fiber length (15–24%

TABLE 6.1.   Some properties of juvenile wood compared with mature wood.

Wood property	Juvenile wood	Mature wood
Specific gravity (green)	0.42[a]	0.48
	0.40	0.53[b]
Density (kg m^{-3})	427.2[a]	489.2
Fiber length (mm)	2.98[a]	4.28
	1.28[c]	2.68
Cell wall thickness (μm)	3.88[a]	8.04
Lumen size (μm)	42.25[a]	32.78
Cell diameter (μm)	50.01[a]	48.86
S$_2$ layer fibril angle (deg)	55[d]	20
	28[e]	10
	37[f]	7
Longitudinal shrinkage, green to 12% moisture content (% of green dimension)	0.57[g]	<0.10
	0.9[c]	<0.10
Breaking strength or MOR (psi)	7770[b]	10660
	4924[b]	9147[b]
Stiffness index or MOE (10^6 psi)	1.12[b]	1.75
	0.594[b]	1.549[b]
Compression strength parallel to the grain index	100[i]	124

[a] Data for 11-year-old (juvenile) versus 30-year-old (mature) loblolly pine from Zobel and Kellison (1972) as presented by Bendtsen (1978).
[b] Data from test of juvenile and adult wood of 36-year-old loblolly pine, McAlister and Clark (1991).
[c] Based upon tests of Caribbean pine by Boone and Chudnoff (1972).
[d] Information from Dadswell (1958) for coniferous woods.
[e] Information from Dadswell (1958) for hardwoods.
[f] Figures for Douglas-fir as reported by Erickson and Arima (1974).
[g] Data for loblolly pine from Pearson and Gilmore (1971).
[h] Figures for Douglas-fir from Senft et al. (1986).
[i] Data based upon tests of plantation-grown conifers (Olson et al. 1947) versus published figures for forest-grown trees of the same species. The comparison here is between small, fast-grown trees (with strength values assumed to be 100) and wood of the same specific gravity from larger, forest-grown trees.

shorter) and a lower proportion of long fibers (16–58% less). Jackson and Mcgraw (1986) commented on the reasons for different strength properties in juvenile wood pulps. They pointed out that the thinner cell walls of juvenile wood result in tighter packing of fiber in a paper sheet, with more contact between adjacent fibers. The result is higher sheet density and higher tensile and burst strength. Tear strength, on the other hand, is directly and negatively influenced by a short fiber length and thin cell wall.

Many problems associated with use of juvenile wood in pulp and paper manufacture develop because the juvenile material is processed under conditions designed for mature wood. Crist (1976) commented on this, recalling an observation that "you can't cook a steak from an old bull with one from a young bull and expect both to be cooked to a 'T'." He went on to say that juvenile wood is more "tender" than mature wood. Cooking refers to the practice in chemical pulping operations of placing wood in a chemical solution and subjecting the mixture to a combination of heat and pressure. This process softens or dissolves lignin, facilitating the separation of cells.

When juvenile wood is cooked under the severe conditions necessary for mature wood, pulp yield and strength suffer. Yet when cooked alone under conditions tailored for it, the pulp yield and strength from juvenile wood improve. Numerous studies support this notion (Barker 1974; Bella and Hunt 1973; Hunt and Keays 1973; Jett and Zobel 1975), with findings that indicate little difference in properties of juvenile and mature wood pulps when each are produced under ideal conditions. Ideal treatments for juvenile wood are, incidentally, generally less energy intensive than traditional ones, because cooking times required for juvenile wood are shorter and energy requirements for beating are lower (Einspahr 1976; Laundrie and Berbee 1972). *Beating* involves the forced movement of softened wood fibers through a narrow gap between an apparatus like a paddle wheel and a bedplate. As they pass through the gap, the fibers are subjected to a beating or pounding action that flattens and unravels them, increasing the potential for fiber-to-fiber bonding in a paper sheet. A more recent report (Hatton and Cook 1992) recommended the chipping of juvenile and mature wood separately, with chips then combined in different ratios to produce kraft pulps with a wide range of properties. The difference between this and previous approaches is that rather than seeking ways to make juvenile and mature wood pulp the same, Hatton and Cook acknowledge differences between juvenile and mature wood pulps and seek to take advantage of these differences in designing various types of paper.

Although separation of juvenile and mature wood has been recommended as a way to improve yields and pulp quality, a growing number of studies have found little difference in pulp quality obtained from young and mature trees as long as trees are not extremely young at the time of harvest. Goyal et al. (1999), for instance, examined the fiber length and tensile and tear strength of handsheets made from kraft pulp from short rotation tree crops and concluded that after seven to eight years of tree growth, papermaking properties may not change appreciably.

All of the preceding discussion refers to the use of juvenile wood in paper made from chemical pulps. When mechanical pulps are considered (see Chapter 15), the situation is somewhat similar. Hatton (1997) found differences in mechanical pulp made from juvenile and adult wood of Douglas-fir and jack and lodgepole pine. Mechanical pulping of juvenile wood yielded a higher proportion of fines (6–16% higher), shorter average fiber length (10–20% shorter), and a lower proportion of long fibers (10–11% lower). In addition, the tensile and tear indices of pulps made from juvenile wood were significantly lower (7–12 and 14–16% lower, respectively) than pulps made from mature wood. The light-scattering coefficient of juvenile pulps, however, was found to be 9–12% higher than for pulps made from mature wood. As in previous studies, the specific energy required to mechanically process juvenile wood was also found to be higher (8–11% higher). The greater light-scattering ability of juvenile wood pulps is apparently related to the finer fibers of such pulps, and means that juvenile wood mechanical pulps are "ideally suited to the production of high quality publication grades." This echoes the earlier findings of Carpenter (1984), who indicated that "low density wood containing relatively large amounts of juvenile wood and earlywood is *preferred* to denser wood."

A question that has drawn increasing attention of researchers in recent years is what effect juvenile wood might have on the properties of wood composite products such as particleboard, flakeboard, and fiberboard. Composite products technology offers an opportunity for the production of greater quantities of large-dimension structural materials without the need to use large-size trees as raw material. In such products, juvenile wood has generally been found to be undesirable. Wasniewski (1989) found decreasing strength and increasing linear expansion in randomly oriented flakeboard as the proportion of juvenile wood was progressively increased. The strength of Douglas-fir flakeboard made from 50-year-old wood was 10% greater than that made from juvenile wood formed in the first several years of growth. In contrast, the strength of 50-year-old solid wood from the same tree was 30–40% higher than early-formed juvenile wood. In an extensive study of 144 trees from three locations in Georgia and Arkansas, Pugel et al. (1989, 1990) found flakeboard, standard particleboard, and fiberboard panels made from juvenile wood to have comparable strength and durability to otherwise identical composite panels made from mature wood. However, both thickness swelling and linear expansion – two undesirable properties – were significantly greater in the juvenile wood panels. Similar results were obtained by Geimer et al. (1997), who studied oriented flakeboard and plywood made from juvenile (1–12 years old) and mature (13–35 years old) loblolly pine (*Pinus taeda*). They found significantly greater linear expansion in both plywood and three-layer oriented waferboard made from juvenile wood than in similar panels made from mature wood. Interestingly, no significant difference was found in the linear expansion of randomly oriented single-layer flakeboards made from juvenile and adult wood. A comprehensive examination of the impact of juvenile wood on properties of composites made from southern pine (Pugel and Shupe 2004) resulted in the conclusion that "Although juvenile wood is clearly inferior to mature wood for most applications, the same is not necessarily true for properties of most wood composites." It was noted that in those cases where juvenile wood leads to undesirable properties, such as increased linear expansion, the proportion of juvenile wood should be controlled. Based on these and other studies, it appears that juvenile wood is a generally suitable raw material for wood panel products but that monitoring and control are nonetheless needed to control dimensional stability.

One approach to the problem of increased linear expansion in oriented strandboard panels containing juvenile wood is to pretreat flakes with acetic anhydride. The pretreatment, known as *acetylation*, has been shown to increase dimensional stability under changing moisture conditions (Rowell et al. 1986, 1993). Fortunately, juvenile wood has been demonstrated to more readily accept acetic anhydride treatment than adult wood, with acetyl weight gain increasing with an increasing proportion of juvenile wood. This result suggests a mechanism that automatically ensures maximum effectiveness of acetylation in wood from the juvenile core.

Juvenile wood is difficult to identify by casual observation, especially in softwoods, although several normal characteristics are sometimes modified. In hardwoods, vessels of juvenile wood are often smaller and arranged differently from those in mature wood. Ring-porous woods may, for example, have juvenile wood that tends to be diffuse porous (Priya and Bhat 1999). Another normally consistent feature that becomes variable in juvenile wood is the type of vessel plate perforations (Jane et al. 1970). Scalari-form perforations have been reported in the juvenile wood of species that normally have simple vessel plate perforations.

There is typically no sharp demarcation between juvenile and mature wood. Instead, a gradual transition in properties occurs from the tree center outward. Bendtsen (1978) explained this well when he wrote: "Wood in the first formed rings has the lowest specific gravity, shortest fibers, largest fibril angles, and so forth. In successive rings from the tree center, the specific gravity increases, fibers become longer, and so on. The rate of change in most properties is very rapid in the first few rings; the later rings gradually assume the character of mature wood." Figures 6.2 and 6.3 illustrate this gradual change in wood properties. A notable exception to the progressive change in properties from the pith outward is documented in a recent study of loblolly pine, a distinct-ring softwood. Whereas latewood was found to exhibit increasing specific gravity and strength from the pith outward, earlywood properties remained constant regardless of ring position or distance from the pith (Kretschmann et al. 2006). The fact that juvenile wood is heterogenous, with properties that change from ring to ring, complicates efforts to more effectively use this material.

The substantial amounts of reaction wood often occurring within the juvenile region contribute to the lack of a well-defined juvenile wood zone. One study of southern yellow pine showed the juvenile wood region to contain 42% reaction wood in comparison with only 7% in the surrounding mature wood (North Carolina State College 1957).

Because of the gradual change in wood properties, it is unclear as to where juvenile wood ends and mature wood begins. The location of this boundary depends, furthermore, upon the property or properties used to define the zone (Abdel-Gadir and Krahmer 1993; Bendtsen and Senft 1986; Koubaa et al. 2005). For example, a property such as wood density or cell length may reach maturity before another feature such as cell wall thickness. Bendtsen and Senft report that in pine "wood density and mechanical properties mature beginning at about age thirteen, while cell length matures about age eighteen." They note that fibril angle continues to fall through until age 30. There is some agreement among researchers that wood density is the most appropriate trait to use in determining the juvenile period. There is also general agreement that juvenile wood

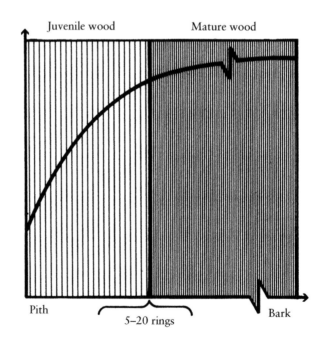

FIGURE 6.2.   Juvenile to mature wood transition in conifers. Many properties show gradual increase. Source: From Bendtsen (1978).

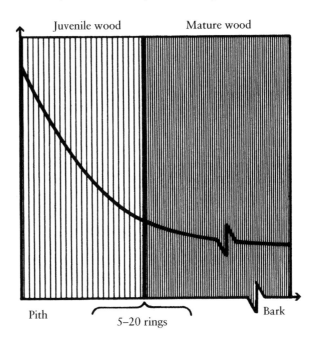

FIGURE 6.3.   Juvenile to mature wood transition in conifers. Some properties show decrease. Source: From Bendtsen (1978).

predominates in the first 5–20 growth rings, with the duration of its formation affected by several factors, including geographic location, site and genetic differences, and silvicultural practices (Peszlen 1995). Kučera (1994) found strong evidence that the period of juvenile wood formation ends at stem-breast height at the point of maximum average height increment. In other words, as soon as the annual increment slows, formation of wood at lower levels in the stem shifts from juvenile wood to mature wood. It should be noted that there is some evidence to indicate that the period of juvenile wood formation is shorter near the upper part of a tree than near the base. Yang et al. (1986) attributed such findings to a relationship between the duration of juvenile wood production and the year of formation of the cambial initials. Some researchers have determined that stimulation of growth (through fertilization, irrigation, or silvicultural treatment) during the period of juvenile wood formation will extend the juvenile period (Larson 1969; Megraw and Nearn 1972). A more recent study shows a direct correlation between growth rate and the juvenile period (Koubaa et al. 2005). Growth acceleration following the juvenile period does not result in reinitiation of juvenile wood formation. The significance of this observation is discussed in Chapter 12.

Reaction Wood

A reaction is a response to a triggering event. Reaction wood was appropriately named. This special kind of wood may be formed if the main stem of a tree is tipped from the vertical (Figure 6.4), and it is known to regulate the orientation or angle of branches

FIGURE 6.4.   A leaning tree such as this one produces considerable reaction wood.

relative to the main stem. It can also arise in response to accelerated growth and is reported to play a role in directing growing stems toward openings in a forest canopy.

Reaction wood formed in hardwoods differs from that formed in softwoods. In softwoods it is termed *compression wood* and in hardwoods, *tension wood*. In both, however, the function of reaction wood is the same: to bring the stem or branch back to the original position.

Compression Wood

If sufficient force is applied to the top of a standing pole, it will bend (Figure 6.5). The side of the pole toward which the top is bent tends to become shorter as the result of an induced compression stress. Conversely, the other side of the pole is stretched slightly as it is subjected to tension stress. In softwoods, reaction wood forms on the compression side (or underside) of a leaning stem, thus the name *compression wood*.

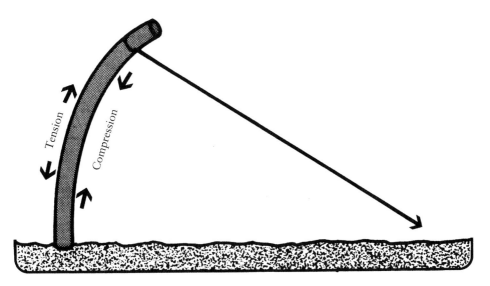

FIGURE 6.5.   Bending a vertical column causes differing stress.

This name, incidentally, refers only to the position in which softwood reaction wood is formed and does not imply that it forms as a result of compression stress. Compression wood also forms almost universally in branches of softwoods, where it functions to maintain branch angle. An exception is a species with drooping branches, such as spruce, in which there is a conspicuous absence of compression wood (Timell 1973c).

Properties.   Compression wood is of interest to the forest products technologist because its properties are considerably different – and in virtually every case less desirable – than those of normal mature wood. Compression wood tracheids, for example, are about 30% shorter than normal (Dinwoodie 1961; Spurr and Hyvärinen 1954). In addition, compression wood contains about 10% less cellulose and 8–9% more lignin and hemicellulose than normal wood (Côté et al. 1966). These factors reduce the desirability of compression wood for pulp and paper manufacture. Watson and Dadswell (1957) obtained a 27% lower sulfite pulp yield and a 17% lower sulfate pulp yield from compression wood than from normal wood. Dadswell and Wardrop (1960) indicated that compression wood not only yields less cellulose but also produces low-strength pulp, especially when subjected to the sulfite chemical pulping process. Sulfite pulp from compression wood was said by Timell (1973c) to be clearly less desirable than normal wood pulp, but use of the kraft (sulfate) process with compression wood was reported to yield an only slightly inferior product. Timell further noted that satisfactory groundwood pulp cannot be made from compression wood, apparently because of the high lignin content as well as an abnormally high proportion of latewood. (Ground wood, sulfite, and kraft pulping methods are described in Chapter 15.) Barefoot et al. (1964) acknowledged the adverse effects of compression wood on pulp quality but tempered this observation by pointing out that compression wood fibers vary from a mild to a severe form. Mild forms of compression wood were found to have a detrimental effect upon tear strength but not upon other paper properties.

FIGURE 6.6.   Longitudinal shrinkage can cause spectacular results.

Compression wood is highly undesirable in lumber or other solid wood products. A major concern when using compression wood in solid form is the longitudinal shrinkage that occurs upon drying. Longitudinal shrinkage is commonly 1–2% (compared with 0.1–0.2% for normal wood) and may be as great as 6–7%. The results of this extreme longitudinal shrinkage are shown in Figure 6.6. Compression wood is higher in density than normal wood of the same species; its density is commonly 10–20% and sometimes as much as 40% higher. Because of the higher density, it might be expected that compression wood also would have higher strength than normal wood. However, compression wood is about equal in strength to normal mature wood of the same species. If material of similar density is compared, the relatively dense compression wood is inferior in most strength properties compared with normal wood. This is a disadvantage for many structural applications, because the most valuable woods are those having a high strength-to-weight ratio; many products such as tool handles, boat masts, and ladders require low weight as well as strength. Understandably, such abnormal properties make it desirable to eliminate compression wood from the raw material going into most solid wood products.

Identification.   Compression wood is relatively conspicuous and can often be identified visually when looking at smooth surfaces. It is especially noticeable in a transverse view.

A stem cross-section containing significant amounts of compression wood often has exceptionally wide growth rings on the lower or compression side of the leaning stem, with much narrower rings to the opposite side of the pith. The pith, as a result, is nearer

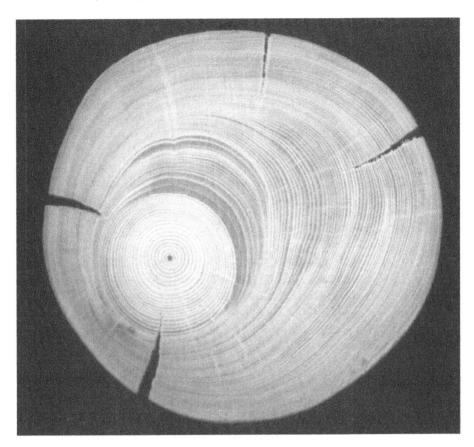

FIGURE 6.7.   Eccentric cross-section of black spruce (*Picea mariana*) exhibiting compression wood. Source: Sample courtesy of K.C. Yang.

to the upper side of the stem, which causes it to be eccentric (Figure 6.7). In addition, the wide growth rings contain a high proportion of latewood, and the contrast between earlywood and latewood is often less distinct than in normal mature wood. This latter characteristic is normally evident on radial and tangential faces of surfaced lumber and is the primary means used to detect compression wood in a mill setting.

Although compression wood may be apparent on a smooth cross-section, detection on the rough-cut end of a log (required if this defect is to be identified in a sawmill) is more difficult. One study, in which 680 logs were sawn after a visual search for compression wood by an expert, showed that visual attempts to identify compression wood in rough-cut logs were unreliable (Hallock 1969). A recent study by Nyström and Kline (2000) demonstrated that compression wood could be detected automatically using a color scanner; the accuracy of this method in a laboratory setting was reported at 87%. Although far from commercialization, this finding raises the possibility that high-speed automated detection of compression wood may someday become a reality.

At the microscopic and submicroscopic levels, more differences between normal and compression wood become evident. Viewed longitudinally, tips of compression wood tracheids are bent and folded. In cross-section, these cells are rounded rather than rectangular and have pronounced intercellular spaces between them. Analysis of the walls of compression wood tracheids shows only S_1 and S_2 layers to the inside of the primary wall, with the slope of microfibrils in both the S_1 and S_2 layers greater than in normal wood. The orientation of microfibrils in the S_1 layer of Norway spruce (*Picea abies*) compression wood has been recorded at 86–90°, or almost perpendicular to the cell axis (Brändström 2004). A number of studies have found the slope of microfibrils in the S_2 layer to be about 45° from the vertical. This large S_2 microfibril angle results in great longitudinal shrinkage (Wardrop and Dadswell 1950). Deep, helically arranged checks extending from the lumen mark this S_2 layer.

Before leaving the topic of compression wood, a few words are in order regarding *opposite wood*, a term used to describe wood formed on the opposite side of the stem from compression wood. This wood is reported to have significantly different characteristics than normal wood (Tanaka and Koshijima 1981; Timell 1973a, b, 1986, pp. 1969–1998; Siripatanadilok and Leney 1985). Some similarities to compression wood have been reported, such as high microfibril angles – in some cases as large as, or larger than, the angles of the compression wood side of the same tree. Opposite wood of softwoods differs from compression wood in having a lower lignin content, higher cellulose content, and cells having thick, highly lignified S-3 layers.

Those interested in a comprehensive discussion of compression wood, its formation, structure, properties, chemistry, and uses, will find the three-volume series by Timell (1986) to be enlightening.

Tension Wood

Tension wood is reaction wood of the hardwood species. It forms on the upper or tension side of leaning stems and of branches.

Properties.   As with compression wood, tension wood properties are quite different from those of normal mature wood. They are not all undesirable, however, and the usefulness of tension wood might best be described by a proverbial good news–bad news tale. With respect to pulp and paper manufacture, for example, the bad news is that tension wood has long been considered a less desirable raw material than normal wood. It requires special care in pulping, and pulp containing large amounts of tension wood produces weaker paper than normal pulp (Jayme and Harders-Steinhäuser 1953). Tensile and burst strengths appear to be most affected (Dadswell et al. 1958). The good news is that tension wood pulp strength compares favorably with that of normal wood after it is subjected to a refining treatment (Isebrands and Parham 1974; Parham et al. 1976). The good news gets better: The cellulose content of tension wood is higher than normal. The higher cellulose content, together with a 5–10% increase in density over normal wood, results in slightly improved chemical pulp yields (Casperson et al. 1968).

Tension wood is especially well suited for both dissolving and mechanical pulps. It is desirable for dissolving pulps because it gives high pulp yields. *Dissolving pulp* is a very pure pulp made by removing residual hemicellulose and lignin from a chemical pulp. It is used in making cellulose products such as cellophane, rayon, and nitrocellulose. For this use, individual fiber strength is unimportant. Compared with normal wood, *mechanical pulping* of tension wood yields higher-strength pulp and is easier to accomplish because the proportion of lignin in tension wood is lower (Scaramuzzi and Vecchi 1968). Recall that compression wood is difficult to pulp mechanically because of a high lignin content.

In the manufacture of solid wood products, there is little good news associated with the presence of tension wood. Tension wood tends to produce a fuzzy surface upon sawing or surfacing, particularly when processed green (Figure 6.8). This causes saws to overheat and makes satisfactory finishing difficult. Satisfactory machining is also more difficult to achieve with tension wood (Schumann and Pillow 1969). Upon drying, tension wood shows a decided tendency to collapse irreversibly (Dadswell 1958; Jane et al. 1970). *Collapse* is the cave-in or flattening of wood cells during drying, often resulting in severely distorted wood surfaces. (More information about this drying defect can be found in Chapter 8.) Tension wood also shrinks excessively along the grain, although to a lesser extent than compression wood. Longitudinal shrinkage of tension wood is usually 1% or less. This degree of longitudinal shrinkage may seem insignificant, but any amount of dimensional change along the grain can create problems. For instance, a change of only 0.5% can mean about 0.6 cm (0.25 in.) of shrinkage for each 0.8 m (4 ft) of length. Warp or twist sometimes results when tension wood is present along only one side or edge of a board.

The strength of tension wood generally compares unfavorably with that of normal mature wood. Most measures of strength are less than in normal wood of similar density, and this is particularly true of compression strength parallel to the grain. In an air-dry condition, tension wood is slightly higher in impact bending strength than normal wood (Panshin and de Zeeuw 1980). Because of limited research comparing

FIGURE 6.8. Sawing through tension wood zones can result in formation of fuzzy surfaces.

tension wood and normal wood strength and conflicting findings, Tsoumis (1968) indicated that caution is advisable where strength is important. He said that because of erratic strength properties "tension wood in wooden structures should be viewed with concern analogous to compression wood, especially if strength of the structure is of primary importance."

Identification.   Tension wood is not easy to detect visually, making removal during wood-processing operations difficult. Unlike compression wood, tension wood usually is not different in color than mature wood. A clue to its presence in a log is the shape of the cross-section, the arrangement of rings within it, or both. As in compression wood, stems containing tension wood often have wider rings in the reaction wood zone than on the opposite side of the pith, which results in an elliptical shape (Figure 6.9). Other indications of the presence of tension wood are the fuzzy surfaces produced during sawing or planing, a lustrous sheen that appears on machined surfaces of some species, and a darker color that characterizes the tension wood of some tropical hardwoods. Unfortunately, none of these are totally reliable indicators of tension wood. Detection is further complicated because tension wood zones are seldom totally composed of tension wood tissue. Such tissue occurs in a mixture with normal cells, with the proportion of tension wood depending upon the degree of lean in a stem. Efforts to find better ways of detecting tension wood continue.

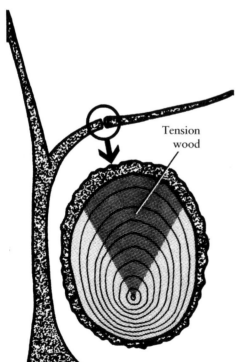

FIGURE 6.9.   Elliptical hardwood stems often indicate tension wood.

Positive identification of tension wood is possible when laboratory methods are available. Examination under a microscope reveals that tension wood contains fewer and smaller vessels and fewer rays than normal wood (Scurfield 1973). Compared with normal mature wood (Figure 6.10a), tension wood fiber walls are often quite thick, with very small lumens, and secondary wall layers are commonly only loosely connected to the primary cell wall. In Figure 6.10b, thick-walled tension wood fibers of *Populus* surround a vessel; loosely attached secondary walls, dislodged in cutting, protrude from cell lumens. The loosely connected secondary wall provides positive evidence of tension wood and also at least partially explains the reason for several tension wood properties. Cells with loosely attached secondary walls have also been reported in the opposite wood of hardwoods (Hamilton et al. 1985). Thick and loosely attached secondary walls of fibers are responsible for the heating of saws during processing and for the fuzzy surfaces that often remain. The soft walls are difficult to cut cleanly, and cutting forces can result in tear-out of fiber bundles from below the surface being formed. Thick fiber walls are also one cause of lower strength in paper made from tension wood fiber. These rigid cells do not easily bend and flatten within the paper and thus fiber-to-fiber bonding is hindered.

The thick and loosely attached secondary wall of tension wood fibers is almost pure cellulose of highly crystalline organization. Because this layer contains little lignin, it is pliable and gelatin-like, rather than stiff like other wall layers, and thus is called a gelatinous (or G) layer. In addition to being almost pure cellulose, the G-layer is composed of microfibrils arranged nearly parallel to the cell axis, varying only about 5° (Côté and Day 1965; Preston and Ranganathan 1947). A highly magnified view of a G-layer is shown in Figure 6.11. The highly cellulosic gelatinous layer is the reason for high yield upon chemical pulping of tension wood.

Careful analysis of tension wood fiber walls shows variability in the sequence of layering. In some cells, the S_2 and S_3 portions of the cell wall are missing and the G-layer lies to the inside of the primary (P) and S_1 cell wall layers, giving a P, S_1, G configuration. Other tension wood fibers are ordered P, S_1, S_2, G or P, S_1, S_2, S_3, G (Dadswell and Wardrop 1955). Apparently, the cell wall configuration is dependent upon the stage of development of a particular cell at the time of stem tilting (Scurfield 1973). Cells that have formed S_1 and S_2 layers of the secondary wall will immediately stop normal development if the stem leans and will shift to G-layer development.

Compression wood shrinks longitudinally because of the large microfibril angle in the S_2 wall layers of longitudinal tracheids. This is not the case in tension wood, which exhibits a normal S_2 orientation in cells having this layer but has longitudinal orientation of microfibrils in the thick G-layer. The explanation for longitudinal shrinkage has long been that tension wood shrinks longitudinally because the loosely attached G-layer does not provide shrinkage restraint as the S_2 layer in normal wood does. Wood shrinks when microfibrils move more closely together as water molecules leave the cell wall. A cell wall layer in which most of the microfibrils are oriented perpendicularly to the cell axis, such as the S_1 layer, will tend to shrink longitudinally as water is lost. This tendency is counteracted in a normal cell by the thick and firmly attached S_2 layer, which does not shrink longitudinally. (For a more complete discussion of why and how wood shrinks, see Chapter 8.)

Results of a study published in 2001 suggest there may be more to tension wood longitudinal shrinkage than simply the lack of an S_2 layer. Clair and Thibaut (2001) found a high degree of longitudinal shrinkage in the gelatinous layer (4.7%) and also

(a)

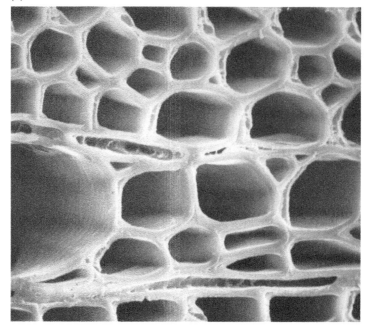

(b)

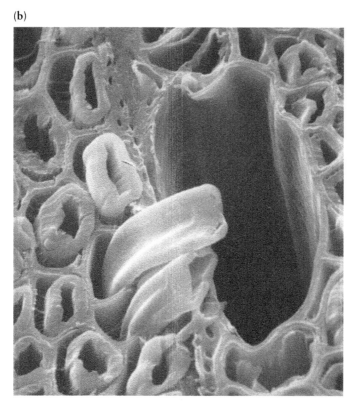

FIGURE 6.10. Normal hardwood fibers versus tension wood fibers. (a) Normal poplar (*Populus* spp.) in cross-section. ×80. (Scanning electron micrographs by Judson G. Isebrands and Russell A. Parham.) (b) Tension wood fibers surround a vessel in xylem of poplar (*Populus* spp.). ×1600.

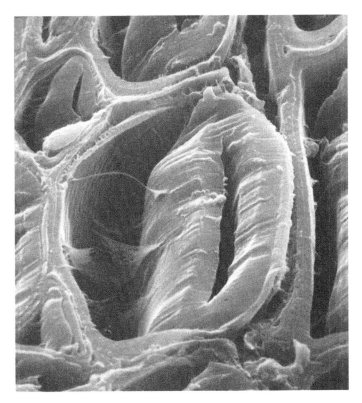

FIGURE 6.11.   Thick gelatinous layer of tension wood fiber detached from the underlying secondary wall. ×2800. Note the transition of the flat microfibril angle in the secondary wall to the longitudinal orientation in the gelatinous layer. Source: Scanning electron micrographs by Judson G. Isebrands and Russell A. Parham.

found that the G-layer is tightly connected to the underlying layer early in drying. These findings point to shrinkage in the G-layer as a contributing factor to higher-than-normal longitudinal shrinkage in tension wood.

Formation of Reaction Wood

A number of investigators have considered how reaction wood brings about redirection of a leaning stem. There are at least two major factors to consider: (i) The sequence of events that triggers the development of reaction wood; and (ii) the mechanics of stem straightening. Regarding mechanics, it appears that stem-straightening forces are brought about by a combination of rapidly developing reaction wood on one side of the stem, which pushes the stem the opposite direction, and cell expansion or contraction of the reaction wood itself resulting from higher or lower lignin content, respectively (Timell 1986).

The question of what triggers reaction wood formation is not yet solved. Scientists who first looked into stem adjustment believed that reaction wood formed as a result of induced compression or tension stress (Metzker 1908), and much of the early experimentation supported this theory (Wardrop 1964). Mechanical stress as a cause of reaction wood formation was later discounted by studies indicating that gravity played a decisive role. This was demonstrated effectively in an experiment in which young stems were bent into complete loops prior to the growing season and then examined after a period of growth (Ewart and Mason-Jones 1906; Jaccard 1938). Growth of softwood stems under these conditions resulted in the formation of compression wood along the lower edges of the uppermost part of the loops where there was indeed compression stress (Figure 6.12a). At the bottoms of the loops, however, compression wood formed on the lower, or tension, side. Precisely the opposite results were obtained in tests with hardwoods (Figure 6.12b). In a related experiment, pine stems were bent, but tipped sideways so that all parts of the stems were oriented horizontally (Burns 1920). These circumstances led to formation of compression wood at neither the compression nor the tension face of the bent stem but on the lower side (Figure 6.13). These results reinforced the earlier view that the force of gravity plays an important role in reaction wood formation, a view that was accepted by early researchers. However, as pointed out by Timell (1986), it has long been known that if branches of conifers are bent upward, compression wood forms on the upper side, forcing them down again. Thus, it appears that mechanisms leading to reaction wood formation are quite complex; as yet, these are not understood.

The addition of a little more uncertainty to the debate about causes of reaction wood formation is provided by two relatively recent papers focused on the relationship of compression growth stresses and lignin content in compression wood zones (Okuyama et al. 1998), and on tensile growth stress and lignin distribution in cell walls of tension wood (Yoshida et al. 2002). Okuyama et al. found that the greater the compressive growth stress in compression wood was, the greater the lignin content of the secondary cell wall. Similarly, Yoshida et al. determined that the percentage of the cross-sectional area of tension wood cells occupied by the G-layer increased as tensile growth stress increased. Both studies reintroduced compression and tension stresses as factors that may play a role in reaction wood formation, or that at least help to shape the properties of this type of wood.

Clues as to how reaction wood formation is induced are provided by experimentation with growth regulators. It has been shown that artificially induced auxins such as indole-acetic acid and gibberellic acid regulate the formation of compression wood (Fraser 1952; Wardrop and Davies 1964; Wershing and Bailey 1942) as well as of tension wood (Sheng et al. 2004). It has also been shown that injection of indole-acetic acid in one side of a vertical softwood stem causes it to tilt away from the injection site (Wardrop and Davies 1964). These findings, coupled with the observations that leaning softwood stems have higher concentrations of auxins at the lower side than at the upper side (Onaka 1949), indicate that auxin concentrations may play a role in compression wood formation. Něcesaný (1958) and Westing (1968) also found that auxin concentrations are associated with tension wood formation. They noted that concentrations of auxin served to inhibit the formation of tension wood.

Reaction wood has been found in nonleaning stems of loblolly pine (Zobel and Haught 1962), yellow poplar (Taylor 1968), and several species of *Populus* (Krempl 1975). Tsoumis (1968) explained this by pointing out that very young trees may be tipped, form tension wood, and recover to the vertical position. He also indicated that reaction wood can form in trees that are tilted by wind action but not permanently

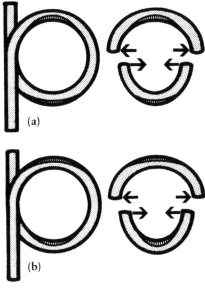

Consistent formation on
lower (CW)/upper (TW) sides

FIGURE 6.12.   Reaction wood formation in growing looped stems: (a) softwood; (b) hardwood.

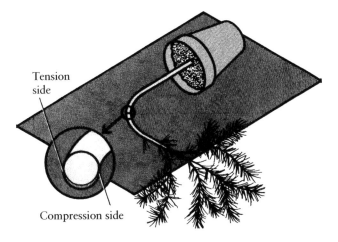

FIGURE 6.13.   Compression wood formation in horizontally oriented stem.

displaced. Thus, gravity appears to play a key role in reaction wood formation even in nonleaning trees. This phenomenon can occur on a large scale after major wind events such as tropical storms or hurricanes, after which reaction wood can persist in the wood supply for decades. Gravity may not, however, be the only environmental factor leading to reaction wood formation. Observations of tropical trees suggest that tension wood may serve to direct crowns toward openings in a dense jungle canopy (Panshin and de Zeeuw 1980). There is also mounting evidence that reaction wood formation is associated with fast growth (Crist et al. 1977; Isebrands and Bensend 1972; Isebrands and Parham 1974; Timell 1986, p. 1696; Tsoumis 1952).

What triggers auxin production upon tipping of stems and how these substances move to specific locations in the stem are other unsolved riddles. Regardless of the trigger/transport mechanisms, it is clear that the system is sensitive and that events occur quickly. Stem displacements as small as 2° can cause compression wood formation, with the amount of reaction wood formed directly related to the angle of lean. Studies have also shown reaction wood to begin developing in as short a period as 24 hours (Kennedy and Farrar 1965). In a more recent study (Yoshizawa et al. 1992), changes in cell structure were not observed until the fourth day after stem inclination.

Branchwood

Ever greater demand for wood and wood products has stimulated interest in finding new sources of wood. The total-tree (or whole-tree) concept arose in the 1960s and early 1970s, led by Hakkila (1971), Keays (1971a, b, c), Koch (1973), Young (1964, 1974), and others. Whereas traditional methods of harvest involve removal of only the main stem, which is trimmed of the top and branches, *total-tree harvest* is characterized by the gathering of main stems, branches, twigs, leaves, and even roots. Tables 6.2 and 6.3 indicate increases in yield of 60–100% or more over conventional methods of harvest through removal of total trees.

Harvesting of all above-ground segments of trees became a reality in the late 1960s and early 1970s with the development in the United States of mobile units with capabilities of on-site chipping of entire trees. This system provided for the collection of chips in semitrailers for movement to processing centers. Tractor-mounted shearing devices, developed at about the same time, ensured maximum removal of above-ground tree parts; this equipment produced stumps only 50–150 mm (2–6 in.) in height.

TABLE 6.2.   Weight distribution (oven-dry basis) of above- and below-ground parts of three 22-year-old, 0.20 m (7.7 in.) diameter, unthinned, plantation-grown slash pine trees cut in central Louisiana.

	Weight fraction (%)		
Portion of tree	Total tree	Above-ground parts	Bark-free stem of 10 cm (4 in.) top (DOB)[a]
Bark-free stem	58.5	70.2	100.0
Roots and stump [roots to a 0.9 m (3 ft) radius]	16.5	19.8	28.2
Stem bark to 10 cm (4 in.) top	12.5	15.0	21.4
Top (with bark)	5.0	6.0	8.5
Needles	4.0	4.7	6.7
Branches (with bark)	3.5	4.2	5.9
Total	100.0		

Source: Koch (1973).
[a] DOB = diameter outside bark.

TABLE 6.3.   Weight distribution (ovendry basis) of above- and below-ground parts of 0.15–0.25 m (6–10 in.) diameter, naturally grown red maple.

Portion of tree	Total tree (%)	Above-ground parts (%)	Bark-free stem of 10 cm (4 in.) top (DOB)[a] (%)
Bark-free stem	50.1	64.6	100.0
Roots and stump (with bark)	22.4	28.9	44.8
Stem bark to 10 cm (4 in.) top	7.7	9.9	15.3
Top (with bark)	5.3	6.8	10.6
Branches (with bark and leaves)	14.5	18.7	29.0
Total	100.00		

Source: Adapted from Young et al. (1963, 1965).
[a] DOB = diameter outside bark.

TABLE 6.4.   Bark as a proportion of the dry weight of various tree parts.

Portion of tree	White pine (%)	Red maple (%)
Merchantable bole	18.7	13.3
Unmerchantable top	16.8	16.6
Branches larger than 2.5 cm (1 in.)	21.1	20.7
Branches smaller than 2.5 cm (1 in.)	71.6	71.2
Stump	20.8	11.9
Roots larger than 10 cm (4 in.)	17.5	10.5
Roots 2.5–10 cm (1–4 in.)	18.5	19.9
Roots less than 2.5 cm (1 in.)	54.4	50.7

Source: Adapted from Young et al. (1965).
Note: For trees ~0.20 m (8 in.) diameter at breast height (1.4 m or 4.5 ft from ground level).

More recent interest in the use of tops, branches, and full trees is based on the need to thin vast areas of forests in the western United States in order to reduce risks of catastrophic fire. In this case interests are focused on finding viable uses for material for which there is currently no market.

Although yield can be greatly increased by whole-tree operations, there has been concern about the nature or quality of material harvested. The properties of branchwood differ from those of wood from the main stem, so manufacturing processes may need to be modified to accommodate these variable components. The use of small branches and twigs as a papermaking raw material became widely practiced in the United States in the 1970s, but was largely abandoned within a decade following experience with problems related to inorganic matter (grit) and excess bark. Such material has never found commercial acceptance in European mills.

From a utilization viewpoint, one of the most significant differences between material from branches and from that of the main stem is that branches have a much higher proportion of bark. This is especially true of those less than 2.5 cm (1 in.) in diameter (Table 6.4). Because bark has markedly different properties than wood (see Chapter 7) and often picks up considerable dirt during harvest, the use of branches for fiber or particle products requires caution. Considerable process modification may be necessary. Processes to separate wood and bark are not yet perfected or widely used.

Aside from the bark, *branchwood* itself differs from wood of the bole. This can make wood identification as well as utilization difficult. Some kinds of cells are more abundant in branchwood than they are in the wood of the main stem. In hardwood branches, vessels and rays are more numerous than in the bole, with fibers present in lesser numbers. Softwoods that normally have resin canals also exhibit this feature in branchwood, although canals in branches are both smaller in diameter and more numerous. Softwood branches also characteristically have a higher than normal ray volume.

Narrow growth rings typify both hardwood and softwood branch material, and longitudinal cells are generally both shorter and of lesser diameter than those in the main stem (Tsoumis 1968). In studies of a number of hardwoods, fiber length in branchwood was found to be on average 25–35% less than that of wood from the main stem (Manwiller 1974; Phelps et al. 1982; Taylor 1977); similar results have been obtained for softwoods (Brunden 1964; Lee 1971).

Early literature indicated that branchwood is generally higher in specific gravity than *stemwood* (Fegel 1941; Jane et al. 1970). Recent work suggests, however, that this relationship is somewhat variable. For instance, although several studies found branches of softwoods to be 5–20% lower in specific gravity than bolewood (Brunden 1964; Phillips et al. 1976; Pugel et al. 1989), McAlister et al. (2000) found the specific gravity of branchwood of loblolly pine to be generally higher than that of stemwood (10% higher on average). Similarly, specific gravity of hardwood branches has been found to be higher in some species (Hamilton et al. 1976; Taylor 1977) and lower or the same in others (Taylor 1977).

Products made from branchwood have different properties than those made from main stemwood. For example, particleboard made from Douglas-fir branchwood was shown to have lower stiffness and breaking strength in bending and lower strength retention after aging than board made from bolewood (Lehmann and Geimer 1974). Particleboard is a product composed of wood particles that have been sprayed or mixed with an adhesive, at 2–8% of the dry weight, and compressed to a desired thickness. Stiffness and breaking strength of boards made entirely from branch material were only 30–65% of those of composite panels made from wood of the bole, depending upon whether branches were small – less than 25 mm (1 in.) diameter – or large. Another study (Pugel et al. 1989) using loblolly pine branches showed the stiffness of flakeboard, standard particleboard, and fiberboard panels to be 10–20% less than that of identical panels made of main stemwood. However, with the exception of flakeboard, breaking strength and internal bond (a measure of the tensile strength of the panel core) were equal or better in the panels made from branchwood. Thickness swelling is also affected by the presence of branchwood. Nemli et al. (2004) found significantly increased thickness swelling in experimental particleboards made from 100% black locust topwood than in panels made from wood from the main stem. Strength properties of the topwood panels were also significantly lower. Particleboard manufactured from logging slash (including branches, twigs, needles, and cones) was found to have considerably lower strength than board made from conventional materials when using the same percentage of adhesive (Boehner and Gertjejansen 1975).

A number of investigations have assessed the suitability of branch material for wood fiber products. Timell (1986, p. 1876) noted that the major disadvantage of using treetops as a source of pulp is that they consist almost entirely of juvenile wood and, in the case of softwoods, have a high content of compression wood as well. Kraft pulp from branchwood requires less beating time and has lower strength than pulp made from wood of the merchantable bole (McKee 1960; Worster and Vinje 1968). One study of western hemlock showed that paper from kraft branch pulp had 20–25% lower tear strength and 40–45% lower burst and tensile strength than paper made from wood of the main stem (Keays and Hatton 1971). In contrast, more recent research led to the conclusion that small-diameter trees and tops and submerchantable logs are suitable raw material sources for kraft pulp mills, based on the finding that there are only small differences in kraft pulping and papermaking performance between the material tested and wood commonly used in kraft mills (Kumar et al. 2004). Little work has been done to determine the suitability of branchwood for pulping processes other than kraft. Studies do suggest, however, that relatively good-quality mechanical or chemi-mechanical pulp can be obtained from this material (Keays 1971a).

Low pulp strength is not the only reason why branchwood is looked upon with disfavor by papermakers. Branchwood is also a less desirable raw material because of the nonuniformity of chip sizes and high proportions of bark. The configuration of branches and twigs, along with higher bark fractions, results in decreased yield and a higher proportion of rejects in the form of oversized material such as slivers, chunks, or uncooked knots that remain after the pulp is passed through a series of screens (Hunt and Hatton 1975). Tests of hardwoods showed pulp yields from branch material to be virtually the same as yields from bolewood when yield was expressed in terms of pulp weight divided by the weight of chips entering the digester. Yield from branch material was far lower, however, when unacceptable chips produced in the chipping operation were considered. More than 50% of branchwood chips were rejects, compared with about 22% for chips from the main stem (Young and Chase 1966). Rejection in this case was based upon nonuniform chip size.

One wood fiber product for which branchwood appears to be an entirely suitable raw material is *hardboard*, a high-density product made from mechanically ground pulp by compressing a fiber mat under heat and pressure. A number of mills in the United States currently use whole-tree chips, including branches mixed with normal wood, for hardboard manufacture.

In summary, branchwood can substantially increase the quantity of wood fiber per area of forest harvested in comparison with the use of the main stem only. However, branch material has several disadvantages, including a high percentage of bark, a significantly lower wood-specific gravity, and short fiber length. These factors markedly limit commercial uses of branchwood. One development that could greatly enhance further use of topwood is the perfection of technologies for economically separating bark from small branches.

Bark

Although the focus of this textbook is on wood and its utilization, several other components of the tree are both abundant and increasingly valuable. One of these materials is bark, a by-product of wood products manufacture. In 2005, ~55 million m³ (or about 20 million metric tons, dry weight basis) of bark was delivered to mills in the United States as part of sawlogs, pulpwood bolts, and wood chips. Worldwide, the volume of bark generated as a result of the harvest of industrial wood approximated 60 million metric tons in 2001. Once considered an expensive and irritating disposal problem, bark is now widely used as an industrial fuel, soil amendment, and ground cover, and is a possible source of chemical feedstocks. Most pulp and paper mills, sawmills, and plywood plants today burn all of their bark to produce energy to run the plant. This is discussed in Chapter 16. More information about bark can be found online.

References and Supplemental Reading

Abdel-Gadir, A.Y. and Krahmer, R.L. (1993). Estimating the age of demarcation of juvenile and mature wood in Douglas-fir. *Wood Fiber Sci.* 25 (3): 242–249.

Andersson, S., Hansen, R., Jonsson, Y., and Nylinder, M. (1978). Harvesting systems for stumps and roots – a review and evaluation of Scandinavian techniques. In: *Complete Tree Utilization of Southern Pine* (ed. C.W. McMillin), 130–145. Madison, WI: Forest Products Research Society.

Barefoot, A.C., Hitchings, R.G., and Ellwood, E.L. (1964). Wood characteristics and kraft paper properties of four selected loblolly pines. I. Effect of fiber morphology under identical cooking conditions. *TAPPI* 47 (6): 343–356.

Barker, R.G. (1974). Papermaking properties of young hardwoods. *TAPPI* 57 (8): 107–111.

Bella, I.E. and Hunt, K. (1973). Kraft pulping of young trembling aspen from Manitoba. *Can. J. For. Res.* 3 (3): 359–366.

Bendtsen, B.A. (1978). Properties of wood from improved and intensively managed trees. *For. Prod. J.* 28 (10): 61–72.

Bendtsen, B.A. and Senft, J. (1986). Mechanical and anatomical properties in individual growth rings of plantation-grown eastern cottonwood and loblolly pine. *Wood Fiber Sci.* 18 (1): 23–38.

Biblis, E.J. and Carino, H.F. (1999). Flexural properties of lumber from 50-year-old loblolly pine plantation. *Wood Fiber Sci.* 31 (2): 200–203.

Biblis, E.J., Carino, H.F., Brinker, R., and McKee, C.W. (1995). Effect of stand density on flexural properties of lumber from two 35-year-old loblolly pine plantations. *Wood Fiber Sci.* 27 (1): 25–33.

Boehner, A.W. and Gertjejansen, R.O. (1975). Effect of three species of logging slash on properties of aspen planer shavings particleboard. *For. Prod. J.* 25 (12): 36–42.

Boone, R.S., and Chudnoff, M. (1972). Compression wood formation and other characteristics of plantation grown *Pinus caribaea*. USDA For. Serv. Res. Pap. ITF-13.

Brändström, J. (2004). Microfibril angle of the S-1 cell wall layer of Norway spruce compression wood tracheids. *IAWA J.* 25 (4): 415–423.

Brunden, M.N. (1964). Specific gravity and fiber length in crown-formed and stem-formed wood. *For. Prod. J.* 14 (1): 13–17.

Burns, G.P. 1920. Eccentric growth and the formation of red wood in the main stem of conifers. VT Agric. Exp. Stn Bull. 219.

Carpenter, C.H. (1984). The mechanical pulping of southern pine containing relatively large amounts of spring and juvenile fiber. In Utilization of the Changing Wood Resource in the Southern United States. North Carolina State Univ., pp.124–146.

Casperson, G., Jacopian, V., and Phillipp, B. (1968). Influence of different cooking processes on the ultrastructure of poplar reactionwood. *Sven. Pap.* 71 (13/14): 482–487. (Ger. Abstr. in Weiner, J., and Roth, L. 1970. Biblio. Ser. 184, Suppl. 2. Inst. Pap. Chem.).

Chang, H., Kadla, J., Li, B., and Sederoff, R. (2005). Exploiting genetic variation of fiber components and morphology in juvenile loblolly pine. North Carolina State Univ./US Dept of Energy, 27 June.

Clair, B. and Thibault, B. (2001). Shrinkage of the gelatinous layer of poplar and beech tension wood. *IAWA J.* 22 (2): 121–131.

Côté, W.A. Jr. and Day, A.C. (1965). Anatomy and ultrastructure of reaction wood. In: *Cellular Ultrastructure of Woody Plants* (ed. W.A. Côté Jr.), 391–418. Syracuse, NY: Syracuse University Press.

Côté, W.A. Jr., Day, A.C., Simson, B.W., and Timell, T.E. (1966). Studies of larch arabinogalactin. I. The distribution of arabinogalactin in larch wood. *Holzforschung* 20 (6): 178–192.

Crist, J.B. (1976). Utilization advantages of material produced in maximum fiber yield plantations. Intensive plantation culture – Five years research. USDA For. Serv. Gen. Tech. Rep. NC-21.

Crist, J.B., Dawson, D.H., and Nelson, J.A. (1977). Wood and bark quality of juvenile jack pine and eastern larch grown under intensive culture. Proc. TAPPI Biol. – Wood Chem. Conf., pp. 211–216.

Cutler, D.E. (1976). Variation in root wood anatomy. In: *Wood Structure and Biological and Technological Research* (ed. P. Baas, A.J. Bolton and D.M. Catling), 143–156. Leiden: Leiden University Press.

Dadswell, H.E. (1958). Wood structure variations occurring during tree growth and their influence on properties. *J. Inst. Wood Sci.* 1: 11–33.

Dadswell, H.E. and Wardrop, A.B. (1955). The structure and properties of tension wood. *Holzforschung* 9 (4): 97–104.

Dadswell, H.E., and Wardrop, A.B. 1960. Recent progress in research on cell wall structure. Proc. 5th World For. Congr., vol. 2, pp. 1279–1288.

Dadswell, H.E., Wardrop, A.B., and Watson, A.J. (1958). The morphology, chemistry and pulping characteristics of reaction wood. In: *Fundamentals of Papermaking Fibers* (ed. K. Bolam), 187–229. London: British Paper and Board Makers Association.

Davis, B.M. and Hurley, D.W. (1978). Fiber from a southern pine root system? In: *Complete Tree Utilization of Southern Pine* (ed. C.W. McMillin), 274–276. Madison, WI: Forest Products Research Society.

DeBell, D., Singleton, R., Harrington, C., and Gartner, B. (2002). Wood density and fiber length in young populus stems: relation to clone, age, growth rate, and pruning. *Wood Fiber Sci.* 34 (4): 529–539.

Deresse, T., Shepard, R., and Shaler, S. (2003). Microfibril angle variation in red pine (*Pinus resinosa* Ait.) and its relation to the strength and stiffness of early juvenile wood. *For. Prod. J.* 53 (7/8): 34–40.

Dinwoodie, M.J. (1961). Tracheid and fibre length in timber – a review of literature. *Forestry* 34 (2): 125–144.

Dyer, R.E. (1967). Fresh and dry weight, nutrient elements, and pulping characteristics of northern white cedar. ME Agric. Exp. Stn Tech. Bull. 27.

Einspahr, D.W. (1976). The influence of short-rotation forestry on pulp and paper quality. II. Short-rotation hardwoods. *TAPPI* 59 (11): 63–66.

Erickson, H.D. and Arima, T. (1974). Douglas-fir wood quality studies. II. Effects of age and stimulated growth on fibril angle and chemical constituents. *Wood Sci. Technol.* 8 (4): 255–265.

Eskilsson, S. (1969). Fiber properties in the spruce root system. *Cell. Chem. Technol.* 3 (4): 409–416.

Ewart, A.C.J. and Mason-Jones, A. (1906). The formation of red wood in conifers. *Ann. Bot. Lond.* 20: 201–203.

Fayle, D.C.F. (1968). Radial growth in tree roots. *Univ. Toronto Fac. For. Tech. Rep* 9: 183.

Fegel, A.C. (1941). Comparative anatomy and varying physical properties of trunk-, branch-, and rootwood of certain northeastern trees. NY State Col. For., Syracuse Univ. Tech. Publ. 55.

Fight, R., Snellgrove, T., Curtis, R., and Debell, D. (1986). Bringing timber quality considerations into forest management decisions: a conceptual approach. In: *Douglas-fir: Stand Management for the Future* (ed. C. Oliver, D. Hanley and J. Johnson), 20–25. Seattle: University of Washington Press.

Foran, C.D. (1984). Wood quality, a pulp mill perspective: case studies of the impact of juvenile wood usage on by-products recovery. In Utilization of the Changing Wood Resource in the Southern United States. North Carolina State Univ., pp. 231–242.

Fraser, D.A. (1952). Initiation of cambial activity in some forest trees in Ontario. *Ecology* 33 (2): 259–273.

Geimer, R., Herian, V., and Xu, D. (1997). Influence of juvenile wood on dimensional stability and tensile properties of flakeboard. *Wood Fiber Sci.* 29 (2): 103–119.

Gerry, E. (1915). Fiber measurement studies: length variations; where they occur and their relation to the strength and uses of wood. *Science* 61 (1048): 179.

Gooding, J.W., and Smith, W.H. (1972). Effects of fertilization on stem, wood properties, and pulping characteristics of slash pine (*Pinus elliottii* var. elliottii Engelm.). Proc. Symp. Effect of Growth Acceleration on the Properties of Wood, Madison, Wisconsin, E1–19.

Goyal, G.C., Fisher, J.J., Krohn, M.J. et al. (1999). Variability in pulping and fiber characteristics of hybrid poplar trees due to their genetic makeup, environmental factors, and tree age. *TAPPI* 82 (5): 141–147.

Groom, L., Shaler, S., and Mott, L. (2002). Mechanical properties of individual southern pine fibers. Part III. Global relationships between fiber properties and fiber location within an individual tree. *Wood Fiber Sci.* 34 (2): 238–250.

Hakkila, P. (1971). Branches, stumps, and roots as future raw material source. In: *Forest Biomass Studies* (ed. H. Young). Orono, ME: University of Maine Press.

Hallock, H. (1969). Sawing to reduce warp of lodgepole pine studs. USDA For. Serv. Res. Pap. 102.

Hamilton, J.R., Cech, F.C., and Shurtliffe, C.E. (1976). Estimating bole specific gravity from limbs of mature black cherry and northern red oak trees. *Wood Fiber* 7 (4): 281–286.

Hamilton, J.R., Thomas, C.K., and Carvell, K.L. (1985). Tension wood formation following release of upland oak advance reproduction. *Wood Fiber Sci.* 17 (3): 382–390.

Hatton, J.V. (1993). Kraft pulping of second-growth jack pine. *TAPPI* 76 (5): 105–113.

Hatton, J.V. (1997). Pulping and papermaking properties of managed second-growth softwoods. *TAPPI* 80 (1): 178–184.

Hatton, J.V. and Cook, J. (1992). Kraft pulps from second-growth Douglas-fir: relationships between wood, fiber, pulp, and handsheet properties. *TAPPI* 75 (1): 137–144.

Hatton, J.V. and Gee, W.Y. (1994). Kraft pulping of second-growth lodgepole pine. *TAPPI* 77 (6): 91–102.

Hon, D.N.-S. and Bangi, A.P. (1996). Chemical modification of juvenile wood. Part 1. Juvenility and response of southern pine OSB flakes to acetylation. *For. Prod. J.* 46 (7/8): 73–78.

Howard, E.T. (1973). Physical and chemical properties of slash pine tree parts. *Wood Sci.* 5 (4): 312–317.

Howard, E.T. (1974). Slash pine rootwood in flakeboard. *For. Prod. J.* 24 (6): 29–35.

Hunt, K. and Hatton, J.V. (1975). Full forest utilization. II. Quality and kraft pulp yield of eastern Canadian hardwoods. *Pulp Pap. Mag. Can.* 76 (1): 97–102.

Hunt, K. and Keays, J.L. (1973). Short-rotation trembling aspen trees (*Populus tremuloides* Michx.) for kraft pulp. *Can. J. For. Res.* 3 (2): 180–184.

Isebrands, J. and Bensend, D. (1972). Incidence and structure of gelatinous fibers within rapid-growing eastern cottonwood. *Wood Fiber* 4 (2): 61–71.

Isebrands, J. and Parham, R.A. (1974). Tension wood anatomy of short rotation *Populus* spp. before and after kraft pulping. *Wood Sci.* 6 (3): 256–265.

Jaccard, P. (1938). Eccentric increment and anatomical–histological differentiation of wood. *Berl. Schweiz. Bot. Ges.* 48: 491–537.

Jackson, M., and Megraw, R.A. (1986). Impact of juvenile wood on pulp and paper products. In Juvenile Wood – What Does It Mean to Forest Management and Forest Products? For. Prod. Res. Soc., Proc. 47309.

Jane, F.W., Wilson, K., and White, O.J.B. (1970). *The Structure of Wood*. London: Adam & Charles Black.

Jayme, G. and Harders-Steinhäuser, M. (1953). Tension wood and its effect in poplar and willow wood. *Holzforschung* 7 (213): 39–43.

Jett, J.B. and Zobel, B.J. (1975). Wood and pulping properties of young hardwoods. *TAPPI* 58 (1): 92–96.

Keays, J.L. (1971a). Complete-tree utilization – an analysis of the literature. III. Branches. Can. Dep. Fish. For. Inf. Rep. VP-X-71.

Keays, J.L. (1971b). Complete-tree utilization – an analysis of the literature. IV. Crown and slash. Can. Dep. Fish. For. Inf. Rep. VP-X-77.

Keays, J.L. (1971c). Complete-tree utilization – an analysis of the literature. V. Roots and stump-root system. Can. Dep. Fish. For. Inf. Rep. VP-X-79.

Keays, J.L. and Hatton, J.V. (1971). Complete-tree utilization studies: the yield and quality of kraft pulp from the components of Tsuga heterophylla. *TAPPI* 54 (I): 99–104.

Kellogg, R.M. and Kennedy, R.W. (1986). Practical applications of wood quality relative to end use. In: *Douglas-fir: Stand Management for the Future* (ed. C.D. Oliver, D. Hanley and J. Johnson), 103–107. Seattle: University of Washington.

Kennedy, R.W. and Farrar, J.L. (1965). Tracheid development in tilted seedlings. In: *Cellular Ultrastructure of Woody Plants* (ed. W.A. Côté Jr.), 419–453. Syracuse, NY: Syracuse University Press.

Koch, P. (1972). Utilization of the Southern Pines, Vol. 1. USDA For. Serv. Agric. Handb. 420, pp. 535–574.

Koch, P. (1973). Whole tree utilization of southern pine advanced by developments in mechanical conversion. *For. Prod. J.* 23 (10): 30–33.

Koch, P. (1974). Harvesting southern pine with taproots can extend pulpwood resource significantly. *J. For.* 72 (5): 266–268.

Koch, P. (1976). Harvesting southern pine with taproots can extend pulpwood resource significantly. *Proc. Appl. Polymer Symp.* 28: 403–420.

Koch, P. and Coughran, S.J. (1975). Development of a puller-buncher for harvesting southern pines with taproot attached. *For. Prod. J.* 25 (4): 23–30.

Koubaa, A., Hernandez, R., Beaudoin, M., and Poliquin, J. (1998). Interclonal, intraclonal, and within-tree variation in fiber length of poplar hybrid clones. *Wood Fiber Sci.* 30 (1): 40–47.

Koubaa, A., Isabel, N., Zhang, S. et al. (2005). Transition from juvenile to mature wood in black spruce (*Picea mariana* [Mill.] B.S.P.). *Wood Fiber Sci.* 37 (3): 445–455.

Kozlowski, T.T. and Pallardy, S.G. (1997). *Physiology of Woody Plants*, 2e, 68–69. San Diego: Academic Press.

Krahmer, R.L. (1986). Fundamental anatomy of juvenile wood and mature wood. In Juvenile Wood – What Does It Mean to Forest Management and Forest Products? For. Prod. Res. Soc. Proc. 47309.

Krempl, H. (1975). Differences in proportion of tension wood in various poplar species. *Holzforsch. Holzverwert.* 27 (6): 131–137.

Kretschmann, D., Cramer, S., Lakes, R., and Schmidt, T. (2006). Selected mesostructure properties in loblolly pine from Arkansas plantations. In: *Characterization of the Cellulosic Cell Wall* (ed. D. Stokke and L. Groom), 149–170. Ames, IA: Blackwell Press.

Kučera, B. (1994). An hypothesis relating current annual height increment to juvenile wood formation in Norway spruce. *Wood Fiber Sci.* 26 (1): 152–167.

Kumar, S., Barbour, R., and Gustafson, R. (2004). Kraft pulping response and paper properties of wood from densely stocked small-diameter stands. *For. Prod. J.* 54 (5): 50–56.

Larson, P.R. (1969). Wood formation and the concept of wood quality. Yale Univ. Sch. For. Bull. 74.

Larson, P.R., Kretschmann, D.E., Clark, A.I., and Isebrands, J.D. (2001). Formation and properties of juvenile wood in southern pines: A synopsis. USDA For. Serv. Gen. Tech. Rep. FPL-GTR-129.

Laundrie, J.E, and Berbee, J.G. (1972). High yields of kraft pulp from rapid-growth hybrid poplar trees. USDA For. Serv. Res. Pap. FPL-186.

Lee, P.W. (1971). Physical properties of the stem, branch, root, and topwood of pitch pines grown in Korea. *Seoul Natl Univ. For. Bull.* 8: 35–45.

Lehmann, W.F. and Geimer, R.L. (1974). Properties of structural particleboards from Douglas-fir residues. *For. Prod. J.* 24 (10): 17–25.

MacPeak, M.D., Burkart, L.F., and Weldon, D. (1990). Comparison of grade, yield, and mechanical properties of lumber produced from young fast-grown and older slow-grown planted slash pine. *For. Prod. J.* 40 (1): 11–14.

Manwiller, E.G. (1972). Tracheid dimensions in rootwood of southern pine. *Wood Sci.* 5 (2): 122–124.

Manwiller, E.G. (1974). Fiber lengths in stems and branches of small hardwoods on southern pine sites. *Wood Sci.* 7 (2): 130–132.

McAlister, R.H. and Clark, A. (1991). Effect of geographic location and seed source on the bending properties of juvenile and mature loblolly pine. *For. Prod. J.* 41 (9): 39–42.

McAlister, R.H. and Clark, A. (1992). Shrinkage of juvenile and mature wood of loblolly pine from three locations. *For. Prod. J.* 42 (7/8): 25–28.

McAlister, R.H., Powers, H.R., and Pepper, W.D. (2000). Mechanical properties of stemwood and limbwood of seed orchard loblolly pine. *For. Prod. J.* 50 (10): 91–94.

McKee, J.C. (1960). The kraft pulping of small diameter slash pines. *TAPPI* 43 (6): 202A–204A.

McKee, J.C. (1984). The impact of high volumes of juvenile wood on pulp mill operations and operating costs. In Utilization of the Changing Wood Resource in the Southern United States. North Carolina State Univ., pp. 178–182.

Megraw, R.A., and Nearn, W.T. (1972). Detailed DBH density profiles of several trees from Douglas-fir fertilizer/thinning plots. Proc. Symp. Effect of Growth Acceleration on the Properties of Wood, Madison, WI, G1–24.

Metzker, K. (1908). Konstruktionsprinzip des sekundaren Holzkorpers. *Naturwiss. Z. For. Land wirtsch.* 6: 249–274.

Něcesaňy, V. (1958). Effect of b-indolacetic acid on the formation of reaction wood. *Phyton* 11: 117–127.

Nemli, G., Hiziroglu, S., Usta, M. et al. (2004). Effect of residue type and tannin content on properties of particleboard manufactured from black locust. *For. Prod. J.* 54 (2): 36–40.

North Carolina State College. (1957). First annual report. NC State – Ind. Coop. For. Tree Improv. Program.

Noskowiak, A.F. (1963). Spiral grain in trees – a review. *For. Prod. J.* 13 (7): 266–277.

Nyström, J. and Kline, D.E. (2000). Automatic classification of compression wood in green southern yellow pine. *Wood Fiber Sci.* 32 (3): 301–310.

Okuyama, T., Takeda, H., Yamamoto, H., and Yoshida, M. (1998). Relation between growth stress and lignin concentration in the cell wall: ultraviolet microscopic spectral analysis. *J. Wood Sci.* 44 (1): 83–89.

Olson, R.A., Poletika, N.V., and Hicock, H.W. (1947). Strength properties of plantation-grown coniferous woods. Conn. Agric. Exp. Stn Bull. 511.

Onaka, F. (1949). Studies on compression and tension wood. Wood Res. Kyoto no. 1.

Page, D.H., El-Hosseiny, F., Winkler, K., and Bain, R. (1972). The mechanical properties of single woodpulp fibers. I. A new approach. *Pulp Pap. Mag. Can.* 73 (8): 72–76.

Panshin, A.J. and de Zeeuw, C. (1980). *Textbook of Wood Technology*, 4e. New York: McGraw-Hill.

Panshin, A.J., Harrar, E.S., Bethel, J.S., and Baker, W.J. (1962). *Forest Products – Their Sources, Production, and Utilization.* New York: McGraw-Hill.

Parham, R.A., Robinson, K.W., and Isebrands, J.G. (1976). Effects of tension wood on kraft paper from a short-rotation hardwood (*Populus* "Tristis #1"). Inst. Pap. Chem. Tech. Pap. Ser. 40.

Pearson, R.G. and Gilmore, R.G. (1971). Characterization of the strength of juvenile wood of loblolly pine (*Pinus taeda* L.). *For. Prod. J.* 21 (1): 23–30.

Peszlen, I. (1995). Juvenile wood characteristics of plantation wood species. Abstract XX IUFRO World Congress, Finland. *IAWA J.* 16 (1): 14.

Phelps, J.E., Isebrands, J.G., and Jewett, D. (1982). Raw material quality of short-rotation, intensively cultured *Populus* clones. I. A comparison of stem and branch properties at three spacings. *IAWA Bull.* 3 (3/4): 193–200.

Phillips, D.R., Clark, A. III, and Taras, M.A. (1976). Wood and bark properties of southern pine branches. *Wood Sci.* 8 (3): 164–169.

Posey, C.E. (1964). The effects of fertilization upon wood properties of loblolly pine (*Pinus taeda* L.). NC State Coll. Sch. For. Tech. Rep. 22.

Preston, R.D. and Ranganathan, V. (1947). The fine structure of the fibers of normal and tension wood in beech (*Fagus sylvatica* L.) as revealed by X-rays. *Forestry* 11 (1): 92–97.

Priya, P.B. and Bhat, K.M. (1999). Influence of rainfall, irrigation, and age in the growth periodicity and wood structure in teak (*Tectona grandis*). *IAWA J.* 20 (2): 181–192.

Pugel, A., Price, E., and Hse, C. (1989). Composites from southern pine juvenile wood. Part I. *For. Prod. J.* 40 (1): 29–33.

Pugel, A., Price, E., and Hse, C. (1990). Composites from southern pine juvenile wood. Part II. *For. Prod. J.* 40 (3): 57–61.

Pugel, A. and Shupe, T. (2004). Composites from southern pine juvenile wood. Part III. *For. Prod. J.* 54 (1): 47–52.

Rendle, B.J. (1960). Juvenile and adult wood. *J. Inst. Wood Sci.* 5: 58–61.

Rowell, R.M., Lichtenberg, R.S., and Larsson, P. (1993). Stability of acetylated wood to environmental changes. *Wood Fiber Sci.* 25 (4): 359–364.

Rowell, R.M., Tillman, S.M., and Simonsen, R. (1986). A simplified procedure for the acetylation of hardwood and softwood flakes for flakeboard production. *J. Wood Chem. Technol.* 6 (3): 427–448.

Scaramuzzi, G. and Vecchi, E. (1968). Characteristics of mechanical pulp from poplar tension wood. *Cellul. Carta* 19 (2): 3–12.

Schumann, D.R., and Pillow, M.Y. (1969). Effect of tension wood on hard maple used for manufactured parts. USDA For. Serv. Res. Pap. FPL–108.

Scurfield, G. (1973). Reaction wood: its structure and function. *Science* 179: 647–655.

Semke, L.K. (1984). Effect of juvenile pine fibers on kraft paper properties. In Utilization of the Changing Wood Resource in the Southern United States. North Carolina State Univ., pp. 160–177.

Senft, J.F., Bendtsen, B.A., and Galligan, W.L. (1985). Weak wood: fast grown trees make problem lumber. *J. For.* 83 (8): 477–484.

Senft, J.F, Quanci, M.J., and Bendtsen, B.A. (1986). Property profile of 60-year-old Douglas-fir. In Juvenile Wood – What Does It Mean to Forest Management and Forest Products? For. Prod. Res. Soc. Proc. 47309.

Sheng, D., Uno, H., and Yamamoto, F. (2004). Roles of auxin and gibberellin in gravity-induced tension wood formation in *Aesculus turbinata* seedlings. *IAWA J.* 25 (3): 337–347.

Siripatanadilok, S. and Leney, L. (1985). Compression wood in western hemlock (*Tsuga heterophylla*). *Wood Fiber Sci.* 17 (2): 254–265.

Spurr, S.H. and Hyvärinen, M.J. (1954). Wood fiber length as related to position in the tree and growth. *Bot. Rev.* 20 (9): 561–575.

Tanaka, E. and Koshijima, T. (1981). Characterization of cellulose in compression and opposite woods of a *Pinus densiflora* tree grown under the influence of strong wind. *Wood Sci. Technol.* 15 (4): 265–273.

Tang, Y. and Pearson, R.G. (1992). Effect of juvenile wood and choice of parametric property distributions on reliability-based beam design. *Wood Fiber Sci.* 24 (2): 216–224.

Taylor, F.W. (1968). Variation of wood elements in yellow poplar. *Wood Sci. Technol.* 2 (3): 153–165.

Taylor, F.W. (1977). A note on the relationship between branch- and stemwood properties of selected hardwoods growing in the mid South. *Wood Fiber* 8 (4): 257–261.

Timell, T.E. (1973a). Studies on opposite wood in conifers. I. Chemical composition. *Wood Sci. Technol.* 7 (1): 1–5.

Timell, T.E. (1973b). Studies on opposite wood in conifers. II. Histology and ultrastructure. *Wood Sci. Technol.* 7 (2): 79–91.

Timell, T.E. (1973c). Ultrastructure of the dormant and active cambial zones and the dormant phloem associated with formation of normal and compression wood in *Picea abies* (L.) Karst. State Univ. NY Tech. Publ. 96.

Timell, T.E. (1986). *Compression Wood in Gymnosperms*, vol. I, II, and III. Berlin: Springer.

Tsoumis, G. (1952). *Properties and Effects of the Abnormal Wood Produced by Leaning Hardwoods.* New Haven, CT: Yale School of Forestry.

Tsoumis, G. (1968). *Wood as Raw Material.* New York: Pergamon Press.

Turner, L.M. (1936). Root growth of seedlings of *Pinus echinata* and *Pinus taeda. J. Agric. Res.* 53: 145–149.

Wardrop, A.B. (1964). The reaction anatomy of arborescent angiosperms. In: *The Formation of Wood in Forest Trees* (ed. M.H. Zimmermann), 405–456. New York: Academic Press.

Wardrop, A.B. and Dadswell, H.E. (1950). The nature of reaction wood. II. The cell wall organization of compression wood tracheids. *Aust. J. Sci. Res.* 5B (1): 1–13.

Wardrop, A.B. and Davies, G.W. (1964). The nature of reaction wood. VIII. The structure and differentiation of compression wood. *Aust. J. Bot.* 12 (1): 24–38.

Wasniewski, J. (1989). Evaluation of juvenile wood and its effect on Douglas-fir structural composite panels. In Proc., 23rd Particleboard and Compos. Materials Symp. Pullman, WA: Washington State Univ.

Watson, A.J. and Dadswell, H.R. (1957). Papermaking properties of compression wood from *Pinus radiata. Appita* 11: 56–70.

Wershing, H.F. and Bailey, I.W. (1942). Seedlings as experimental material in the study of "redwood" in conifers. *J. For.* 40 (5): 411–414.

Westing, A.H. (1968). Formation and function of compression wood in gymnosperms. II. *Bot. Rev.* 34 (1): 51–78.

Worster, H.E. and Vinje, M.G. (1968). Kraft pulping of western hemlock tree tops and branches. *Pulp Pap. Mag. Can.* 69 (14): 57–60.

Yang, K.C.; and Benson, C.A. 1997. Formation, distribution and its criteria for determining the juvenile-mature wood transition zone. Pages IX-7 in Proc. CTIA/IUFRO International Wood Quality Workshop; 18–22 August, Quebec.

Yang, K.C., Benson, C.A., and Wong, J.K. (1986). Distribution of juvenile wood in two stems of *Larix laricina. Can. J. For. Res.* 16 (5): 1041–1049.

Yang, K.C., Chen, Y.S., and Benson, C.A. (1994). Vertical and radial variation of nuclear elongation index of living sapwood ray parenchyma cells in a plantation tree of *Cryptomeria japonica. IAWA Journal* 15 (3): 323–327.

Yoshida, M., Ohta, H., and Okuyama, T. (2002). Tensile growth stress and lignin distribution in the cell walls of black locust (*Robinia pseudoacacia*). *J. Wood Sci.* 48 (2): 99–105.

Yoshizawa, N., Satoh, I., Yokota, S., and Idei, T. (1992). Response of differentiating tracheids to stem inclination in young trees of *Taxus cuspidata. IAWA Bull.* 13 (2): 187–194.

Young, H.E. (1964). The complete tree concept – A challenge and an opportunity. Proc. Soc. Am. For., pp. 231–233.

Young, H.E. (1974). Complete tree concept: 1964–1974. *For. Prod. J.* 24 (12): 13–16.

Young, H.E. and Chase, A.J. (1966). Pulping hardwoods? Try sulfate process on branches, roots. *Pulp Pap.* 40 (27): 29–31.

Young, H.E., Gammon, C., and Hoar, L.E. (1963). Potential fiber from red spruce and red maple logging residues. *TAPPI* 46 (4): 256–259.

Young, H.E., Hoar, L.E., and Ashley, M. (1965). Weight of wood substance for components of seven tree species. *TAPPI* 48 (8): 466–469.

Zobel, B.J., and Haught, A.E., Jr. (1962). Effect of bole straightness on compression wood of loblolly pine. NC State Coll. Sch. For. Tech. Rep. 15.

Zobel, B.J., Jett, J.B., and Hutto, R. (1978). Improving wood density of short-rotation southern pine. *TAPPI* 61 (3): 41–44.

Zobel, B.J. and Kellison, R.C. (1972). Short rotation forestry in the southeast. *TAPPI* 55 (8): 1205–1208.

Zobel, B.J., Kellison, R.C., and Kirk, D.G. (1972). Wood properties of young loblolly and slash pines. Proc. Symp. Effect of Growth Acceleration on the Properties of Wood, Madison, WI, M1–22.

Zobel, B.J. and Sprague, J.R. (1998). *Juvenile Wood in Forest Trees*. Berlin: Springer.

7

Wood and Water

Water is a natural and necessary constituent of all parts of a living tree. In the xylem portion of the stem, water commonly makes up over half the total wet or "green" weight. Stated another way, the weight of water in green wood is commonly equal to or greater than the weight of dry wood substance. When a tree dies or a log is processed into lumber, veneer, chips, etc., the wood immediately begins to lose some of its moisture to the surrounding atmosphere. As drying continues, the dimensions and the physical properties of the wood begin to change. Some water, however, generally remains within the structure of the cell walls even after the wood has been manufactured into lumber or other wood-based products.

The physical and mechanical properties, resistance to biological deterioration, and dimensional stability of any wood-based product are all affected by the amount of water present. Because almost all properties of wood and wood products are affected by water, it is important to understand the nature of water in wood and how it is associated with wood microstructure and properties. This chapter is devoted to this subject and, in addition, covers some practical aspects of wood drying and dimensional change. For satisfactory use of wood as a raw material these relationships must be clearly understood.

Location of Water in Wood

Water in green or freshly harvested wood is located within both the cell wall and the cell lumen. The amount of water within the cell wall structure of a living tree remains essentially constant from season to season, but the amount of water in the lumen may vary. The water in the lumen may contain dissolved food materials produced by photosynthesis as well as organic and inorganic compounds. This solution is commonly referred to as *sap*.

When wood is dried during manufacture, all liquid water in the cell lumen is removed. The cell lumen always contains some water vapor, however. The amount of water remaining in the cell walls of a finished product depends upon the extent of drying during

Forest Products and Wood Science: An Introduction, Seventh Edition. Rubin Shmulsky and P. David Jones.
© 2019 John Wiley & Sons Ltd. Published 2019 by John Wiley & Sons Ltd.
Companion website: www.wiley.com/go/shmulsky

manufacture and the environment into which the product is later placed. Once removed by drying, water will recur in the lumen only if the product is exposed to liquid water. Three common ways for this to occur are installing wood in the ground, placing it where it can be rained on, and locating it where condensation or leaking water occur.

Figure 7.1 may help in visualizing the location of water in a wood cell. As long as there is any liquid water in the lumen, the wall of the cell will be saturated, i.e. it will contain as much water as it can physically adsorb (see the following discussion). Most physical properties of wood (other than weight) are not affected by differences in the amount of water in the cell lumen. For example, if the lumen is one-quarter full of liquid water, the cell (and the wood) will have the same strength as when the lumen is half full.

The green (wet) cell is illustrated in Figure 7.1a. As green wood begins to dry, water is first removed from the lumen. When the wood is dried to the point where all the water in the lumen is removed, water then begins to leave the cell wall. Almost all wood products used in buildings or where there is no contact with the ground contain water in the form shown in Figure 7.1b.

The point at which all the liquid water in the lumen has been removed but the cell wall is still saturated is termed the *fiber saturation point* (FSP). This is a critical level because below this point the properties of wood are altered by changes in moisture content. If dry wood is used where it has no contact with a source of liquid water, the amount of water in the wood will always be less than the FSP.

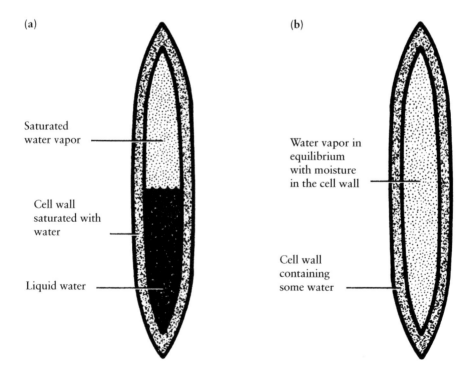

(a)

Saturated
water vapor

Cell wall
saturated with
water

Liquid water

(b)

Water vapor in
equilibrium
with moisture
in the cell wall

Cell wall
containing
some water

FIGURE 7.1.   Water in a cell of green wood (a) and dry wood (b).

Nature of Water in Wood

To simplify discussion, the liquid water found in the lumen of wood is often referred to as *free water* and the water within the cell wall is called *bound water*. This is an appropriate description, because free water is relatively easy to remove and so is the first to be removed when wood is dried. Bound water is held more tightly because of surface adsorption within the wood structure. As moisture content decreases below the FSP, the strength of the wood–water bond increases.

The water within the cell wall (bound) is held by adsorption forces, mainly hydrogen bonds. This is not to be confused with the *absorption* that takes place, for example, when a synthetic sponge soaks up water. *Absorption* results from surface tension and capillary forces, and it results in bulk accumulation of water in the porous wood. *Adsorption*, in contrast, involves the attraction of water molecules to hydrogen-bonding sites present in cellulose, hemicellulose, and lignin. Hydrogen bonding occurs on the hydrogen side of the hydroxyl (OH) groups found throughout the chemical elements of wood. Figure 7.2 illustrates water molecules held to a fragment of hemicellulose by hydrogen bonding. Four water molecules are monomolecularly (one per site) adsorbed onto the wood. At higher moisture contents, additional water molecules first bind to the remaining hydroxyl groups and then begin to polymolecularly adsorb, or piggyback, onto each other. In saturated green wood, as many as 6–10 water molecules may be attracted to each accessible sorption site. It is believed that the OH groups of adjacent cellulose molecules can be mutually bonded, or crosslinked. The location of water molecules in relation to the adjacent molecules is depicted in Figure 7.3.

Moisture Content Calculation

The amount of water in wood or a wood product is usually expressed as the *moisture content* (MC), defined as the weight of the water expressed as a percentage of the moisture-free or *ovendry* (OD) wood weight. The term *weight* rather than mass is employed throughout this book to conform to general usage. Thus:

$$\text{percentage MC} = (\text{weight of water/OD weight}) \times 100$$

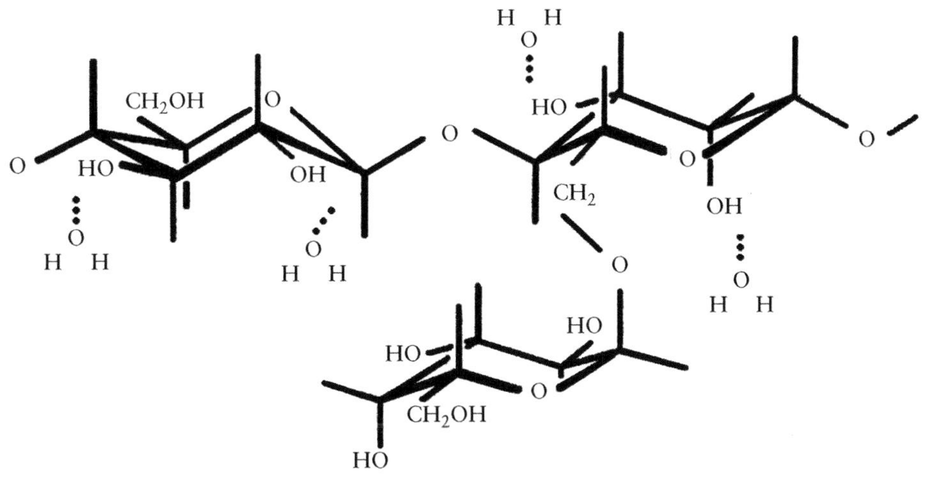

FIGURE 7.2.   Attraction of water to a fragment of galactoglucomannan hemicellulose.

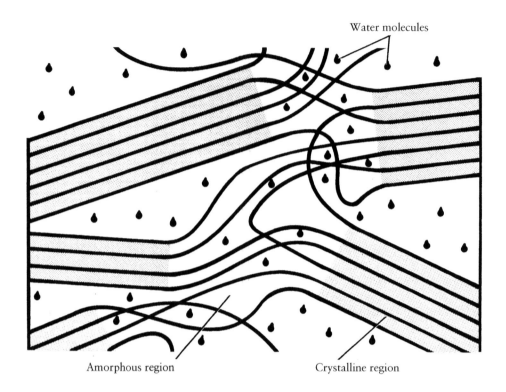

FIGURE 7.3.   Relationship of water to cellulose molecules.

Note that because the denominator is the dry weight, not the total weight, the moisture content as calculated in this way can be over 100%. In the pulp and paper industry and in parts of Europe, other practices are used to calculate moisture content, as will be discussed later.

The most reliable method of determining the moisture content is to weigh the wet sample, dry it in an oven at $103 \pm 2°C$ for 24 hours to drive off all water, and then reweigh it. For samples greater than 25 mm (1 in.) in any dimension the time should be increased and drying should stop when the samples reach a constant weight. The details of this ovendry method are described in American Society for Testing and Materials (ASTM) Standard D4442-92 (2003). Bench-scale microwave ovens can also be used to rapidly dry small samples or wafers, thereby reducing drying time from hours to minutes. Care must be taken, however, to stop the drying process before the wood burns. When using the ovendry method, the moisture content is computed as follows:

$$\text{percentage MC} = \left[(\text{weight with water} - \text{OD weight})/\text{OD weight}\right] \times 100$$

For example: a block of green redwood has a total weight of 970 g. After oven drying, the weight is 390 g. What was the moisture content when the weight was 970 g?

$$\text{percentage MC} = \left[(970-390)/390\right] \times 100 = 149\%$$

If a similar block of wood with the same ovendry weight is partially dried until the weight drops to 540 g, what is the moisture content after partial drying?

$$\text{percentage MC} = \left[(540-390)/390\right] \times 100 = 38\%$$

Note that when calculating the moisture content, the amount of water is expressed as a percentage of the weight of the dry wood. This method of calculation is the accepted standard for all lumber, plywood, particleboard, and fiberboard products in the United States and in most countries. In the pulp and paper industry and when wood is used as a fuel, the amount of moisture is often expressed as a percentage of the total weight, i.e. weight of wood plus water. When the wet weight basis is used in this text, it is indicated as *percentage* MC (wet basis). Such notation is a good practice to follow whenever dealing with MC calculations. The basic equation for moisture content can be manipulated into forms that are convenient to use in other situations. for example, solving the equation for ovendry weight yields:

$$\text{OD weight} = \text{green weight}/\left[1+(\text{percentage MC}/100)\right]$$

This form is useful for estimating the dry weight of green wood when the green weight is known and moisture content has been obtained from a small sample. For example, a load of pulpwood weighs 15 tons (16.5 short tons). The moisture content is found to be 90% by oven drying representative cross-sections from several bolts. An estimate of the dry weight of wood is desired so that the yield of pulp can be predicted.

$$\text{OD weight} = 15 \text{ tons}/(1+0.90) = 7.89 \text{ tons}$$

In another situation, 12 m³ (5080 board feet) of lumber to be exported has a moisture content of 19%. The ovendry weight of a cubic meter of this species (volume measured at 19%) is known to be 680 kg. The total weight of the shipment is desired.

$$\text{Green weight} = \text{OD weight} \times \left[(1+(\text{percentage MC}/100)\right]$$

thus

$$\text{Green weight} = (680 \text{ kg m}^{-3} \times 12 \text{ m}^3)(1+0.19) = 9710 \text{ kg}$$

Consider a situation where a rail car containing 165 m³ (70,000 board feet) of lumber, weighing 730 kg m⁻³ (3.8 lb per board foot) when green, is to be shipped. The shipping cost is $0.10 kg⁻¹ ($4.5 per 100 lb). The lumber is estimated to average 60% MC when green. How much money would be saved in shipping cost if this lumber were dried to 15% MC prior to shipping?

$$\text{OD weight per m}^3 = 730 \text{ kg m}^{-3}/1.60 = 456 \text{ kg}$$

$$\text{Weight at 15\% MC per m}^3 = 456 \text{ kg} \times 1.15 = 525 \text{ kg}$$

Weight savings per m^3 = 730 − 525 kg = 205 kg

Total weight savings = 165 m^3 × 205 kg m^{-3} = 33,825 kg

Savings in shipping cost = 33 825 kg × $0.10 kg^{-1} = $3383 per rail car

Moisture Content Measurement

Determination of moisture content during manufacture and subsequent handling to verify conformance to commercial standards are generally accomplished by the ovendry method, described in the preceding section, or by the use of electrical moisture meters, which have the advantage of being relatively simple and fast. Other methods of determining moisture content are sometimes used for research purposes where high precision is required. Such methods are outlined in ASTM D4442-92 (2003).

The major disadvantages of the ovendry method are that (i) it is a destructive test requiring that a sample be cut from the piece, (ii) it can take several days to complete, and (iii) a few species contain volatile components other than water that can be driven off during drying, thereby resulting in erroneously high moisture indications. However, for most situations the ovendry method is a more reliable indication of moisture content than that obtained by meters or other nondestructive methods. In fact, moisture meters and other nondestructive tests are calibrated using the ovendry method.

A variety of electrical meters are available to measure the moisture content of lumber, chips, particles, and other products. Although meters are generally less precise than the ovendry method, their instant readout, ease of operation, and nondestructive nature make them well suited for many applications.

One of the most commonly used handheld meters for lumber is the *resistance-type moisture meter*, which measures the electrical resistance between pins driven into the wood. This type of meter indicates the moisture content based upon the relationship shown in Figure 7.4. Insulated pins make it possible to measure the resistance between the tips of the pins and therefore to determine the moisture content at different depths. Resistance-type meters are generally reliable in the 6–30% MC range. Because the electrical resistance of wood varies with temperature, corrections must be made if the wood temperature is significantly different from the calibration temperature indicated by the manufacturer. Pin orientation (with or across the grain) must also follow the guidelines of the manufacturer. In addition, corrections for species are often necessary because extractives can influence resistance. Above the FSP, electrical resistance meters give only a qualitative measure of moisture content, so other methods should be used when dealing with green wood.

Some electric meters are based upon the effect that moisture has on the behavior of wood as a capacitor when placed in a high-frequency field. The capacitance of wood varies with the density and moisture content. These meters measure the dielectric constant or power loss of the sample. Such meters must be calibrated for each species to account for density differences. The effective range of ~3–30% MC for capacitance/power-loss meters is only slightly greater than for the resistance-type meters. These meters have plate-type electrodes that may contact the surface of the lumber or veneer but no pins

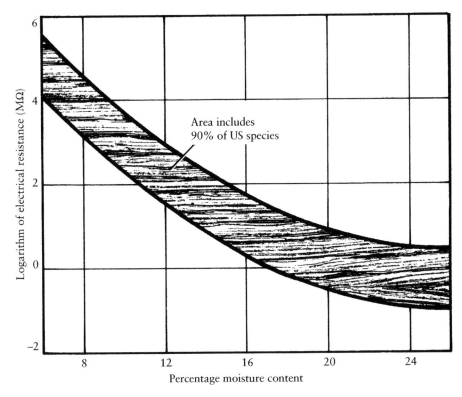

FIGURE 7.4.   Relationship of electrical resistance to MC. Source: USDA (2010).

need be driven, which is a particular advantage in valuable woods or when greater speed of measurement is needed. However, the proper use of species corrections is more critical with these meters than with resistance meters. Several classic types of handheld meters are shown in Figure 7.5. For further discussion on meter accuracy refer to Wengert and Bois (1997) and Wilson (1999).

A variety of meters are used in wood products manufacturing to measure the moisture content of lumber, veneer, strands, particles, or fibers that are bulk-piled or on conveyors. Microwave or radio frequency power absorbtion and attenuation, phase shift, and a combination of neutron and gamma radiation gauges are some of the technologies currently available for moisture measurement. Some of these moisture-measuring systems are suitable for automatically controlled production processes. These technologies can also be mated to pyrometers for temperature detection, x-ray machines or scales for density correction, and laser or light detectors for profile measurement. This combination of sensing technology can then be used to provide information on which to base production decisions. Such systems can provide moisture content information to an archive data file, monitor output of a production line, or provide input to a microprocessor used to control a drying process. Real-time moisture meters are currently used in many applications to determine when the wood material has reached its target moisture level.

FIGURE 7.5.   Several types of handheld moisture meters. Source: Courtesy of McCarthy Products Co. and University of Minnesota.

Relation of Moisture Content to the Environment

Because of the sorptive nature of wood, it has the ability to exchange water vapor with the surrounding air until it obtains moisture equilibrium with the air. Thus, wood is called a *hygroscopic material*. If wood is in equilibrium with the surrounding environment and the air then becomes drier, it will lose water (or desorb) until it again comes into equilibrium. The term *sorption* is applied to the combined or general phenomena of adsorption and desorption.

Below the FSP, the forces holding the bound water to the wood become greater as the moisture content decreases. As wood approaches the dry condition, less polymolecular adsorption and more monomolecular adsorption are involved. Thus, as moisture content decreases, the energy required to remove bound water increases.

The relationship between the relative vapor pressure in the environment and the moisture content of wood in equilibrium with that environment is not linear. This results from the different ways in which bound water is held. The graph of this relationship at a constant temperature is called a *sorption isotherm*. Isotherms developed for white spruce in 1931 by Seborg and Stamm illustrate the shape of the curve that is typical of most species and most wood products (Figure 7.6). Three curves are shown: desorption, adsorption, and cyclic desorption–adsorption. These curves show that if a piece of wood has desorbed to the point of equilibrium it may have a moisture content as much as 3% higher than if it had adsorbed to the point of equilibrium at the same relative vapor pressure. Above a relative vapor pressure of about 0.5, the initial desorption curve of green wood is slightly above that of a previously dried piece.

The moisture content that wood or a wood-based product attains while in an environment of constant temperature and humidity is termed the *equilibrium moisture content* (EMC). At room temperature, 20°C, considerable time is often required before large pieces of wood will achieve equilibrium – i.e. reach their EMC.

The difference between the desorption and adsorption curves is referred to as *hysteresis* or the lag effect. Hysteresis is common to many types of hygroscopic materials. A rather simplistic view of this complex phenomenon, but one that may help to visualize the dynamic nature of the water–wood equilibrium process, follows. In the green condition, the hydroxyl groups of the cellulosic cell wall are satisfied by water molecules, but as drying occurs these groups move closer together, allowing the formation of weak cellulose-to-cellulose bonds. When adsorption of water then occurs, fewer sorption sites are available for water than was the case originally. Those interested in a thorough discussion of sorption, isotherms, and hysteresis should study Stamm (1964) or Skaar (1972).

In situations where a wood product is subjected to alternating high- and low-humidity conditions, the moisture content approaches the middle curve of Figure 7.6. There is also a species effect; some woods vary markedly from the typical values shown in Figure 7.6, but the general shape of the sorption isotherm for all species is similar. The FSP indicated in Figure 7.6 is about 31% MC. Thirty percent is a value often used as a typical FSP for wood. Higgins (1957) demonstrated that the FSP varies widely. He also found that the hysteresis effect was considerably greater in some species than in others.

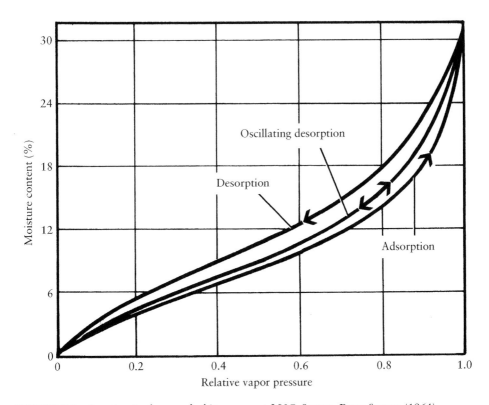

FIGURE 7.6.   Sorption isotherms of white spruce at 25 °C. Source: From Stamm (1964).

TABLE 7.1.   The fiber saturation point of several species.

Species	Fiber saturation point (% MC)
Southern yellow pine	29
Sitka spruce	28
Western redcedar	18
Redwood	22
Teak	18
Rosewood	15

Source: Higgins (1957).

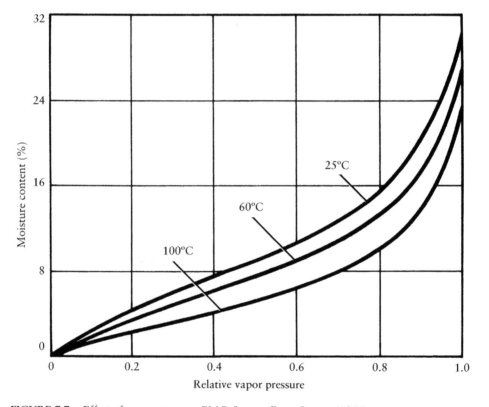

FIGURE 7.7.   Effect of temperature on EMC. Source: From Stamm (1964).

Fiber saturation points for selected species are shown in Table 7.1. Note that the dry-bulb temperature is simply the air temperature. One cause of variation in the FSP is the presence of extractives (Nzokou and Kamdem 2004). Species high in extractives usually have relatively low FSPs. The extractives occupy some sites in the cell wall that would otherwise attract water.

Temperature also has an effect on wood–water relationships. The general relationship is illustrated in Figure 7.7. Note that this temperature effect is relatively small. In addition to this real-time effect, high temperatures also have a permanent effect on the

TABLE 7.2. Percentage MC of wood in equilibrium with dry-bulb (air) temperatures and relative humidity conditions.

Dry-bulb temperature, °C (°F)	Relative humidity							
	20%	30%	40%	50%	60%	70%	80%	90%
21 (30)	4.6	6.3	7.9	9.5	11.3	13.5	16.5	21.0
10 (50)	4.6	6.3	7.9	9.5	11.2	13.4	16.4	20.9
21 (70)	4.5	6.2	7.7	9.2	11.0	13.1	16.0	20.5
32 (90)	4.3	5.9	7.4	8.9	10.5	12.6	15.4	19.8
43 (110)	4.0	5.6	7.0	8.4	10.0	12.0	14.7	19.1
54 (130)	3.7	5.2	6.6	7.9	9.4	11.3	14.0	18.2
66 (150)	3.4	4.8	6.1	7.4	8.8	10.6	13.1	17.2
77 (170)	3.0	4.3	5.6	6.8	8.2	9.9	12.3	16.2

Source: Adapted from Simpson (1991).

wood itself. Wood that has been subjected to temperatures in excess of 100 °C for long periods becomes less hygroscopic; i.e. it equalizes at a lower moisture content than normal wood. This is one of the reasons why products such as fiberboard and particleboard have lower equilibrium moisture content values than solid wood products. These products are often subjected to temperatures in excess of 150 °C during manufacture, the result of which is permanently altered (reduced) moisture isotherms.

It is assumed that for most applications and manufacturing considerations, wood attains the same EMC under a given temperature and relative humidity. A table prepared from data for Sitka spruce and used throughout the world is presented in the Wood Handbook (USFPL 2010) and also in Appendix A1, and is shown here in condensed form (Table 7.2). Note that wood's variability in sorption and FSP characteristics must be acknowledged. Note also that at 21 °C (70 °F) the moisture content of wood subjected to relative humidity from 30 to 70% will vary from 6.2 to 13.1%. This is the range of humidity conditions to which most wood is subjected in use. This is, therefore, the range of MC for normal wood used indoors where it is generally protected from direct contact with water.

Wood products such as plywood and particleboard tend to have slightly lower EMCs than the raw wood from which they are produced. This is because of the heat treatment effect discussed above, densification, and the addition of resins, coatings, and sizing materials, which are usually less hygroscopic than wood. The less the extent of reconstitution involved in the manufacturing process is, the less the effect on the EMC. For example, plywood and laminated wood products generally have EMC characteristics similar to those of wood or lumber whereas fiber- and particle-based products often exhibit considerably different characteristics. Table 7.3 shows EMC values for softwood plywood, particleboard, tempered hardboard, and decorative laminate samples as compared with solid wood. The idea of using heat energy as a means to permanently reduce the EMC and isothermic response of wood has been applied to solid wood products, such as flooring, as a means of stabilization.

If accurate EMC information is needed when dealing with a specific forest product, a laboratory determination should be made. Even products of the same type will vary in EMC because of differences in raw materials used and the specific manufacturing process.

TABLE 7.3.   EMC of typical forest products at 21 °C (70 °F).

Relative humidity (%)	Moisture content (%)				
	Wood	Softwood plywood	Particleboard	Oil-treated hardboard	High-pressure laminate
30	6.0	6.0	6.6	4.0	3.0
42	8.0	7.0	7.5	4.6	3.3
65	12.0	11.0	9.3	6.9	5.1
80	16.1	15.0	11.6	9.5	6.6
90	20.6	19.0	16.6	10.8	9.1

Source: Heebink (1966).

TABLE 7.4.   MC of green wood.

Species	Moisture content (%)	
	Heartwood	Sapwood
Hardwoods		
White ash	46	44
Aspen	95	113
Yellow birch	74	72
American elm	95	92
Sugar maple	65	72
Northern red oak	80	69
White oak	64	78
Sweetgum	79	137
Black walnut	90	73
Softwoods		
Western redcedar	58	249
Douglas fir	37	115
White fir	98	160
Ponderosa pine	40	148
Loblolly pine	33	110
Redwood	86	210
Eastern spruce	34	128
Sitka spruce	41	142

Source: USFPL (2010).

Moisture Content of Green Wood

The moisture content of green wood is important because of its direct relation to the weight of logs and green lumber. Therefore, moisture content is of concern to those who design harvesting and transport equipment, purchase wood on a weight basis, or ship or transport green wood.

   The moisture content of green wood varies considerably among species. Note that among species shown in Table 7.4, the moisture content of heartwood ranges from 33 to 98% and that of sapwood from 44 to 249%. The values in Table 7.4 should be considered as only general indications. Within any species there is considerable variation depending upon the location, age, season of harvest, and tree size. In softwoods, the average green moisture content tends to decrease as a tree grows older. Koch (1972) reported a 30% difference in the moisture content of southern pines over 45 years of age compared with trees under 25 years of age.

Within a single tree there is typically considerable variation in moisture content. The differences between sapwood and heartwood (Table 7.4) are one source of such variation. When wood in a tree changes from sapwood to heartwood, the amount of moisture in the cell wall often decreases, with the drop in moisture content preceded or accompanied by the deposition of extractives. These extractives tend to take the place of water molecules associated with cellulose and hemicellulose. Some extractives may also be left in solution or suspended in the free water in the lumina of the heartwood cells.

Hardwoods generally have smaller differences in moisture content between sapwood and heartwood. This contrasts markedly with softwoods, where the moisture content of sapwood is usually much higher than that in heartwood, often by a factor of 3–4. If you are lost in the forest and forced to build a fire from green wood, try burning the heartwood of a softwood, not a hardwood!

For most design or total weight estimates, the moisture contents of green wood given in Table 7.4 should be adequate. A more complete listing can be found in Table 2.1. Many trade associations and railroads publish lists of weights, by species, for green and dried products such as lumber, poles, and ties. Such information is useful, but if the actual moisture content can be measured, a more accurate estimate of green weight can be made with the procedures described in Chapter 8.

If green weight is to be used as the basis for purchasing logs or pulpwood it is advisable to conduct an on-site study of the green moisture content. The effect of log or bolt size and the season of the year should be determined. The goal of most weight-scaling procedures is to pay a fixed amount per unit of dry wood. Occasionally, adjustments to the price per unit weight may be made for defects, for small diameters, or for the degree of seasoning of the wood. Little information has been published regarding seasonal variation of moisture content, although many firms and associations have compiled such data. Koch (1972) reported that in southern pine the moisture content of increment cores is higher during midwinter than in summer. In Minnesota, where aspen pulpwood is often purchased by weight, some mills use a weight-to-volume ratio (density) of ~905 kg m^{-3} (4800 lb cord^{-1}) during the winter and 870 kg m^{-3} (4600 lb cord^{-1}) as the conversion factor during the summer, which reflects a higher moisture content in the winter. For many species, a higher moisture content is reported in the summer months.

Shrinking and Swelling

As wood dries below the FSP – i.e. loses bound water – it shrinks. Conversely, as water enters the cell wall structure, the wood swells. This dimensional change is a completely reversible process in small pieces of stress-free wood. In wood panel products such as fiberboard and particleboard, however, the process is often not completely reversible. This results, in part, from the compression that wood fibers or particles undergo during the manufacturing process. In large pieces of solid wood, dimensional change may not be completely reversible because of internal drying stresses.

Shrinking of the cell wall, and therefore of the whole wood, occurs as bound-water molecules escape from between the hemicellulose and long-chain cellulose molecules. These chain molecules then move closer together. The amount of shrinkage that occurs is proportional to the amount of water removed from the cell wall. Swelling is simply the reverse of this process. Because the S_2 layer of the cell wall is generally thicker than the other layers combined, the molecular orientation in this layer largely determines how shrinking occurs. In the S_2 layer most of the chain molecules are oriented almost parallel to the long axis of the cell. Thus, both transverse dimensions decrease as these molecules move closer together. For the same reason, the length of the cell is not greatly affected as the cell wall substance shrinks or swells. Interestingly, cell lumen diameter remains approximately constant while the cell wall shrinks and swells.

In reaction wood and other abnormal wood, the orientation of microfibrils in the S_2 layer is often at a steeper angle from the cell axis. Therefore, as the wood dries there is a measurable shortening of the cell; consequently, longitudinal shrinking occurs. Longitudinal shrinkage in such abnormal wood can be as great as 6–7% when going from the FSP to the ovendry condition. A 40×90 mm (2×4 in.) stud, 2.44 m (8 ft) long for the wall of a home, would shrink almost 75 mm (3 in.) in length during drying from its FSP to EMC condition if it were manufactured from such material. Shrinking and swelling are expressed as a percentage of dimension before the change occurred. Thus:

percentage shrinkage = (decrease in dimension or volume/original dimension or volume) * 100

percentage swelling = (increase in dimension or volume /original dimension or volume) * 100

The longitudinal shrinkage of normal wood is negligible for most practical purposes. This is one of the characteristics that makes lumber and wood products such useful building materials. If this were not so, the consequences of moisture content change could be disastrous. Some longitudinal shrinkage does usually occur in drying from the green to the ovendry condition, but this amounts to only 0.1–0.2% for most species and rarely exceeds 0.4%.

From an idealized "soda-straw" concept of wood cells, one might visualize that the radial and tangential dimensions would shrink or swell by the same amount. Tangential shrinkage, however, is greater than radial shrinkage by a factor of 1.5–3.0. Several anatomical traits are believed to be responsible for this differential, including the presence of ray tissue, frequent pitting on radial walls, domination of summerwood in driving moisture-induced shrinking and swelling in the tangential direction, and differences in the amount of cell wall material radially versus tangentially. The average transverse shrinkage values of a number of domestic and imported species are shown in Table 7.5. These values are reasonable guidelines for estimating dimensional behavior; however, the actual shrinkage of individual pieces in service may vary significantly from these averages.

Variation in the shrinkage of different samples of the same species under the same conditions results primarily from three factors:

- *The size and shape of the piece.* This affects the grain orientation in the piece and the uniformity of moisture through the thickness.
- *The density of the sample.* The higher the density of the sample is, the more it will tend to shrink (or swell).

TABLE 7.5. Shrinkage values of wood from green to ovendry MC.

Species	Shrinkage[a] (%)		
	Radial	Tangential	Volumetric
Domestic hardwoods			
White ash	4.9	7.8	13.3
Quaking aspen	3.5	6.7	11.5
Yellow birch	7.3	9.5	16.8
American elm	4.2	9.5	14.6
Sugar maple	4.8	9.9	14.7
Northern red oak	4.0	8.6	13.7
Black walnut	5.5	7.8	12.8
Imported hardwoods			
Apitong	5.2	10.9	—
Balsa	3.0	7.6	—
Mahogany	3.0	4.1	—
Teak	2.5	5.8	—
Khaya (African mahogany)	2.5	4.5	—
Softwoods			
Western redcedar	2.4	5.0	6.8
Coast Douglas fir	4.8	7.6	12.4
White fir	3.3	7.0	9.8
Western hemlock	4.2	7.8	12.4
Loblolly pine	4.8	7.4	12.3
Sitka spruce	4.3	7.5	11.5

Source: USFPL (2010).
[a] Percentage shrinkage = change in dimension/green dimension × 100.

- *The rate at which the sample is dried.* Under rapid drying conditions, internal stresses are set up because of differential shrinking. This often results in less final shrinkage than would otherwise occur. In contrast, however, some species shrink more than normal when dried rapidly under high-temperature conditions owing to the collapse of some cells.

Shrinking of lumber during manufacturing is significant and must be considered when determining the green target size of the piece to saw from a log. For example, if a nominal 50×250 mm (2×10 in.) Douglas fir plank is to be sawn from a log, it must be expected that it will shrink about 1.3 mm (0.05 in.) in thickness and 10 mm (0.4 in.) in width when it is dried to 15% MC. Because the final dry and surfaced size of that piece must be at least 38×235 mm (1.5×9.25 in.) to meet most United States softwood structural lumber standards, adequate shrinkage allowance must be provided. Allowance must also be made for sawing variation, pattern (quarter versus flat), and planing. Thus, softwood sawmills must use green target sawing sizes close to the nominal dimensions when manufacturing lumber despite the fact that actual finished sizes are notably smaller.

The amount of shrinkage is generally proportional to the amount of water removed from the cell wall, which suggests that higher-density species with more cell wall material should shrink more per unit moisture content change than lower-density species. This is indeed usually the case. Note that high-density woods also lose a greater amount of water per percentage moisture content change. As an example, sugar pine contains about

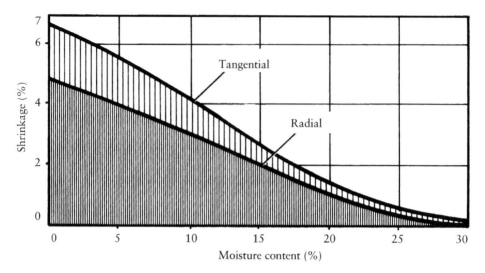

FIGURE 7.8.   Relationship between MC and shrinkage (southern pine). Source: From Peck (1947).

0.34 g ovendry wood substance cm^{-3}, whereas the same volume of longleaf pine contains about 0.54 g dry wood substance. If each of these woods loses 10% MC, the sugar pine will lose 0.034 g water cm^{-3} but longleaf pine will lose 0.054 g cm^{-3}. The normal volumetric shrinkage, green to ovendry, for sugar pine is 7.9%, whereas that for longleaf pine is 12.2%. In this example, there is a close relationship between the amount of water lost per unit of moisture content change and the resulting shrinkage.

Often, however the relationship between the mass of water removed and the resulting shrinkage is not direct. One would expect, based upon the density difference, that black walnut containing 0.55 g dry wood substance cm^{-3} would shrink more than eastern cottonwood containing 0.40 g wood cm^{-3}. However, the average green to ovendry volumetric shrinkage of walnut is only 12.8% compared with 13.9% for cottonwood. A major factor that tends to mask the effect that density has upon shrinking and swelling is the presence of extractives, which tend to lower the FSP and increase the bulk of the cell wall. Thus, the heartwood of some species, such as walnut, is more dimensionally stable than the sapwood and more stable than density might suggest.

The relationship between shrinkage and moisture content is essentially linear. Figure 7.8 shows the shape of this relationship for southern pine. This near linearity makes it relatively simple to estimate shrinkage between any two moisture contents if the green to ovendry shrinkage values for the wood are known. The rationale for such a calculation is shown in Figure 7.9, which illustrates the shrinkage expected in the tangential dimension of loblolly pine when dried from 15 to 8% MC. Assuming that the FSP is 30%, the rate of tangential shrinkage for this species is 7.4/30, or 0.25% shrinkage per percentage moisture content change. Because the moisture content change is 7%, the total shrinkage expected is about 7×0.25, or 1.75%. If this piece were dried from the FSP (30%) to 8% MC, the predicted shrinkage would be 22×0.25, or 5.5%. Note that when drying from a higher moisture content, such as 50–8%, the predicted shrinkage would also be 5.5% because it is assumed that shrinkage does not commence until the wood is dried to the FSP. The estimate of radial shrinkage from 15 to 8% MC equals 4.8/30×7, or 1.1%.

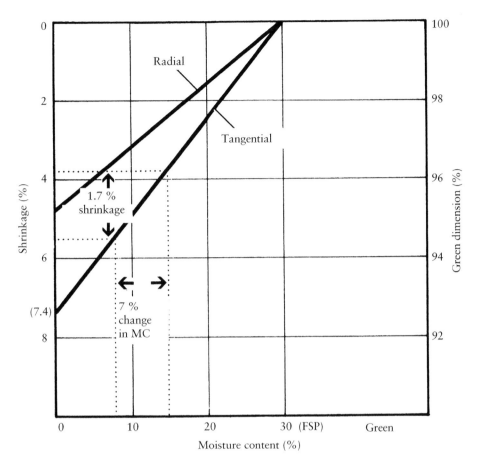

FIGURE 7.9.   Estimating shrinkage and change in size for loblolly pine with 4.8% radial and 7.4% tangential shrinkage, from green to ovendry.

In actual practice, wood shrinkage may commence before the average moisture content drops below the FSP. This is a result of drying and shrinking in the surface layers of wood while the core is still wet. However, this need not be considered in general practice when estimating shrinkage as described above.

Dimensional Changes and Environmental Changes

It follows from the previous discussion that wood used where the humidity fluctuates, as it does in almost all situations, will continually change in moisture content and therefore size. If humidity changes are small, these dimensional changes will not be noticed and will have no impact on satisfactory use. Even large fluctuations in humidity may have little effect if these conditions last for only short periods (hours or days) and the wood does not have time to come to the new EMC. Problems can arise, however, when a wood product is used under humidity and temperature conditions that cycle over long periods of time if the user of that product has not anticipated the changes in dimension.

It is often said that the solutions to most wood moisture problems are highly elementary, an observation that can be summarized in the following statement. To avoid 90% of the problems associated with wood products the following three tongue-in-cheek rules should be followed: (i) keep wood dry; (ii) don't let wood get wet; and (iii) maintain a separation between water and wood (Suddarth 2001). For the most trouble-free use of wood, the goal should be to fabricate products at the predicted in-service moisture content, work to reduce moisture changes, and design products to cope with dimensional change. Sometimes a compromise is made between attainment of the ideal moisture content and practical considerations in manufacturing and shipping. Framing lumber for light-frame wood buildings, as an example, is commonly manufactured at 15–19% MC, a moisture content that is 5–10% above that to which it will eventually equilibrate in most areas of the United States. This is usually not a problem, because small changes in the dimensions of studs, rafters, and floor joists are not noticeable. The United States standards for the manufacture of softwood lumber (ALSC 2010) specify a small percentage of shrinkage in use, which is to be considered acceptable. If green framing lumber is used, however, the dimensional changes may be large enough to cause problems such as cracked plaster, nail pops in gypsum walls, noisy floors, and distortion of wall surfaces. Also, if high proportions of juvenile wood are present in the wood, such as in lumber from small-diameter trees, the associated abnormal shrinkage can cause severe warp.

For interior uses such as furniture and millwork it is much more critical for satisfactory performance to use lumber at the proper moisture content. The objective when seasoning wood for interior products is to dry to a moisture content that is within the range of the EMCs experienced in service. Heated spaces in winter may experience EMC conditions as low as 3%; the same spaces in summer may have EMC conditions of 13%. As a result, the recommended moisture content for wood to be used indoors is 6–8%. For demanding situations, where dimensional changes could obviously cause problems, the designer or user should carefully consider the moisture content of the wood, the species, the conditions of use, and the amount of expected dimensional fluctuation. The following examples illustrate the types of cases deserving careful consideration. These will also serve to show procedures that can be used for estimating dimensional changes.

Example 7.1

A gymnasium floor was constructed with wood strip flooring nailed tightly such that essentially no cracks or gaps were left between the strips. The size of the gymnasium was 15.2 × 36.6 m (50 × 120 ft), with the strip floor laid parallel to the long axis of the gym. The flooring was hard maple (sugar maple) dried to a uniform 6% MC. The flooring was nailed to 38 × 89 mm (2 × 4 in.) wood nailer strips attached by mastic to a concrete slab. Less than 25 mm (1 in.) of space for expansion was left between the wood strip floor and the concrete block wall. Unfortunately, the wood floor was nailed before the concrete floor had completely dried. Under these conditions, the moisture content of the maple increased to about 9%.

Within a short time after the floor was installed, it began to buckle; i.e. ridges began to develop where the strip flooring rose and was pulled off the nailer strips. What was the source of the problem and how could it have been avoided? The buckling was the result of transverse swelling in the floor that was restrained at the edges so that it could not remain flat. The only way for the flooring to accommodate swelling was to move upward or buckle. The contractor could have anticipated this problem by estimating the amount of swelling that might develop. Assuming that the flatsawn and quartersawn wood would swell by the average of the radial and tangential values (see sugar maple, Table 7.5), it would swell (4.8 + 9.9%)/2, or 7.4%, in going from an ovendry condition to the FSP. In this situation, the moisture content changed by only 3%, so the anticipated swelling would be 3/30 or 10% of the total possible swelling. This change would amount to 0.1 × 7.4, or 0.74%. The total swelling that could be anticipated across the 15.2 m (600 in.) wide gymnasium is 113 mm (15.2 m × 0.74%) or 4.4 in. Because this potential swelling is considerably greater than the expansion space available, a problem of this type should have been foreseen. The problem could have been prevented by allowing the concrete floor to adequately dry before installing the wood floor.

Example 7.2

Rough-sawn western hemlock lumber 19 mm (0.75 in.) thick by 185 mm (7.25 in.) wide, with square edges, was to be used as paneling in an office building. The lumber was stored in the basement of the building for 2.5 months during the summer prior to installation. During that time it equilibrated to the 21 °C and 70% relative humidity (RH) conditions. When installed, the pieces were nailed as close together as possible. Cracks between the individual pieces were not over 1.6 mm (0.063 in.) and were not considered objectionable in view of the naturally rough appearance of the wall. After installation, the conditions in the heated building during the winter averaged 21 °C and 20% RH. If the paneling was flatsawn, what width of gap could be expected between each piece after shrinkage had occurred?

> EMC at 21 °C and 70% RH = 13.1% MC (see Table 7.2)
> EMC at 21 °C and 20% RH = 4.5% MC
> Percentage MC change = 13.1 − 4.5 = 8.6%
> Shrinkage green to OD = 7.8% (see Table 8.5)
> Percentage shrinkage expected = 8.6/30 × 7.8 = 2.2%
> Shrinkage per piece (average width of cracks) = 2.2% × 185 mm = 4.1 mm (0.16 in.)

In the real-life situation, the cracks averaged about 3.1 mm (1/8 in.). The difference between the 3.1 mm shrinkage actually encountered and the 4.1 mm estimate was probably because some of the pieces were not flatsawn. Therefore, some radial as well as tangential shrinkage was experienced. These gaps were, however, considered unacceptable and a large claim was filed against the building materials supplier. In this case, the fault lay with the contractor, who stored the material in a humid environment prior to installation.

Example 7.3

A large Douglas-fir timber (203 × 305 mm actual size) was used as a mantel over a fireplace. The timber extended across the entire end wall of the room and the ends of the timber were plastered into the adjacent walls. Large, solid-sawn timbers can only rarely be purchased dry, and in this case, the timber was green (actually about 45% MC) when installed. The 305 mm dimension was the radial face of the timber, and the 203 mm dimension was the tangential surface of the wood. During a normal year, the conditions in the house averaged 21 °C and 30% RH. How much could it be expected that this timber would shrink in use, i.e. how big a gap in the plaster would develop where the mantel joined the plaster?

EMC at 70 °F and 30% RH = 6.2% MC
Percentage MC change through which shrinkage would occur = 30 − 6.2 = 23.8%
Percentage radial shrinkage green to OD = 4.8%
Percentage tangential shrinkage green to OD = 7.6%
Percentage radial shrinkage green to 6.2% MC = 23.8/30 × 4.8 = 3.8%
Percentage tangential shrinkage green to 6.2% = 23.8/30 × 7.6 = 6.0%
Radial shrinkage = 3.8% × 305 mm = 11.6 mm
Tangential shrinkage = 6.0% × 203 mm = 12.2 mm

Thus a gap about 12 mm (0.5 in.) wide could be expected to develop at each end of the mantel where it was plastered into the wall. In the actual situation these cracks measured about 9.5 mm (0.375 in.). The moisture gradient and drying stresses in this large timber restricted the total shrinkage.

These examples illustrate some practical uses for knowledge of the EMC and shrinkage behavior of wood. Although these procedures give only an estimate of what will actually occur, this is often sufficient to avoid costly and wasteful problems in manufacture and use.

Dimensional Changes in Veneer-, Fiber-, and Particle-based Panel Products

The shrinking and swelling characteristics of wood shown in Table 7.5 and in the examples above are determined from measurements on small samples 2.5 cm square by 10 cm long. Such samples are cut such that the radial or tangential dimensions are along the 10 cm dimension of the specimen. Because of the small specimen size and the fact that it is only 2.5 cm long in the longitudinal direction, drying occurs rapidly and relatively uniformly throughout the sample. This process avoids large moisture content gradients from surface to center that could cause internal stresses. Therefore, the shrinking/swelling values obtained are those exhibited under unrestrained conditions.

The dimensional stability of most lumber products corresponds closely to unrestrained values for wood. Products such as solid wood furniture, millwork, laminated beams, and construction lumber all behave in a similar way with regard to *radial, tangential*, and *longitudinal shrinkage*. Forest products produced from veneer, particles, and fiber, in contrast, have unique dimensional behaviors under moisture change. These differences result from three basic causes: (i) the degree of restraint to swelling provided by one element in the product to other elements in the product; (ii) the degree of compression or crushing that the wood elements (veneer, particle, or individual fibers) undergo during the manufacture of the product; and (iii) the effect that adhesives and other additives have on the ability of the wood elements to respond dimensionally to moisture change. In some cases, additives bulk the cell walls to some degree, thus lowering the EMC of the wood.

Plywood is produced by face-laminating veneer, generally 4.2 mm (0.17 in.) or less in thickness, in such a way that in each layer (veneer) the longitudinal direction is at 90° to the adjacent layer. Oriented strandboard (OSB) has an analogous construction made with strands. Construction of three-ply plywood is shown in Figure 7.10. If the veneers are not glued together they can shrink or swell in the same way as normal wood. However, when glued into plywood the face veneers restrain swelling of the core veneer in its transverse direction, and the core restrains the swelling of the faces in their transverse direction. As a result, plywood is relatively dimensionally stable along the panel plane. OSB exhibits dimensional stability for the same reasons. Small, but measurable dimensional changes in plywood and OSB do occur. In general, these products exhibit much less dimensional change in either radial or tangential direction than the normal characteristics of their constituent species. However, they exhibit greater dimensional change in the

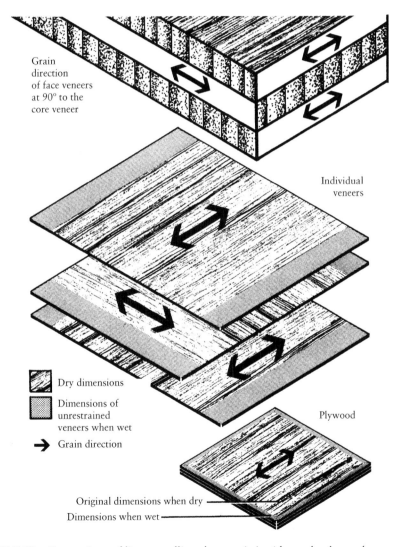

Grain
direction
of face veneers
at 90° to the
core veneer

Individual
veneers

Dry dimensions

Dimensions of
unrestrained
veneers when wet

Plywood

→ Grain direction

Original dimensions when dry
Dimensions when wet

FIGURE 7.10.   Comparison of linear swelling characteristics (three-ply plywood versus unrestrained veneer).

longitudinal direction, and therefore it is advisable to leave a gap between adjacent 1.22 × 2.44 m (4 × 8 ft) sheets when covering a roof, wall, or floor. Manufacturers supply specific instructions to this end.

The second factor that affects the dimensional change of wood-based panel products is the amount of compression that the product undergoes during manufacture. The thickness swelling or shrinking of plywood with moisture change is about the same as that of normal solid wood because little compression occurs. However, in some cases, thickness swelling in plywood may be slightly more than that in normal wood if excessively high pressures occurred during the pressing process. Wood that is compressed tends to partially recover its original dimension when rewet. This behavior is exemplified with OSB.

A dent in wood furniture can often be removed by steaming. The crushed wood tends to recover its original shape. Much the same thing can happen to an entire panel of OSB or particleboard. In the manufacture of particleboard and OSB, small shavings, flakes, strands, or wafers of wood are sprayed with droplets of a synthetic resin adhesive. These wood elements are compressed between 1.2 and 2.0 times their original density and simultaneously the resin is cured. If such a product is subjected to steaming or other moisture content increases, the wood swells normally and in addition, the crushed particles tend to return to their original thickness. For this reason, compressed wood-based products often exhibit greater thickness swelling than normal wood.

The third factor is product additives, of which synthetic resin adhesives and waxes are the most common. The wax (or sizing) is intended to provide resistance to liquid water pick-up. Wax does not bulk the cell wall or change the ultimate EMC; rather, it helps the product shed liquid water, making it water repellent. Synthetic adhesives can, however, alter the recovery of the crushed particles or fibers. Generally, the greater the amount of adhesives used to manufacture a composite product, the less the thickness-swell response to moisture pick-up is.

Not only are the wood elements in a product held more tightly when more resin is used, but some resin penetrates into the cell walls and provides a degree of bulking, or replacement of water molecules. Figure 7.11 shows this effect in a wafer-type particleboard. Board A was made with 3% phenolic resin, and boards B and C were made with 10% phenolic resin. In the case of board B, 7% of the total 10% resin was applied to green particles so that the resin could more easily enter the cell wall structure. In board C, all of the resin was applied to dry particles. Note the difference in thickness between board types when wet and also how much of the swelling was irreversible, i.e. remaining after the panels were redried.

Figure 7.12 shows the dimensional change in the plane of the panel in the same experiment. Note that after moisture cycling, the panel was actually smaller in the plane of the panel than it had been. This type of behavior is often found in particle- and fiberboard products. Generally, this effect is so small that it is not noticeable.

Most fiber and particle products are manufactured under commercial or industry standards, which place limits on the swelling properties. Specific property limitations in the standards vary depending on use of the product. For example, in the commercial standards for OSB, limits on *linear expansion* (in the plane of the panel) range from 0.25 to 0.55%. Thickness swell is restricted to 15–25%. In the product standard for *hardboard* (a high-density wood-fiber product), limits for thickness swell range from 8 to 30% but linear swelling is not specified. Prudent users of wood-based products should obtain the manufacturer's data on dimensional characteristics. Products of the same type

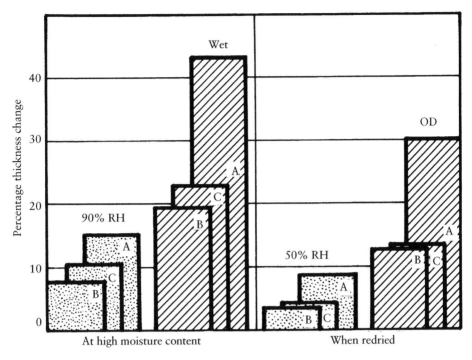

FIGURE 7.11. Thickness swelling of a particleboard (experimental waferboards exposed to wetting–redrying and to EMC cycled from 50 to 90 to 50% relative humidity). Source: From Haygreen and Gertjejansen (1972).

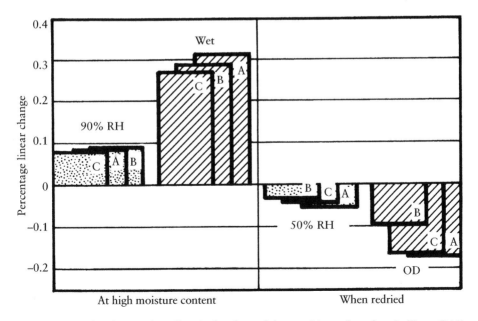

FIGURE 7.12. Shrinking and swelling in the plane of the panel (same boards as in Figure 7.11).

but from different manufacturers can vary considerably in this regard. Dimensional change can almost always be accommodated by proper design, but if dimensional changes are not anticipated, problems can arise.

Because plywood is a relatively stable product and its dimensional characteristics cannot be easily altered by manufacturing variables, there are few specifications in the plywood product standard as to dimensional stability. O'Halloran (1975) found that about 0.2% shrinking or swelling occurs for each 10% change in the relative humidity. Hygroscopic expansion of plywood consists of a uniform percentage of swelling or shrinking across the width or length or the panel plus an edge effect, which is independent of panel size.

Moisture Movement During Drying

The movement of water in wood during drying takes place as mass movement of liquid water or water vapor, or diffusion of individual water molecules. *Diffusion* is a phenomenon that occurs as water moves from areas of higher concentration to those of lower concentration. Thus, for diffusion to occur there must be a moisture concentration gradient or a vapor pressure gradient across the cell walls. The rate of diffusion is related to the temperature, the steepness of the moisture gradient across the cells, and the characteristics of species that determine the ease with which diffusion can occur. The rate of diffusion in a species can be expressed as the diffusion coefficient. Diffusion through individual cells is noticeable only below the FSP. Above the FSP, free water moves out of wood as a result of surface drying, vapor diffusion, and capillary forces. At that stage of drying, wood can be thought of as a series of partially filled tubes, with water evaporating from its ends.

The rate at which lumber dries is determined by the rate at which water is removed from the surfaces and the rates of internal liquid and vapor diffusion. In the initial drying stages the rate is often controlled by surface evaporation and, in the later stages, by the diffusion characteristics. For highly permeable woods such as the southern pines, surface evaporation is the primary limiting factor that controls the drying rate.

In some species, the structure of wood inhibits the mass movement of liquid water. Such woods are referred to as *impermeable* or *refractory*. Tyloses, aspirated pits, and the deposition of extractives on pit membranes are examples of wood features that inhibit the movement of water. In woods with these characteristics, the movement of water must be principally by diffusion; thus, drying is an extremely slow process. Redwood, white oak, and walnut are a few of the species having relatively impermeable heartwood. Sapwood is generally *permeable* in all species. A few species, such as western hemlock and aspen, contain pockets or localized zones that are impermeable. After drying, these latter woods may still contain wet spots. These impermeable wet areas are subject to drying downgrade, that is loss of quality, yield, or other value, if extreme care is not exercised in the drying process. Green sorting of lumber, veneer, or other wood elements is often economically justified when drying these obstinate species.

Using an analogy between electrical conduction and diffusion, Stamm (1964) developed the theoretical transverse drying diffusion coefficients shown in Figure 7.13. Note the more than 10-fold increase in the rate of diffusion by raising the temperature from 50 to 120 °C.

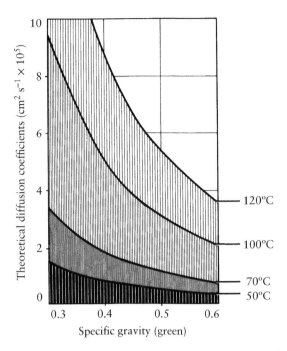

FIGURE 7.13. Theoretical drying diffusion coefficients for softwoods in radial direction. Source: From Stamm (1964).

A variety of treatments to increase the movement of water through wood – i.e. to increase permeability or diffusion – have been developed, but few have found wide acceptance. Erickson et al. (1966) found that freezing redwood lumber prior to drying improved drying performance. Application of hygroscopic chemicals such as urea, sodium chloride, and calcium chloride alters the moisture gradient and permits an increased rate of drying for some species. However, when so treated the wood retains a hygroscopic surface layer that can cause problems in use. Solvent replacement can improve permeability by eliminating pit aspiration during drying. Incising or predrilling improves moisture transport by exposing additional end-grain area. Presteaming of wood has been found to be beneficial in some cases. Unfortunately, a universally effective means of improving liquid water movement and/or diffusion has not yet been found.

Methods of Drying Lumber and Other Solid Wood Products

Most lumber, whether hardwood or softwood, is dried in some type of dry kiln. Modern kilns provide a controlled temperature and humidity environment and are equipped with fans to force air circulation and ventilation. Conventional kilns operate at temperatures up to about 100 °C, whereas high-temperature kilns used to dry softwood lumber operate above the boiling point of water. These latter kilns can dry some species up to 50 mm (2 in.) thick in less than 24 hours. The lumber in such a kiln is dried in air that has been heated by steam coils or directly by the combustion gases from a gas-, oil-, or wood residue-fired burner. Figure 7.14 illustrates the main elements in a typical crossventilated kiln heated by steam.

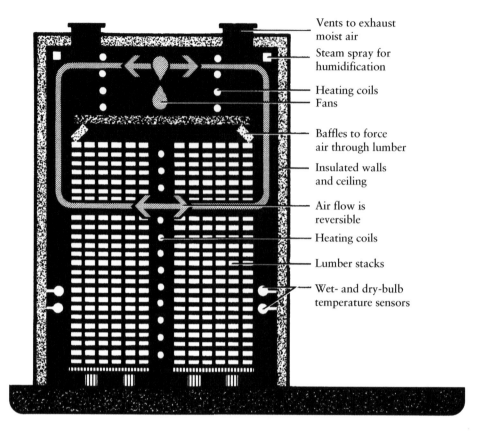

Vents to exhaust
moist air

Steam spray for
humidification

Heating coils
Fans

Baffles to force
air through lumber

Insulated walls
and ceiling

Air flow is
reversible

Heating coils

Lumber stacks

Wet- and dry-bulb
temperature sensors

FIGURE 7.14.   The main components of a modern steam-heated dry kiln (cross-sectional view).

Drying in a kiln progresses via a series of temperature and relative humidity steps designed to dry the wood gently while it is at a high moisture content. After the free water has been removed and wood strength increases, more severe drying conditions are imposed to maintain an adequate rate of drying. The series of temperature and humidity conditions imposed during drying is referred to as a *kiln schedule*. Most hardwood schedules are controlled according to the moisture content of the lumber; thus, changes in drying conditions are made when the moisture content drops to predetermined levels. Softwoods are more frequently dried by a time schedule; i.e. drying conditions are changed at predetermined times. A high temperature kiln schedule and resulting moisture content response for 50 mm (2 in.) thick yellow pine is shown in Figure 7.15. From this figure, it can be seen that after the initial warmup, the lumber dries at a relatively constant rate until the wood surface goes below the FSP. From that point on, between about 540 and 600 min, the drying rate decreases.

Two contrasting moisture content schedules for drying 25 mm (1 in.) thick hardwood lumber are shown in Table 7.6. Note the differences in temperature and relative humidity at each moisture content level for the difficult-to-dry white oak as compared with the easily dried basswood. These schedules are from recommendations in the *Dry Kiln Operator's Manual* (Simpson 1991), which is widely used as the guide to kiln drying.

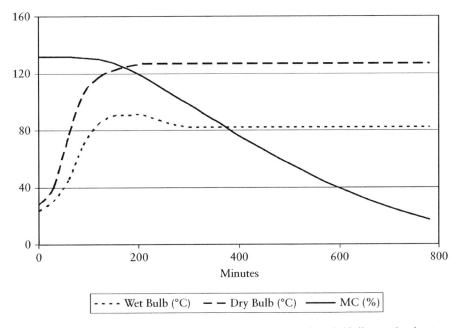

FIGURE 7.15. Moisture content (%) response of 50mm (2 in.) thick loblolly pine lumber in a high-temperature kiln. Wet- and dry-bulb temperatures are in degrees Celsius. Lumber reached target moisture content in ~13h.

TABLE 7.6. Kiln drying schedules for drying 25 mm (1 in.) thick basswood and white oak lumber.

	Basswood			White oak	
Moisture content (%)	Temperature, °C (°F)	Relative humidity (%)	Moisture content (%)	Temperature, °C (°F)	Relative humidity (%)
>60	71 (160)	58			
70	71 (160)	43			
50	71 (160)	31	>40	43 (110)	87
40	71 (160)	21	40	43 (110)	84
35	71 (160)	21	35	43 (110)	75
30	77 (170)	24	30	49 (120)	62
25	77 (170)	24	25	54 (130)	35
20	82 (180)	26	20	60 (140)	25
15	82 (180)	26	15	82 (180)	26

Source: Simpson (1991).

More recent information related to hardwood drying, *Drying Hardwood Lumber* (Denig et al. 2000), is also available. Kiln-drying schedules for southern pine are provided by Denig et al. (1993). Table 7.7 from this latter publication shows high-temperature drying times for two thicknesses of southern pine. After gaining experience with the species and thicknesses that are commonly dried, many firms develop their own drying schedules.

TABLE 7.7.   Sample schedules for high-temperature drying of southern yellow pine.

			Dry-bulb temperature	Wet-bulb temperature
Thickness, mm (in.)	Width, mm (in.)	Time (hours)	°C (°F)	°C (°F)
25 (1)	100–250 (4–10)	14–18	110 (230)	82 (180)
45 (1.75)	100–250 (4–10)	22–26	116 (240)	82 (180)

Source: Adapted from Denig et al. (1993).

Because an entire kiln load of lumber can become checked, split, honeycombed, and otherwise ruined by drying it too aggressively, care must be employed when developing new schedules. Drying too slowly can also cause economic loss, decreased throughput, and degrading by mold and staining. Means of optimizing drying operations, minimizing degrading, and maximizing the rate of drying are discussed by numerous authors including Holmes (1989), Ziegler (1988), Rice et al. (1994), and Carter and Sprague (1989). Some industry trends in the drying practices for hardwoods and softwoods are summarized by Armstrong and Pahl (1994).

Some lumber (such as thick timbers or railroad ties) is still marketed air-dried, but more commonly air-drying is used as a preliminary step to kiln drying. This can greatly lower drying costs, particularly for difficult-to-dry species that may take many weeks to kiln dry. For refractory species of hardwoods – i.e. those that are difficult to dry without degrading – kiln drying may require over 2900 MJ m^{-3} of lumber (6.5 million BTU per thousand board feet). This amounts to about 70% of all the energy used in the manufacture of this lumber. Also, lumber is air-dried at small mills that do not operate kilns and in regions where the ambient humidity is normally very low, such as the Southwest and the Rocky Mountain regions of the United States. Air-drying may be satisfactory if adequate time is available and the drying specifications are not too stringent. From the user's standpoint, a disadvantage of air-drying is that it may not achieve the lower moisture content required for some applications. From the manufacturer's standpoint, air-drying has the disadvantages of requiring costly inventory to spend time in the drying yard, a substantial amount of yard space, and little control of degrade. Air-drying times vary from a month under ideal conditions for easy-to-dry woods to a year or more for refractory woods dried under more difficult conditions.

The amount of energy required to evaporate water from wood is shown in Figure 7.16. Although the heat of vaporization varies slightly with the temperature in the range of common drying temperatures, it is about 2.3 MJ kg^{-1} water evaporated. To accomplish drying below the FSP, the heat of wetting must be supplied as well as the heat of vaporization. The total actual energy required to dry wood is much greater than the sum of the heats of wetting and vaporization. Heat losses in the kiln from the venting of water vapor, air leakage, radiation, conductive losses through the kiln walls, and the use of steam to provide humidity control are also involved. The total amount of energy consumed in conventional kilns generally varies between 3.7 and 7.0 MJ kg^{-1} water evaporated.

Dehumidification drying has gained significant popularity in North America. It was first used commercially in Europe; to date it has been used principally for the drying of hardwoods. The primary difference between this and conventional drying is that water is

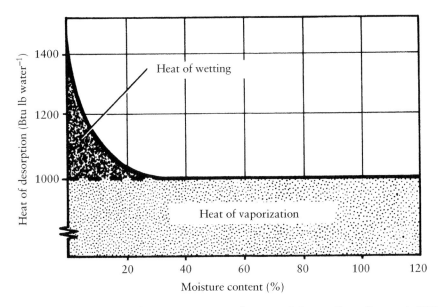

FIGURE 7.16.   Energy required to evaporate water from wood. Source: From Comstock (1975).

removed as condensate on refrigerated coils rather than by venting the moist air to the atmosphere. Because energy loss from venting is eliminated and heat from the refrigeration unit compressor heats the kiln, the dehumidification drying method offers great efficiency with respect to energy consumption. Another potential benefit of dehumidification drying is that airborne emissions are reduced or eliminated, which is a key factor for permitting pollution control. The emissions are condensed and can be captured or treated as a liquid, which is easier and less costly than the treatment of airborne volatiles. Early installations of dehumidification kilns had the disadvantage of being slower than conventional dry kilns, but technical improvements, including the use of high-temperature refrigerants, have reduced this limitation. Because the form of energy required for this type of drying system is electricity (to run the compressor), it does not adapt itself to the utilization of direct or indirect heat generated from the combustion of mill residues.

A number of other methods have been used experimentally, and in a few instances commercially, for drying solid wood products. These may have advantages for special situations but are generally more expensive and less predictable than kiln drying. Wood scientists and industry engineers are actively engaged in improving drying practices. Some nonconventional drying methods have been studied. They include the following:

1. *Immersing wood in a heated organic liquid.* Liquids such as fuel oil or perchloroethylene are heated, raising the temperature of the wood above the boiling point of water, thus driving off moisture.
2. *Vapor drying.* This method uses an organic liquid with a boiling point above 100 °C. The drying chamber contains the wood and the organic vapor. The condensation of these vapors on the wood heats it rapidly, driving out the water, which is then separated from the solvent vapor in a condenser and separator.

3. *Radiofrequency/dielectric heating.* This approach involves placing wood between two electrodes and subjecting the material to an electric field oscillating at high frequency. Being polar, water molecules in wood rotate in the alternating field, thus generating heat. In woods that dry easily, the internal temperature tends to rise only slightly above the boiling point until the free water is gone. However, in the case of impermeable woods, the wood temperature may rise to destructive levels. A variation of this process is the combination of microwave energy with hot-air impingement drying. This technology has been used to a limited extent commercially for high-value thick stock and thin materials such as veneers.

4. *Combination of radio frequency heating and vacuum drying.* Lumber is heated dielectrically at 7–9 MHz while it is in a chamber with a partial vacuum. In this case, water boils at a lower temperature, thereby accelerating the drying process. Such kilns, capable of drying 24 m³ (10,000 board feet) of lumber, have been built, and good results are reported (Wengert and Lamb 1982).

5. *Press drying.* Wood is press dried by placing it between two heated platens. Heat transfer is by conduction from metal to wood and therefore is very rapid. This technique works well on thin pieces of easy-to-dry species. It is used commercially for high-quality veneer, but only one commercial application for lumber has been reported. This drying method can be combined with high pressures to produce a densified product. Figure 7.17 shows experimental results of simultaneously drying and densifying two species of pine. In one of the samples shown, phenolic resin was added to increase dimensional stability.

6. *Solar drying.* Solar-heated kilns have proven to be an effective means of small-scale commercial lumber drying. Many low-capital-cost designs have been

FIGURE 7.17.   Normal and densified loblolly and Norway pine. The specific gravity of the samples is indicated.

developed and are used semi-commercially. Although there is less control of drying parameters and no steam-spray is available, much high-quality lumber is dried by solar methods. Because drying rates are low, it is often possible to dry lumber without appreciable stress gradients in solar kilns.

Veneer, Particle, and Fiber Drying

The primary difference between drying veneer or strands compared with lumber is that veneer and strands, being very thin, develop smaller moisture-stress gradients. The drying stresses and impermeable zones that cause problems in the drying of lumber are usually not limiting factors in veneer and strand drying. Other considerations, however, such as the glueability of the surface, may dictate how fast and at what temperature veneer and strands can be satisfactorily dried.

Veneer dryers employ conveyors to move the veneer through a heated chamber where temperatures range from 150 to 260 °C. In many hardwood veneer dryers, air is circulated in a manner similar to that in a dry kiln. Modern structural veneer plants, and some hardwood plants, however, utilize jet dryers. These are also called *impingement driers* because a curtain of air at velocities of 10–20 m s^{-1} is directed against the surface of the veneer. The high velocity produces turbulent air on the surface of the veneer. This eliminates the laminar boundary layer that slows down heat and moisture transfer under ordinary drying conditions.

Most strand-, particle- and fiberboard plants utilize a high-speed drying system of some sort because of the large amounts of material to be dried; commercial plants in the United States dry up to 900 tons (1000 short tons) of wood per day. Three principal types are commonly used: drum dryers, tube dryers, and screen bed dryers. *Rotating drum dryers* are the most common type used in composite board plants. The wood elements make one, two, three, or four passes from one end of the dryer to the other and then are discharged. Inlet temperatures of such dryers can be as high as 870 °C when wet furnish is being dried. Inlet temperatures are reduced to about 260 °C or lower to reduce the fire hazard if dry furnish such as planer shavings is involved.

Wood particles can be dried at air temperatures above the burning point (about 230 °C) as long as moisture is present in the wood to evaporatively cool the surface. Dryer control systems must be designed to ensure that dried wood is not present in the preliminary high-temperature stages of the dryer. Wood element movement through these dryers is controlled by air velocity. The finer particles, which dry faster, are blown more rapidly through the drum and therefore are exhausted before reaching the combustion point.

The drying of fibers for dry-process fiberboard production can be accomplished in *tube dryers*. The fibers are introduced into a stream of gas heated from 200 to 320 °C. These dryers may have a second stage operating at a lower temperature. Moisture is often flashed off in a few seconds; thus, effective feed and temperature control systems are critical.

Screen bed dryers are used in high-quality strand-drying operations. Here, long flat strands are deposited on flat screens and passed through heated dryer sections. These dryers are also appropriate for long-strand operations, such as laminated-strand lumber. These dryers are reported to produce less curling of strands during drying, thereby improving subsequent adhesive blending and pressing, and have lower air emissions.

All types of dryers must be cleaned regularly to remove chips, dust, and broken pieces of wood debris that inevitably build up along the floor. Failure to address this detail invariably leads to hazardous and costly dryer fires.

The moisture content to which strand-, fiber-, or particleboard elements are dried for the manufacture of panel products depends on the specific product, the amount of water added with the resin and wax size, and the pressing cycle. Generally, the wood furnish is dried to between 2 and 8% MC. Precise control is necessary because a moisture content 2% higher than desired can generate excess steam and cause blows or internal explosions in panels when the press is opened. A moisture content 2% below the desired level can cause poor bonds and thereby reduce mechanical properties.

References and Supplemental Reading

ALSC. American Softwood Lumber Standard. PS 20-10. American Society for Testing and Materials. 2010.

Armstrong, J.P. and Pahl, T.L. (1994). Changes in the kiln-drying practice of the forest products industry in West Virginia: 1982 to 1992. *For. Prod. J.* 44 (9): 54–56.

ASTM. D4442-92 (2003). *Standard Test Methods for Direct Moisture Content Measurement of Wood and Wood-Base Materials.* Conshohocken, PA: American Society for Testing and Materials.

Carter, L. and Sprague, M. (1989). Improve lumber drying by analyzing kiln environment. *For. Ind.* 116 (7): 12–15.

Comstock, G. (1975). Energy requirements for drying. Proc. For. Prod. Res. Soc. Energy Symp. Denver, CO.

Denig, J., Hanover, S.J., and Hart, C.A. (1993). Kiln drying southern pine lumber. NC. State Univ. Coop. Ext. Ser. Pub.

Denig, J., Wengert, E.M., and Simpson, W.T. (2000). Drying Hardwood Lumber. USDA For. Ser. FPL-GTR-118.

Erickson, R.W., Haygreen, J.G., and Hossfeld, R. (1966). Drying prefrozen redwood. *For. Prod. J.* 16 (8): 57–65.

Haygreen, J.G. and Gertjejansen, R.O. (1972). Influence of the amount and type of phenolic resin on the properties of a wafer-type particleboard. *For. Prod. J.* 22 (12): 30–34.

Heebink, B.G. (1966). Thoughts on the term EMC. Soc. Wood Sci. Tech. Log. (Nov.):1–3.

Higgins, N.C. (1957). The EMC of selected native and foreign woods. *For. Prod. J.* 7 (10): 371–377.

Holmes, S. (1989). Optimizing grade recovery in lumber drying. *For. Ind.* 116 (7): 15–20.

Koch, P. (1972). Utilization of Southern Pines, Vols 1 and 2. USDA For. Serv. Agric. Handb. 420, Chap. 8, 20.

Lewis, D.C. (1990). Dehumidification kilns let small mills dry lumber. *For. Ind.* 117 (5): 39–40.

McMillen, J.M., and Wengert, E.M. (1978). Drying eastern hardwood lumber. USDA For. Serv. Agric. Handb. 528.

Nzokou, P. and Kamdem, D.P. (2004). Influence of wood extractives on moisture sorption and wettability of red oak (*Quercus rubra*), black cherry (*Prunus serotina*), and red pine (*Pinus resinosa*). *Wood and Fiber Sci.* 36 (4): 483–492.

O'Halloran, M.R. (1975). Plywood in hostile environments. Am. Plywood Assoc. Res. Rep. 132.

Peck, E.G. (1947). Shrinkage of wood. USDA For. Prod. Lab. Rept. 1650.

Peck, E.G. (1955). Moisture content of wood in use. USDA For. Prod. Lab. Rept. 1655.

Rice, R.W., Howe, J.L., Boone, R.S., and Tschernitz, J.L. (1994). A survey of firms kiln drying lumber in the U.S. *For. Prod. J.* 44 (7/8): 55–62.

Seborg, C.O. and Stamm, A.J. (1931). Sorption of water by paper making materials. *Ind. Eng. Chem.* 23: 1271.

Simpson, W.T. (1991). Dry Kiln Operator's Manual. USDA For. Serv. Agric. Handb. 188.

Skaar, C. (1972). *Water in Wood*. Syracuse, NY: Syracuse University Press.

Spalt, H.A. (1958). Fundamentals of water vapor sorption by wood. *For. Prod. J.* 8 (10): 288–295.

Stamm, A.J. (1957). Adsorption in swelling vs. non-swelling systems. *TAPPI J.* 40 (9): 761–764.

Stamm, A.J. (1964). *Wood and Cellulose Science*. New York: Ronald Press.

Suddarth, S. (2001). Reflections of a fruit fly: The value of experience in research. 2nd International Advanced Engineered Wood Composites Conference, Bethel Maine.

USDA 2010. Wood Handbook: Wood as an Engineering Material: Chapter 4. Centennial edn. General technical report FPL; GTR-190. Madison, WI: USDA For. Serv., For. Prod. Lab., pp. 4.1–4.19.

US Forest Products Laboratory. (2010). Wood Handbook: Wood as an Engineering Material. USDA For. Ser. FPL-GTR-190.

Virginia Polytechnic Institute and State University. (1979). Proc. Symp. Wood Moisture Content, Temperature and Humidity Relationships.

Wengert, E.M. and Bois, P. (1997). Evaluation of electric moisture meters on kiln-dried lumber. *For. Prod. J.* 47 (6): 60–62.

Wengert, E.M., and Lamb, F.M. (1982). Hardwood drying tests new methods. For. Ind. (Jan.): 21–22.

Wilson, P.J. (1999). Accuracy of a capacitance-type and three resistance-type pin meters for measuring wood moisture content. *For. Prod. J.* 49 (9): 29–32.

Ziegler, G.A. (1988). Addressing the basics of successful lumber drying. For. Ind. (Jul.):6–10.

8

Density and Specific Gravity

The specific gravity of wood is its single most important physical property. Most mechanical and physical properties of wood are closely correlated to specific gravity and density. In general discussions, the terms *specific gravity* and *density* are often used interchangeably. These terms have distinct definitions even though they refer to the same characteristic. The strength of wood, as well as its stiffness, increases with increasing specific gravity. The yield of pulp per unit volume is directly related to the specific gravity. The heat transmission of wood increases with the specific gravity as well as the heat per unit volume produced in combustion. The shrinking and swelling behavior of wood is also affected, although the relationship is not as direct as with the strength properties. It is possible to learn more about the nature of a wood sample by determining its specific gravity than by any other single measurement. Perhaps it is for this reason that density was the first wood property to be scientifically investigated.

The cellular structure of wood gives it many of its unique properties. The density of the wood substance in the cell wall is about $1520 \, kg \, m^{-3}$. Density is inversely related to its porosity, that is, to the proportion of the *void volume*. A piece of sugar pine with a density of $380 \, kg \, m^{-3}$ dry wood substance m^{-3} includes about 25% cell wall material and 75% voids (principally lumen space) by volume. In contrast, white oak with a density of $750 \, kg$ dry wood substance m^{-3} has a void volume of about 50%. When considering the density of wood, it can be helpful to visualize the void volume to which it corresponds. It is easy to see why a block containing 50% void volume will resist crushing to a much greater extent than a block with 75% void volume.

The physicomechanical properties of wood are mainly determined by three characteristics: (i) the porosity or proportion of void volume, which can be estimated by measuring the density; (ii) the organization of the cell structure, which includes the microstructure of the cell walls and the variety and proportion of cell types (the

Forest Products and Wood Science: An Introduction, Seventh Edition. Rubin Shmulsky and P. David Jones.
© 2019 John Wiley & Sons Ltd. Published 2019 by John Wiley & Sons Ltd.
Companion website: www.wiley.com/go/shmulsky

organization of the cell structure is principally a function of species); and (iii) the moisture content. The effect of bound water on the properties of wood was discussed in Chapter 7. In the engineering and use of wood materials, it is important to keep these three characteristics in mind. Density and specific gravity are the two physical properties commonly used to describe the mass of a material per unit volume. These properties are commonly used in connection with all types of materials.

Density (D) is defined as the mass or weight per unit of volume. It is usually expressed in kilograms per cubic meter ($kg\,m^{-3}$), grams per cubic centimeter ($g\,cm^{-3}$), or pounds per cubic foot ($lb\,ft^{-3}$). A word of caution is in order when discussing wood density. There is no universally accepted procedure for calculating the density of wood. For instance, although density is frequently expressed in terms of green weight and green volume when calculating weights for transportation or construction, this is not always the case. It is, therefore, important to be sure of the basis of the calculation when discussing wood density. It is good practice to calculate density (the mass per unit volume) by determining the mass and the volume at the same moisture content. The moisture content at which the density is determined should then be noted.

Specific gravity (SG), or *relative density*, is the ratio of the density of a material to the density of water. With many materials the weight and the volume are determined under the same conditions. However, a dilemma exists for hygroscopic materials such as wood. Because both the weight and the volume vary with changes in moisture content, the specific gravity of wood is always calculated using oven-dry (OD) weight or mass. Volume is determined for the appropriate moisture content. Because of this recognized standard procedure, confusion about the basis of the calculation is avoided. Specific gravity is defined in physics as *the ratio of the density of a material to the density of water at 4 °C*. Water has a density of $1\,g\,cm^{-3}$ or $1000\,kg\,m^{-3}$ at that standard temperature. A modification of this definition for wood, as mentioned above, is that the mass is always determined in the OD condition. Specific gravity has no units because it is a direct ratio. In the metric system, specific gravity can be visualized by thinking of it as grams of dry wood substance per cubic centimeter. For example, OD (0% moisture content, MC) wood with a specific gravity of 0.50 (SG at 0% MC, SG_{od}) has 0.50g of dry wood substance per cm^3 or $500\,kg\,m^{-3}$. In the Imperial system, water has a density of $62.4\,lb\,ft^{-3}$, and therefore the density of wood with a specific gravity of 0.50 (SG at 0% MC) is $0.50 \times 62.4\,lb\,ft^{-3}$, or $31.2\,lb\,ft^{-3}$.

An advantage of using the metric system is that the calculation of specific gravity is simplified because $1\,cm^3$ of water weighs exactly $1\,g$. Specific gravity can thus be calculated directly by dividing the OD weight in grams by the volume in cubic centimeters. When using these two metric units for OD wood, the density and the specific gravity are numerically the same. This is not however the case for other moisture contents.

$$D = mass/volume\,(at\,any\,given\,moisture\,content)$$

$$SG = (OD\,mass/volume)/density\,of\,water$$

$$Density\,of\,water = 1\,g\,cm^{-3}, 1000\,kg\,m^{-3}, 62.4\,lb\,ft^{-3}$$

The following example illustrates the difference between density and specific gravity shown in both the Imperial and the metric systems. Suppose that a block of wood has the following measurements at three moisture contents:

	Green	At 12% MC	Oven-dry
Volume, metric:	$0.00220\,m^3$	$0.00209\,m^3$	$0.00200\,m^3$
Volume, imperial	$0.0777\,ft^3$	$0.0738\,ft^3$	$0.0706\,ft^3$
Mass, metric	$1.80\,kg$	$1.12\,kg$	$1.00\,kg$
Mass, imperial	$3.97\,lb$	$2.47\,lb$	$2.20\,lb$

The density and specific gravity of that block can be calculated and expressed correctly in any of the following ways:

$$D(\text{OD weight and volume}) = 1.000\,kg/0.00200\,m^3 = 500\,kg\,m^{-3}$$

or

$$= 2.20\,lb/0.0706\,ft^3 = 31.2\,lb\,ft^{-3}$$

$$D(\text{green weight and volume}) = 1.800\,kg/0.00220\,m^3 = 818\,kg\,m^{-3}$$

or

$$= 3.97\,lb/0.0777\,ft^3 = 51.1\,lb\,ft^{-3}$$

$$D(\text{weight and volume at 12\%}) = 1.120\,kg/0.00209\,m^3 = 536\,kg\,m^{-3}$$

or

$$= 2.47\,lb/0.0738\,ft^3 = 33.5\,lb\,ft^{-3}$$

$$SG_g(\text{green}) = \left(1.000\,kg/0.00220\,m^3\right)/1000\,kg\,m^{-3} = 0.454$$

or

$$= \left(2.20\,lb/0.0777\,ft^3\right)/62.4\,lb\,ft^{-3} = 0.454$$

$$SG_{12}(\text{12\% MC}) = \left(1.000\,kg/0.00209\,m^3\right)/1000\,kg\,m^{-3} = 0.478$$

or

$$= \left(2.20\,lb/0.0738\,ft^3\right)/62.4\,lb\,ft^{-3} = 0.478$$

$$SG_{od}(OD) = \left(1.000\,kg/0.00200\,m^3\right)/1000\,kg\,m^{-3} = -0.500$$

or

$$= \left(2.20\,lb/0.706\,ft^3\right)/62.4\,lb\,ft^{-3} = 0.500$$

The *total weight* of a wood product is, of course, the sum of the weight of the wood substance plus the moisture. As explained above, when density is calculated, this total weight should be used as the numerator in the equation; when specific gravity is calculated, it is standard practice to use the OD weight as the numerator. This difference is particularly significant when dealing with green wood. Consider this example: a cubic meter of green lumber has a weight of 675 kg at 50% MC, a weight of 855 kg at 90% MC, and an OD weight of 450 kg. The density at 50% MC is 675 and at 90% MC it is 855 kg m^{-3}; the specific gravity is 0.45 at both moisture contents. The specific gravity is the same at both moisture levels because the volume is constant above the fiber saturation point (FSP), and the OD weight is used in both cases. This is an important distinction in the use of specific gravity rather than density when studying, reporting, or otherwise working with wood properties.

Effects of Moisture Content

Density and specific gravity can be calculated at any moisture content. Density decreases as moisture content decreases, but below the fiber saturation point (FSP) the specific gravity increases as the moisture content decreases. This occurs because the dry weight remains constant while the volume decreases during drying. The greater the volumetric shrinkage, the greater the difference between the green and the OD specific gravities.

The following example illustrates the effect of volumetric change. Sample "A" has a green weight and volume of 800 g and 1000 cm³, respectively. When OD, the weight is 500 g. This sample undergoes 12% volumetric shrinkage from green to OD, and thus the dry volume is 880 cm³. For sample A:

$$SG_g = \left(500g/1000\,cm^3\right)/\left(1g\,cm^{-3}\right) = 0.50$$

$$SG_{od} = \left(500g/880\,cm^3\right)/\left(1g\,cm^{-3}\right) = 0.57$$

Sample "B" has the same green weight and volume and the same OD weight as sample A. It shrinks only 6% in volume when drying, so the OD volume is 940 cm³. Therefore, for sample B:

$$SG_g = \left(500g/1000\,cm^3\right)/\left(1g\,cm^{-3}\right) = 0.50$$

$$SG_{od} = \left(500g/940\,cm^3\right)/\left(1g\,cm^{-3}\right) = 0.53$$

Although the values of green specific gravity of A and B are the same, the OD specific gravity differs by 0.04. This amount is not trivial, especially considering that the specific gravity range for commercial North American species is ~0.35–0.70.

Figure 9.1 shows the typical relationship between specific gravity and moisture content. This figure makes it possible to convert from green specific gravity to specific gravity at any moisture content. This figure is based on an average volumetric shrinkage value (%S_v) for each density. Alternatively, specific gravity can be calculated based on the following equation:

$$SG_{n\%MC} = SG_g / \left(1 - \%\text{volume shrinkage to}\,n\%MC/100\right)$$

In the United States, for wood that is above the FSP, specific gravity is commonly computed in either the green or 12% MC condition. For computation of SG_g, the MC at FSP should be used: 28 or 30%. The use of SG_g as a fundamental property has the advantage of being reliably reproducible because the wood is fully swollen and in its natural unstressed condition. Green specific gravity is often used in studies of silvicultural practices, when estimating shipping weights of logs and pulpwood, and for estimating yields of pulp or other products.

The amount of shrinkage that occurs when a sample of wood is dried is affected somewhat by the size and shape of the piece and the means of drying. Therefore, specific gravity values based on volumes determined below the FSP can be affected.

Cell Wall Density and Porosity

An inverse relationship exists between the void volume of wood (*porosity*) and density because the density of dry cell wall substance is approximately the same for all species. That is, if sections of void-free cell wall material are taken from a low-density species like basswood, tested for specific gravity, and compared with results of a similar test from a dense wood such as hickory, the two specific gravity values will be almost identical. For general purposes it can be assumed that the density of the dry wood cell wall material is ~1.5 g cm^{-3}; that is, the specific gravity is 1.5. If a wood species contained no cell lumina or other voids, as would be the case if it were completely crushed, it would have an OD specific gravity of 1.5. If it were 50% porous, its specific gravity would be 0.75. The approximate void volume of wood can be calculated by the following equation:

$$\text{Percentage void volume} = \left[1 - \left(SG_{od}/1.50\right)\right] \times 100$$

The density of cell wall material has been studied by many wood scientists dating back to the mid 1800s. Observed density values are affected by the measurement technique employed. In determining cell wall density, volume is generally determined by displacement of a fluid. Different fluids vary in their ability to penetrate the voids in the wall and in their physical association with chemical components of wood, and thus these measurements often vary. In most cases, studies have determined the cell wall specific gravity to be between 1.45 and 1.54.

Calculation of Weight and Buoyancy

The weight of wood products can be easily estimated if the moisture content and specific gravity are known. The weight of the product should be calculated using the specific gravity that corresponds as closely as possible to the moisture content of the product at that time.

$$\text{Weight} = \text{volume} \times \text{specific gravity} \times \text{density of water} \times \left[1 + \left(\%MC/100\right)\right]$$

Example: 10 m^3 (353 ft^3) of mahogany (*Khaya* sp.) is shipped at an average moisture content of 40%. The average green specific gravity of African mahogany is known to be 0.43. What is the estimated weight?

$$\text{Weight} = 10\,\text{m}^3 \times 0.43 \times 1000\,\text{kg}\,\text{m}^{-3}\,31.40 = 6020\,\text{kg}$$

or

$$= 353\,\text{ft}^3 \times 0.43 \times 62.4\,\text{lb}\,\text{ft}^{-3} \times 1.40 = 13260\,\text{lb}$$

Similar calculations have been performed for a series of moisture contents and specific gravities and the results are compiled in Table 8.1. The densities in this table are based on the weight and volume at the same moisture content.

The following examples illustrate the application of density calculations to wood and wood products.

TABLE 8.1.   Total density of wood: $\text{kg}\,\text{m}^{-3}$ ($\text{lb}\,\text{ft}^{-3}$).

MC (%)	Specific gravity				
	0.30	0.40	0.50	0.60	0.70
0	300	400	500	600	700
	(18.7)	(25.0)	(31.2)	(37.4)	(43.7)
10	330	440	550	660	770
	(20.6)	(27.5)	(34.3)	(41.1)	(48.1)
20	360	480	600	720	840
	(22.5)	(30.0)	(37.4)	(44.9)	(52.4)
40	420	560	700	840	980
	(26.2)	(34.9)	(43.7)	(52.4)	(61.2)
60	480	640	800	960	1120
	(30.0)	(39.9)	(49.9)	(59.9)	(69.9)
80	540	720	900	1080	1260
	(33.7)	(44.9)	(56.2)	(67.4)	(78.6)
100	600	800	1000	1200	1400
	(37.4)	(49.9)	(62.4)	(74.9)	(87.4)

Note: The total density is based upon the weight and volume at the moisture content indicated.

Example 1

White ash has an OD specific gravity of 0.61 and a green specific gravity of 0.55. What percentage of the total volume of dry wood is made up of cell wall substance?

Percentage solid wood substance by volume $= 0.61/1.50 = 41\%$

What will the weight of $1\,\text{m}^3$ of this wood be if the moisture content is 38%?

$$\text{Weight} = 1\,\text{m}^3 \times 0.55 \times 1000\,\text{kg}\,\text{m}^{-3} \times 1.38 = 759\,\text{kg}$$

Recall that total weight = oven-dry weight 3 [1+(percent MC/100)]. What will the moisture content of this wood be if it is completely saturated with water (all voids filled)?

Percentage void volume $= 100 - \left[(0.61/1.50) \times 100\right] = 59\%$

Void volume per cubic centimeter $= 0.59\,cm^3$

Mass of water per cubic centimeter when filling all voids $= 0.59\,cm^3 \times 1\,g\,cm^{-3} = 0.59\,g$

Saturated MC $= 0.59\,g/0.55\,g \times 100 = 107\%$

At what moisture content will a log of white ash have zero buoyancy (i.e. just float)? Recall Archimedes's principle – the force buoying up a solid body in a liquid is equal to the weight of the liquid displaced by the body. There is also an equal force exerted downward on the liquid by the mass of the body.

Mass of OD wood per cubic centimeter of green wood $= 0.55\,g$

Total mass per cubic centimeter for zero buoyancy $= 1\,g\,cm^{-3}$

Weight of water per cubic centimeter for zero buoyancy $= 1\,g \pm 0.55\,g = 0.45\,g$

Moisture content $= 0.45/0.55 = 82\%$

Example 2

A shipment contains $80\,ft^3$ of ponderosa pine lumber. The moisture content as measured by an electrical meter averages 12%. The average specific gravity of ponderosa pine at 12% MC as indicated in Online Appendix Table A.5 is 0.40. What is the estimated weight of the lumber?

Weight $= 80\,ft^3 \times 0.40 \times 1.12 \times 62.4\,lb\,ft^{-3} = 2240\,lb$

What will the specific gravity of this ponderosa pine be at the OD condition and when green? (Refer to Figure 8.1.)

OD specific gravity $= {\sim}0.42$

Green specific gravity $= {\sim}0.38$

Impact of Extractives and Inorganic Materials on Specific Gravity

Wood often contains measurable quantities of extractives and infiltrates including terpenes, resins, polyphenols such as tannins, sugars, and oils as well as inorganic compounds such as silicates, carbonates, and phosphates. These materials are located within the cell wall, where they are deposited during the maturation of the secondary wall and as remnants in the cell cavity after heartwood formation. Thus, heartwood has higher concentrations of these materials than sapwood; therefore, the density of heartwood is often slightly higher than that of sapwood.

The amount of extractives in wood varies from less than 3 to over 30% of the OD weight. Obviously, the presence of these materials can have a major effect on the density. In some species, including pine, it has been shown that the presence of extractives

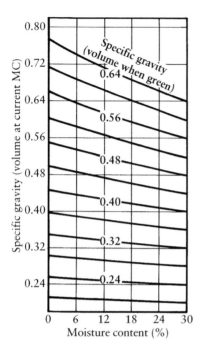

FIGURE 8.1.   Relation of specific gravity and MC.

contributes significantly to observed specific gravity variation. The specific gravity of wood from which extractives have been removed tends to be more uniform than where the weight of extractives is included. In research work, it is often desirable to determine the density of the wood without extractives. Both water and organic solvents are used for extraction when determining extractive-free density. Some of the bulking effect is lost when extractives are removed, so in addition to weight loss, samples tend to show greater dimensional change with moisture fluctuations.

Methods of Determining Specific Gravity

In most cases, the dry weight is calculated by oven drying the sample, as would be done for moisture content calculation. Because high temperatures may drive off some of the extractives in addition to the water, it is sometimes desirable to determine the moisture content by distillation, a process that involves condensing and weighing the collected vapor.

Sample volume may be obtained in a variety of ways. For a piece that is regular in shape, such as a section of lumber, the simplest method is to measure the dimensions as accurately as possible and calculate the volume. If the sample is irregular in shape, such as a tree cross-section or a wood chip, the volume can be obtained by the displacement method. Equipment to do this is illustrated in Figure 8.2. The scale or balance records the weight of the fluid displaced. This value can then be converted to a volume by dividing the weight change by the density of the fluid used. The displacement procedure using water as the fluid works well with green material because little water is adsorbed by the wood. When dry wood is immersed, the sample should be coated with paraffin wax or a similar substance so that water will not penetrate the block and produce an erroneously low volume determination. Use of a high-surface-tension, nonwetting fluid avoids the

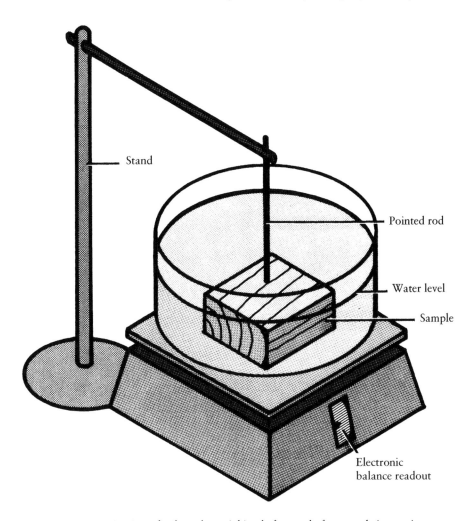

FIGURE 8.2.   Determination of volume by weighing before and after sample immersion.

wetting problem with dry samples, but because many such fluids (e.g. mercury) present safety hazards, the use of water and a waxy coating is usually more convenient.

A third method of determining volume involves the use of a graduated cylinder, as shown in Figure 8.3. In this case the volume is simply the difference between the fluid level before and after immersion. This is a quick and simple technique but involves the same problem for dry samples as the displacement method. The details of these and other methods of determining the specific gravity of wood-based materials are discussed in ASTM Standard D 2395 (American Society of Testing and Materials, 2001).

Relationship of Density to Growth Rate

Intuition or common sense might suggest that wood density should decrease as the rate of tree growth increases. This is not necessarily the case. When wood density is related to growth rate, the response depends on the species and the range of growth rates involved.

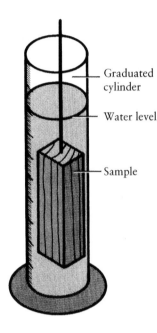

Graduated cylinder

Water level

Sample

FIGURE 8.3.   Determining volume by the difference in the water level (read cylinder before and after immersion).

Other factors such as age, tree vitality, site, and location of wood in the tree are more closely correlated to density than is the rate of growth. The only species group in which density is closely related to growth rate is the ring-porous hardwoods. In these species the density tends to increase as the growth rate increases (see Figure 11.2). Softwoods and diffuse-porous hardwoods do not exhibit a consistent relationship between wood density and growth rate.

The rate of growth is used to estimate density (and thus strength) in the grading of several types of lumber products. Hickory tool handles are often graded based on the average number of growth rings per centimeter or rings per inch. The fewer the growth rings per unit length, the higher the density and, therefore, the higher the grade. Ash shovel and hoe handles and implement parts can also be graded based on the number of growth rings per centimeter (or inch). A ring-porous handle with 12 or more annual rings per centimeter (30 or more rings per inch) cannot be relied on to perform satisfactorily under heavy use.

Although the density of softwoods is generally not related to growth rate, density is directly related to the percentage of latewood in a growth ring. There is typically a large difference in density between earlywood and latewood. In southern pine, the specific gravity of the earlywood is less than one-half that of latewood, e.g. 0.28 versus 0.70 (Koch 1972). For this reason, relatively wide latewood zones indicate relatively high density. Because the strength of wood increases with density, wide growth rings indicating a low proportion of latewood may be of concern in products where strength is important.

Because of the desirability of a high percentage of latewood, a so-called *density rule* is sometimes applied in the grading of southern pine, Douglas fir, and western larch lumber. Such lumber may be graded as *dense* if it averages six or more growth rings per inch and if it contains one-third or more latewood. The requirement of at least six rings per inch reduces the chance of using juvenile wood or compression wood, both of which have poor strength and dimensional properties. Material graded as *dense* carries engineering strength values 15–30% higher than normal lumber. Owing to changes in timber resources and merchandising, the dense grade is rarely produced and is not widely available. Southern pine structural lumber containing 15% or less latewood is deemed to be exceptionally light and is excluded from the top three structural grades.

Most research directed to the question of the relationship between growth rate (or rings per centimeter) and density has been conducted in conjunction with studies on fertilization, irrigation, genetic improvement, or other intensive silvicultural practices. This topic is discussed in more detail in Chapter 11.

Variability of Density

Wood density varies greatly within any species because of a number of factors. These include location in a tree, geographic location within the range of the species, site condition (soil, water, and slope), and genetic source. The user of a wood or wood-based product, however, often has no control or knowledge of where the particular tree was cut, in what part of the tree the product originated, or the tree's maturity level. The user is concerned primarily with the variability that may be encountered in the density of the product, regardless of its source. This expected variability is important. Such knowledge is needed, for example, (i) in estimating the variability in the strength of a wood product, (ii) in establishing a procedure for purchase of wood on a weight basis, and (iii) in estimating the amount of pulp to be obtained per unit volume of raw wood.

Generally, the specific gravity of most species in North America has a *coefficient of variation* (COV) of about 10%. To estimate the range of wood densities one might normally encounter in any such species, the COV is multiplied by the average density (specific gravity) and by 1.96 (to include 95% of a normally distributed population). This figure is then added to and subtracted from the mean to determine the upper and lower estimates, respectively. For example, the average green of black cherry, as indicated in Online Appendix Table A.5, is 0.47. Therefore, the range of density to be expected in black cherry is ~0.47 ± (10% × 0.47 × 1.96) = 0.47 ± 0.09 = 0.38 to 0.56. Wood users frequently question the quality of material they purchase if it has a specific gravity that differs from the average value published in the *Wood Handbook* (USFPL 2010) (Online Appendix Table A.5) or in trade association literature. It should be understood that such differences are to be expected because of natural variability. Table 8.2 shows the average and range of green specific gravity for some important species in the United States, assuming a 10% COV. Figure 8.4 shows the specific gravity variation of in a Douglas fir glue-laminated beam.

Lists of species and their average specific gravities are found in Online Appendix Tables A.5, A.6, and A.7. These list important woods grown in the United States, Canada, and tropical countries, respectively. Online Appendix Table A.5 indicates the specific gravity both at 12% MC and when green; Tables A.6 and A.7 provide only the green specific gravity.

Historically, relatively limited information has been available in North America on the variability of density in tropical species. In some instances, as with balsa, it has been found that the COV is considerably higher than 10%. Recently, information sharing has

TABLE 8.2.   Range of specific gravity for important species in the United States.

Species	Average green specific gravity	Normal[a] range of specific gravity
Softwoods		
Douglas-fir (coast)	0.45	0.36–0.54
Loblolly pine	0.47	0.38–0.56
Longleaf pine	0.54	0.43–0.65
Ponderosa pine	0.38	0.31–0.45
Western hemlock	0.42	0.34–0.50
Western white pine	0.35	0.28–0.42
White fir	0.37	0.30–0.44
Hardwoods		
Black walnut	0.51	0.41–0.61
Northern red oak	0.56	0.45–0.67
Quaking aspen	0.35	0.28–0.42
Southern red oak	0.52	0.42–0.62
Sugar maple	0.56	0.45–0.67
Sweetgum	0.46	0.37–0.55
White ash	0.60	0.48–0.72
Yellow-poplar	0.42	0.34–0.50

Source: USFPL 2010.
[a] Assuming a 10% COV.

improved dramatically with the development and application of information technology and the internet. Worldwide, information sharing related to forest products research, development, applications, and performance has increased substantially and the future outlook shows further development. This information sharing has given much support to researchers, producers, and users of wood products in the global marketplace.

The variability in density of major softwood species in the United States became of considerable concern in the 1960s and early 1970s. This arose in large part because of questions regarding the properties of structural timbers and plywood produced from smaller second-growth timber. Large-scale wood density surveys were carried out by the USDA Forest Service to determine regional and tree-to-tree variability in southern pines and the major western softwoods. About 25,000 trees were sampled in the western survey alone. The density of sample trees was determined from increment cores taken at "breast height," about 1.4 m (4.5 ft) from the ground level. From these cores and analyses of selected whole trees, regressions were developed to predict average whole-tree densities. These studies provided new insights into the variability of these species and led to updated strength and design values being assigned to lumber of several species.

Some of the results of the southern and western wood density (specific gravity) surveys are shown in Figures 8.5–8.7 and Table 8.3. The three values indicated in Figure 8.5 for each survey region are from data taken from loblolly pine in three diameter classes. Note that these averages varied from 0.45 to 0.57 over the range. The density variation of Douglas-fir (Figure 8.6) is almost as great. This information on Douglas-fir verified the lower strength values assigned to material produced in the interior southern region.

Consistent increases in the specific gravity of Douglas-fir with tree age are shown in Figure 8.7. The western density survey also involved other major western species.

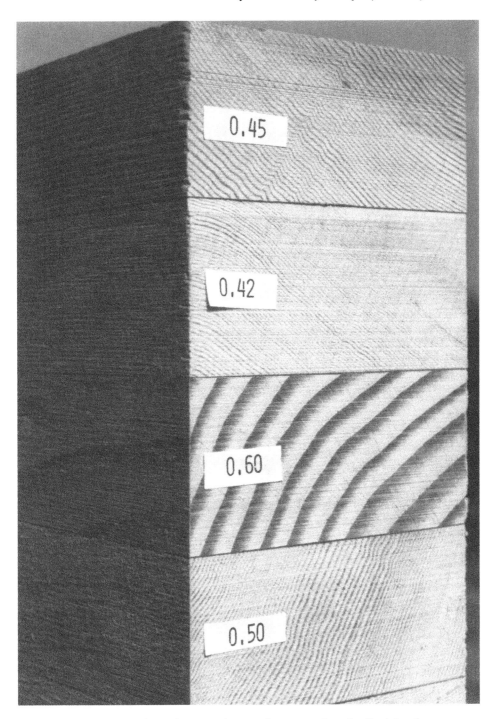

FIGURE 8.4. Variation of specific gravity between lamina in a Douglas-fir glulam beam.

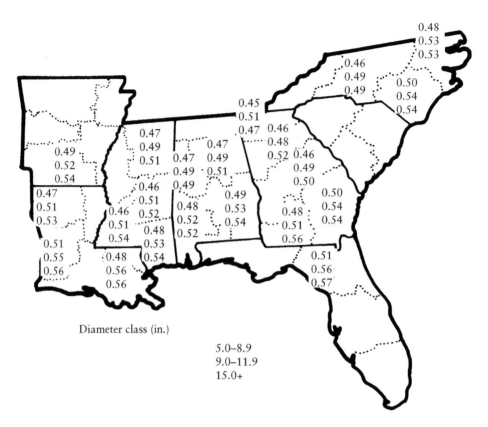

FIGURE 8.5. Variation in specific gravity of loblolly pine by survey units and tree diameter classes (From USDA Forest Service 1965a).

These findings are summarized in Table 8.3. Compare the range of specific gravity measured in this survey with that estimated using the assumption of a 10% COV. Similar results have been recently reported for loblolly pine, one of the principal southern pines (Clark et al. 2006).

Sources of Variation in Specific Gravity

Many factors, including site, climate, geographic location, growth stresses, and species, affect the specific gravity of wood. Because many of these act in combination, it is difficult to separate the independent effects. There is a great deal of scientific literature dealing with these relationships, the inconsistencies of which indicate their complex interactions (Koch 1985).

Site-related factors such as moisture, sunlight, nutrients, wind, and temperature can affect specific gravity. These are determined to a large extent by elevation, aspect, slope, latitude, soil type, stand composition, and spacing. All of these factors can affect the size and wall thickness of the cell and thus the density. Tree species differ greatly in their sensitivity to these factors.

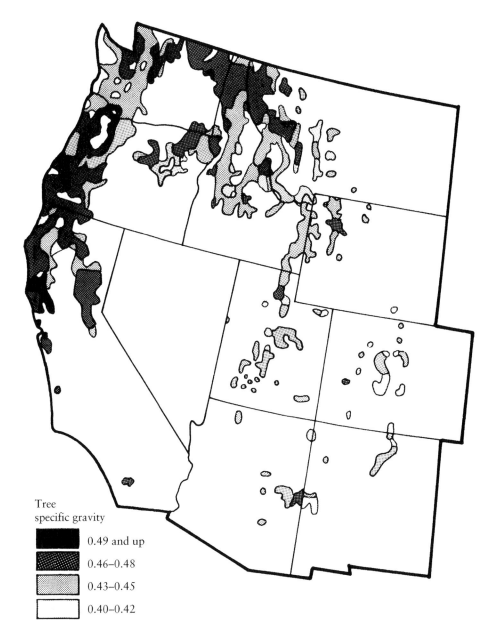

FIGURE 8.6.   Estimated specific gravity of Douglas-fir (From USDA Forest Service 1965b).

It is also common for density to vary significantly within a tree. Figure 8.8 shows the variation found within a young yellow poplar tree in West Virginia. There, density varied from 0.36 to 0.42 at different tree heights and from 0.37 to 0.40 at different distances from the pith at selected heights. In many species, butt logs tend to have a higher density than logs cut from higher in the main stem. However, in some woods, such as tupelo gum and yellow cypress, the wood near the base of the tree may be lighter than normal wood.

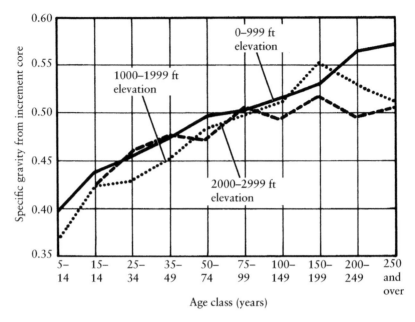

FIGURE 8.7.   Relation between age and specific gravity by elevation classes for Douglas-fir in the Cascade Mountains (From USDA Forest Service 1965b).

TABLE 8.3.   Specific gravity data for 15 western species.

Species	Mean estimated tree specific gravity	Range of estimated tree specific gravity	Number of trees sampled	Wood Handbook average green specific gravity
Douglas fir	0.45	0.33–0.59	9133	0.45
White fir	0.37	0.26–0.54	2150	0.37
California red fir	0.36	0.31–0.46	0840	0.36
Grand fir	0.35	0.24–0.55	0862	0.35
Pacific silver fir	0.40	0.28–0.55	0330	0.40
Noble fir	0.37	0.26–0.44	0158	0.37
Western hemlock	0.42	0.30–0.52	1040	0.42
Western larch	0.48	0.38–0.54	0678	0.48
Black cottonwood	0.31	0.28–0.40	0120	0.31
Ponderosa pine	0.37	0.27–0.54	5337	0.38
Sugar pine	0.34	0.28–0.45	0299	0.34
Western white pine	0.36	0.29–0.45	0292	0.35
Lodgepole pine	0.38	0.26–0.55	3516	0.38
Engelmann spruce	0.35	0.23–0.58	1789	0.33
Western redcedar	0.32	0.27–0.42	0504	0.31

Source: Maeglin and Wahlgren (1972), USFPL (2010).

Generally, in softwoods the density decreases with height in the tree and increases with distance from the pith. In large softwood logs, the density often increases outward from the pith and then reaches a fairly constant level. For the most part, these observations support the old adage "trees grow strongest where the stresses are the highest." At the

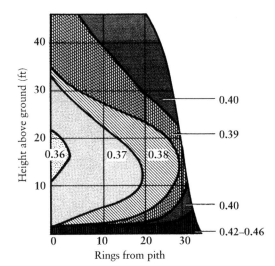

FIGURE 8.8.   Radial and longitudinal specific gravity distribution in yellow poplar (From Koch et al. 1968).

base of the tree, the mechanical bending forces are greatest, so it is reasonable that the densest and strongest wood will be formed there.

Juvenile Wood and Reaction Wood

As explained in Chapter 6, the density of juvenile wood is usually less than that of mature wood, and such wood has a correspondingly lower strength. Low strength is of concern when the portion of a log near the pith is utilized for lumber or veneer. Another problem in the utilization of juvenile wood is the tendency to exhibit abnormally high longitudinal shrinkage. In such lumber, shrinkage results in excessive warp, which renders the product unusable for most applications. This is less serious in cross-laminated panels such as plywood for reasons discussed in Chapter 13.

The specific gravity of compression wood in softwoods, in contrast to juvenile wood, is generally greater, by up to 40%, than that of normal wood. This higher density can often be detected visually because of the higher proportion of latewood. Yet in some cases, the density of compression and normal wood may not differ significantly. Despite its usually higher density, compression wood is generally avoided in lumber products because of its high longitudinal shrinkage and aberrant strength properties. A more complete discussion of reaction wood is provided in Chapter 6. The commercial standards for ladder rails and scaffold planks specifically exclude compression wood because of its unreliable strength and the tendency to exhibit *brash failure* (abrupt, low-energy fracture). Recognizing the influence of juvenility, the American Institute of Timber Construction limits the amount of pith-associated wood permitted in tension laminations of laminated beams. In higher-grade glulam beams, no more than one-eighth of the cross-section can consist of such wood. Tension wood in hardwood, like compression wood in softwoods, often possesses higher than normal specific gravity with erratic strength performance.

Density of Forest Products

The density of veneer-, strand-, particle-, and fiber-based products differs from the density of the wood raw material because of the weight of adhesives and other additives and the compression of the wood that occurs during the manufacturing process. Plywood is ordinarily only slightly more dense than the wood from which it is produced – usually not more than 5–15% greater. The pressure used in making plywood is intended only to provide good contact between the veneers and not to densify the wood, although slight densification occurs. Particleboard and oriented strandboard (OSB), in contrast, are usually produced at densities 20–60% greater than the density of the species used. Particleboard densities in the range of 590–880 kg m^{-3} are common. Of this weight, about 3–12% is the weight of the resin (adhesive) and wax used to impart water repellency. OSB is typically produced at densities ranging from 610 to 700 kg m^{-3}. In OSB, resin and wax typically account for 3–6% of the finished weight. Fiber-based products vary widely in density. Insulation board or cellulosic fiberboard used for wall sheathing is produced at 160–480 kg m^{-3}, medium density board at 500–800 kg m^{-3}, and hardboard at 800–1120 kg m^{-3}. These products may contain from 0.5 to 30% bonding resins and other additives to improve strength and water-resistant properties.

Metrication

The *International System of Units* (SI) is used throughout most of the world and has become commonplace in the United States. This system is a modern version of the meter–kilogram–second–ampere system that was adopted by international treaty in 1935. SI units are recommended for wood science research; most work is reported in SI units. In areas of applied technology such as product sizes, codes and standards, and engineering and structural design, much work in the United States is still based on Imperial units, but in an effort to compete in the global economy this is changing.

Wood scientists and technologists in the United States and around the world need to be able to work and think in terms of either meters and newtons or feet and pounds. This practice facilitates stronger communication and understanding in the global wood and wood-based products arena. An excellent guide to the proper use of metric units is *Standard for Metric Practice*, E 380, published by the American Society for Testing and Materials. The following comments may aid in understanding the metric units commonly encountered in forest products research. Table 8.4 contains conversions for some of the commonly used non-SI units.

One advantage of SI is that only one base unit is used for each physical quantity. Base units include meter (m) for length, second (s) for time, and kilogram (kg, not g) for mass. All mechanical properties are then derived from the base units. Some derived units are given special names such as newton (N) for force, joule (J) for work or energy, and watt (W) for power. The SI units force, energy, and power are the same regardless of whether the process is mechanical, electrical, chemical, or thermal.

The use of the centimeter (cm) is discouraged in SI and should be avoided. The use of the centimeter for length, the cubic centimeter for volume, and the gram for mass will however undoubtedly persist in wood research.

TABLE 8.4. Common non-SI to SI conversions.

Non-SI unit	SI unit equivalent
Length	
foot	3.048×10^{-1} m (meters)
inch	2.540×10^{-2} m
mile (US statute)	1.609×10^{3} m
Area	
acre	4.046×10^{3} m^2
hectare	1.000×10^{4} m^2
ft^2	9.290×10^{-2} m^2
in^2	6.452×10^{-4} m^2
Volume	
ft^3	2.832×10^{-2} m^3
in^3	1.639×10^{-5} m^3
board foot	2.360×10^{-3} m^3
Force	
kilogram-force	9.807 N (newtons)
kilopond	9.807 N
lbf	4.445 N
Mass	
pound (avoirdupois)	4.536×10^{-1} kg (kilograms)
ton (short)	9.072×10^{2} kg
ton (metric)	1.000×10^{3} kg
Moment	
lbf·in	1.130×10^{-1} Nm (newton meters)
lbf·ft	1.356 Nm
Energy and work	
Btu	1.055×10^{3} J (joules)
ft·lbf	1.356 J
Density (mass per unit volume)	
lbf ft^{-3}	1.602×10 kg m^{-3}
g cm^{-3}	1.000×10^{3} kg m^{-3}
Pressure or stress	
lbf in^{-2} (psi)	6.895×10^{3} Pa (pascals)
kgf m^{-2}	9.807 Pa
Other SI symbols	
s	second
k	kilo or 10^{3}
M	mega or 10^{6}

Confusion about the definition of a *pound* is common when using the Imperial system. This unit is used both as a mass and as a force unit, although properly it should be indicated as *pound-force* (*lbf*) when used in the latter sense. When used to indicate mass, the pound can be converted to kilograms. When indicating force, the pound should be converted to newtons (N), the unit of force in SI. A *newton* is the force resulting from the acceleration of a kilogram, i.e. kg m s^{-2}.

For large quantities of material, the ton is often used in commerce as the unit of mass. A *long ton* is 2240 lb, a *short ton* 2000 lb, and a *metric ton* (tonne, ton, or t) is 1000 kg. None of these terms are SI, and their definition should be given when they are used to avoid confusion. Note that the term *weight*, as commonly used, generally applies to mass. Because of the uncertainty as to whether mass or force is indicated, however, the use of the term weight is discouraged by SI. Nonetheless, it too will undoubtedly remain in common practice and is used in this text as a unit of mass.

Many wood science references in the European literature use *kilogram-force* (*kgf*) or *kilopond* (*kp*) as the units of force. These have been replaced in SI with the newton (N). The newton is used for other derived units involving force. For example, stress is expressed as newtons per square meter (N m^{-2}) and called a pascal (Pa); the unit of energy, a *newton meter* (Nm), is termed a *joule* (J); and the unit of power, called a *watt* (W), is Nm s^{-1}.

For engineering purposes it is generally more convenient to work with *kilopascals* (kPa = 1000 Pa) than with pascals. It may be helpful to remember that 1 pound per square inch (psi) is ~7 kPa. In some cases the use of *megapascals* (MPa) is more convenient. Note that in SI the prefix M equals 10^6. In the Imperial system M is often used to indicate 10^3, as in *MSF* (1000 ft^2), another source of possible confusion. Also, the use of the terms such as *billion* and *trillion* should be properly defined or avoided. In the United States billion means 10^9, but it means 10^{12} in many other countries.

The standard method for metric measurement of lumber and wood-composites utilizes the actual length, width, and thickness to calculate the cubic meter volume. The use of cubic meters facilitates ready comparisons among virtually all forest products industries on a worldwide scale.

References and Supplemental Reading

American Society of Testing and Materials (2001). Standard test methods for specific gravity of wood and wood-based materials. ANSI/ASTM Vol. 4.09; D2395.

Brown, H.P., Panshin, A.J., and Forsaith, C.C. (1952). *Textbook of Wood Technology*, vol. 2, 1–22. New York: McGraw-Hill.

Clark, A., Daniels, R., and Jordan, L. (2006). Effect of tree age, rate of growth, and geographic location on plantation loblolly pine wood stiffness and strength. Presented at Forest Products Society 60th International Convention, Newport Beach, California, June 25–28.

Kellogg, R.M. and Wangaard, F.F. (1969). Variation in the cell wall density of wood. *Wood Fiber* 1: 180–204.

Koch, P. (1972). Utilization of Southern Pines, Vol 1. *USDA For. Serv. Handb.* 420: 235–264.

Koch, P. (1985). Utilization of Hardwoods Growing on Southern Pine Sites, Vol 1. *USDA For. Serv. Handb.* 605: 465–548.

Koch, C.B., Brauner, A., and Kulow, D. (1968). Specific gravity variation within yellow poplar. WV Agric. Exp. Stn Bull. 564T.

Kollmann, F.P. and Côté, W.A. Jr. (1968). *Principles of Wood Science and Technology*, 160–180. New York: Springer.

Maeglin, R.R., and Wahlgren, H.E. (1972). Western wood density survey, Report no. 2. USDA For. Serv. Res. Pap. FPL-183.

Oberg, J.C. (1989). Impact of changing raw material on lumber and panel products. USDA SE For. Exp. Stn Gen. Tech. Rep. SE-63. pp.17–33.

Senft, J.F., Bendtsen, B.A., and Galligan, W.L. (1985). Weak wood-fast grown trees make problem lumber. *J. For.* 83 (8): 477–482.

Shepard, R.K. and Shottafer, J.E. (1990). Effect of early release on the specific gravity of black spruce. *For. Prod. J.* 40 (1): 18–24.

Shepard, R.K. and Shottafer, J.E. (1992). Specific gravity and mechanical property relationships in red pine. *For. Prod. J.* 42 (7/8): 60–67.

Siau, J.F. (1995). *Wood: Influence of Moisture on Physical Properties*, 227. Blacksburg, VA: VPI&SU Press.

Stamm, A.J. and Sanders, R. (1966). Specific gravity of wood substance of loblolly pine as affected by chemical composition. *TAPPI* 49: 397–400.

Thomas, R.J., and Kellison, R.C. (1989). Impact of changing raw material on paper manufacture. USDA SE For. Exp. Stn Gen. Tech. Rep. SE-63. pp. 33–47.

USDA Forest Service. (1965a). Southern wood density survey, 1965 Status Rep. Res. Pap. FPL-26.

USDA Forest Service. (1965b). Western wood density survey – Report no. 1. Res. Pap. FPL-27.

US Forest Products Laboratory. (2010). Wood Handbook: Wood as an Engineering Material. USDA For. Serv. FPL-GTR-190.

9

Strength and Mechanics

Among its positive attributes, the mechanical properties of wood rank highly. The availability, cost, structural performance, and reliability of wood make it the material of choice for home building. As discussed in the previous chapter, the quality of wood – in the standing tree – influences the structural performance of wood. Density, slope of grain, knots, and other characteristics all impact wood structural performance. This chapter examines how structural properties are assessed. When engineers search for design solutions, they are led to wood based primarily on its availability, cost, and performance. For example, for county road bridges that have relatively short spans, high loads, and intermittent traffic, wood is a highly competitive material. For highway bridges that may be miles long, steel and concrete are more cost-effective choices. Beyond growing trees with good form, large diameter, and great length, it is important for foresters to keep in mind that the structural properties of wood are important. These are measurable, and two logs that might be of similar appearance and specifications may be worth vastly different amounts of money based on their quality.

The strength and resistance to deformation of a material are referred to as its *mechanical properties*. *Strength* is the ability of a material to carry applied loads or forces. Resistance to deformation, or *stiffness*, determines the amount a material is compressed, stretched, bent, or otherwise distorted by an applied load. Changes in shape that take place instantaneously as a load is applied and are recoverable when the load is removed are termed *elastic deformation*. The flexing of a diving board at a swimming pool when someone bounces on it is an example of elastic deformation. If the deformation, on the other hand, develops slowly after the load is applied, it is termed a *rheological* or *time-dependent* property.

Mechanical properties are usually the most important characteristics of wood products for structural applications. A *structural application* is any use where strength is one of the primary criterion for selection of the material. Figure 9.1 shows two structural applications of wood. Structural uses of wood products include, among other things,

Forest Products and Wood Science: An Introduction, Seventh Edition. Rubin Shmulsky and P. David Jones.
© 2019 John Wiley & Sons Ltd. Published 2019 by John Wiley & Sons Ltd.
Companion website: www.wiley.com/go/shmulsky

FIGURE 9.1.   Structural lumber products used as primary building members. Source: Top photo courtesy of the United States Forest Products Laboratory (USFPL) (1974); bottom photo courtesy of the American Institute of Timber construction. During the four decades since these photos were released, building materials and construction techniques have advanced. However, the components shown herein such as studs, floor and roof trusses, and glue laminated arches remain popular and in widespread use throughout North America.

floor joists and rafters in homes, power transmission and distribution poles, structural panel roof and wall sheathing and subflooring, glue-laminated beams and decking in commercial buildings, particleboard flooring in mobile homes, steps and rails of wood ladders, piling, and sailboat masts. Structural applications of wood and wood-based products are omnipresent in modern society. A sound argument can be made that basic human evolution and expansion (hunting, shelter, transportation, protection, etc.) has been largely shaped by humankind's ability to exploit the strength of wood.

The term *strength* is often used in a general sense to refer to all mechanical properties. However, because there are many different types of strength and elastic properties, it is important to be specific about the mechanical property being discussed. A wood species that is relatively strong with respect to one strength property may rank lower for a different property. The type of mechanical property most critical to any application is determined by the nature and type of loading to which that product will be subjected. For example, in a floor joist the *modulus of elasticity* (MOE) is very important because it determines the amount the joist will bend or *deflect* under load and thus how solid the floor will seem. In the case of wood flooring, the surface hardness determines the resistance to denting when under a concentrated load. In general, it is not difficult for an engineer to design a structure, be it a house, bridge, skyscraper, or other that is "strong enough," but it requires honed art and skill to design one that is "just strong enough." Some of the most important mechanical properties of wood products are listed in Table 9.1. To appreciate the meaning of the various strength properties of wood, it is necessary to first understand some basic engineering mechanics. Ultimately, these properties guide designers in building structures that perform as intended and protect the safety of their occupants.

Concepts of Stress, Strain, and Flexure

Two basic concepts used throughout the study of mechanics are stress and strain. *Stress* is a distributed force per unit area. A stress occurs where a load or force acts upon a solid member, such as a load upon a column. The most common true or pure forces in wood are compression, tension, and shear. Stress also occurs internally within a rigid body as loads are applied. Stress is usually expressed in pascals (Pa or Nm^{-2}) or pounds per square inch (lb in.$^{-2}$ or psi). For example, when air is forced into a rubber tire, it exerts an outward pressure or stress, readily measured as Pa or psi. When an external force (such as wind or the weight of a refrigerator), such as compression or tension, is applied to any rigid body (like a wood beam), internal stresses result. These stresses distort and deform the shape and size of the body. The change in length per unit of length in the direction of the stress is called the *strain*. Because strain is expressed in like units of length divided by the length (for example mm^{-1}), the units cancel and strain is thus *unitless*.

Figure 9.2 illustrates stress and strain in a wood test specimen under compression parallel to the grain. When a load of 35 600 N (8000 lb) is applied to the specimen, an internal parallel-to-grain stress of 13.8 MPa (2000 psi) is created. The stress is uniformly distributed at all distances from the end; therefore, the total deformation of 0.183 mm (0.0072 in.) is distributed uniformly along the 152 mm (6 in.) length. Because strain is change in length per unit of length, the resulting strain is 0.0012 (0.183/152 mm). Note that the strain is the same for any system of units: metric, Imperial, etc.

TABLE 9.1.   Important mechanical properties of wood.

Property	How or where this property can be important
Strength properties	
Breaking strength (MOR)	Determines the load a beam will carry before breaking
Compression strength parallel to grain	Determines the load a short post or column will carry
Compression strength perpendicular to grain	Important in the design of the connections between wood members in a building and the supports or bearing areas for beams and joists on sill plates, timber bridge stringers on bearing caps, and beams on timber piles
Tension strength parallel to grain	Important for the bottom member (chord) in a wood truss and in the design of connections between structural members
Shear strength parallel to grain	Often determines the load-carrying capacity in short-span deep beams. Often critical for I-joists and LVL
Toughness	Measure of the amount of work expended in breaking a small specimen in impact testing. Useful in detecting decay at early stages
Resilience	Measure by the amount of energy absorbed when a piece is bent within its elastic range
Side hardness	Relates to the resistance to denting, as for flooring
Work to maximum load	Measure of the energy absorbed by a specimen as it is slowly bent to failure
Elastic properties	
Modulus of elasticity	Measure of the resistance to bending deflection, that is, directly related to the stiffness of a beam; also a factor in the bending strength of a long column such as a utility pole
Modulus of elasticity parallel to grain	Measure of the resistance to elongation or shortening of a specimen under uniform tension or compression
Modulus of resilience	Measure of the energy that a volume of wood can absorb without becoming permanently deformed. Useful in industrial products such as crane and dragline mats, mine timbers, and other temporarily loaded elements

Strain results whenever stress is applied to any solid body. In wood, if the stress does not exceed a level called the *proportional limit*, there is a linear relationship between the amount of stress and the resulting strain. When the stress is removed the strain is completely recovered. The shape of a typical stress–strain curve for wood tested parallel to the grain is shown in Figure 9.3. Below the proportional limit, the ratio of stress to strain, that is, the slope of the linear relationship, is called the MOE. In compression and tensile tests, this ratio is sometimes termed *Young's Modulus* to differentiate it from the MOE as determined by bending. In the example shown in Figure 9.2, the MOE is 11 500 MPa (1.67×10^6 psi). Notice that the greater the level of stress that is required to produce a given strain level, the greater the resistance to deformation, the steeper the line, and the higher the MOE of the material.

The concepts of stress and strain are simple in uniaxial tension and compression but become more complex in beams (bending members). When a beam such as a wood floor joist is loaded, the top half is stressed in axial compression and the bottom half is stressed in axial tension. The maximum stresses develop at the outer top and bottom extreme surfaces of the beam. In simple beam analysis, it is assumed that the stresses vary in a

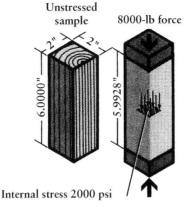

$$\text{Stress} = \frac{8000 \text{ lb}}{4 \text{ in.}^2} = 2000 \text{ psi}$$

$$\text{Strain} = \frac{6.000 - 5.9928}{6} = 0.0012 \text{ in. in.}^{-1}$$

Young's modulus (MOE) = stress ÷ strain = 2000 ÷ 0.0012 = 1.67×10^6 psi

FIGURE 9.2. Illustration of stress and strain in compression parallel to the grain. Trees, studs, wood columns, and certain members in wood trusses are loaded in this type of compression.

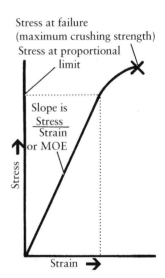

FIGURE 9.3. Relation between stress and strain in a typical compression parallel-to-grain test. In general, the strain below the proportional limit is recoverable while strain above the proportional limit is permanent.

linear manner from the top to bottom surface as shown in Figure 9.4. No tension or compression stresses occur at mid-depth in a rectangular beam of uniform section. This central plane, free from compression or tension, is termed the *neutral axis*.

In reality, the tensile and compressive stress distribution in wood beams is not precisely as idealized as in Figure 9.4, but the equal stress distribution above and below the neutral axis as shown is generally assumed and this is sufficient for most engineering purposes. Actually, clear, straight-grained wood is stronger in tension than in compression,

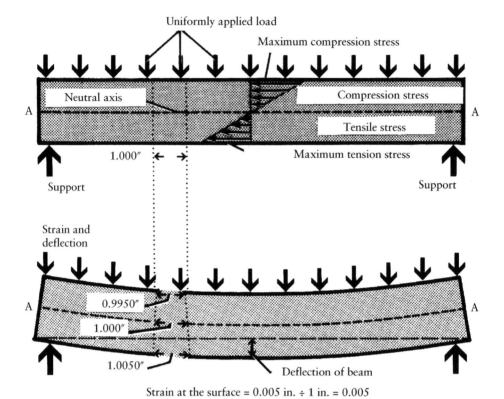

FIGURE 9.4.   Stress/strain in a uniformly loaded beam. Building codes model loads in this manner.

but lumber containing knots and grain deviation is usually stronger in compression than in tension.

Because no tensile or compressive stresses develop at the neutral axis of the beam, the length (AA in Figure 9.4) of the neutral axis remains the same when the beam is loaded. The top surface of the beam is shortened and the bottom surface is lengthened as a result of the compression and tension stresses, respectively. The amount of bending or deviation from initial position is termed *deflection*. In a simply supported uniformly or centrally loaded beam, the maximum deflection occurs at midlength. The amount deflection that occurs when a beam is loaded depends upon the distribution, location and magnitude of the load(s); the length and sectional size of the beam; and the beam's MOE. The higher the MOE is, the less a beam will deflect under a given load.

The MOE of wood materials is often determined with a *static bending* test. Figure 9.5 shows a wood pole during a static test. For testing, the butt of the pole is fixed, the load is applied toward the top, and the pole is tested as a cantilever beam. The figure shows the dramatic deflection that a pole can develop (on the order of several feet) prior to breaking. For most smaller and rectangular beams, such as those from dimension lumber, the test is that a beam is center-point loaded while the load and deflection are measured. From this data, the MOE is calculated using its relationship with the beam size, span,

FIGURE 9.5. Class 4, 12.2 m (40 ft) long, southern pine wood utility pole being tested in static bending. In this case, the butt is fixed in slings while a perpendicular load is applied near the tip. Load and deflection are measured throughout the duration of the test.

load, and deflection. This is often a simpler test to conduct and more closely related to most use situations than the MOE (Young's Modulus) as determined from a tensile or compressive test.

For a test specimen loaded in bending by a concentrated load at the center of its span and supported at its ends, the MOE can be calculated from the following formula:

$$\text{MOE} = \frac{PL^3}{48ID}$$

where

> P = the concentrated center load (newtons, below the proportional limit)
> D = the deflection at midspan (m) resulting from P
> L = the span (m)
> I = moment of inertia, a function of the beam's section – (width $\times$ depth³)/12 for beams with a rectangular cross-section; units are "n" to the fourth power, e.g. m⁴ or in.⁴

If the MOE of the beam is known, this same equation can be solved for D to predict the amount of deflection that would result from a concentrated load applied at midspan.

The MOE for wood ranges from about 3450 to 19 300 MPa (0.5–2.8 $\times$ 10⁶ psi). Engineers and structural designers use the MOE to determine required beam sizes. To do so, they must also consider the acceptable deflection, the span of the beam, and the load that will be carried. Common examples of insufficient design stiffness are floors that squeak or sag and where dishes rattle in cabinets as people walk across the floor. Several examples may help to illustrate how the MOE can be calculated and how knowledge of the MOE can be used by structural engineers. For reference, the reader is directed to Table 8.4 for metric to Imperial conversions.

Example 9.1

A $50 \times 50 \times 762$ mm ($2 \times 2 \times 30$ in.) clear, dry specimen of red oak is supported near each end of its 711 mm (28 in.) long span and loaded at the center in a universal testing machine. A gradually increasing load is applied, and when the load reaches 6680 N (below the proportional limit), the deflection at midspan under the load measures 6.6 mm. The MOE of this specimen is determined as follows:

$$\text{MOE} = \frac{PL^3}{48ID} = \frac{\left[6680\,\text{N}\left(0.711\,\text{m}\right)^3\right]}{\left\{48\left[0.050\,\text{m}\left(0.050\,\text{m}\right)^3 / 12\right]0.0066\,\text{m}\right\}} = 14\,600\,\text{MPa}$$

or, using Imperial units: $[1500\,\text{lb}\,(28\,\text{in.})^3]/\{48\,[1.97\,\text{in}\,(1.97\,\text{in.})^3/12]\,0.260\,\text{in.}\} = 2.10 \times 10^6\,\text{psi}$

Example 9.2

A 0.075 m square piece of red oak of the same type and quality as in Example 9.1 is to be placed between two roof beams and a space heater weighing 1360 kg (13 350 N) is to be hung from the center of the span. The *span* (the distance between the roof beams) is 1.27 m. How much will the 0.075×0.075 m specimen deflect when the load is applied?

$$D = \frac{PL^3}{48(\text{MOE})}I = \frac{\left[13350\,\text{N}\left(1.27\,\text{m}\right)^3\right]}{\left\{48\left(14.6 \times 10^9\,\text{Pa}\right)\left[0.075\,\text{m}\left(0.075\,\text{m}\right)^3 / 12\right]\right\}} = 0.015\,\text{m}$$

Example 9.3

A strip of particleboard 12.5 mm thick and 76 mm wide is loaded at midspan between supports that are 305 mm apart. When the load reaches 89 N, the deflection is measured as 0.0015 m. The MOE of the particleboard is therefore:

$$\text{MOE} = \frac{PL^3}{48ID} = \frac{\left[89\,\text{N}(0.305\,\text{m})^3\right]}{\left\{48\left[0.076\,\text{m}(0.0125\,\text{m})^3 / 12\right]0.0015\right\}} = 2840\,\text{MPa}$$

The maximum bending strength of solid wood and wood-based products is usually expressed in terms of the *modulus of rupture* (MOR). The MOR is calculated from the maximum load (load at failure) in a bending test, using the same static bending testing procedure for determining the MOE. The MOR is calculated from the classic *flexure formula*: bending strength = Mc/I where M is the maximum moment, c is the distance from the neutral axis to the extreme fiber (usually half of the depth), and I is the moment of inertia. When using a center-point-loaded test specimen with a rectangular cross-section, the flexure formula reduces to the following equation:

$$\text{MOR} = \frac{1.5PL}{bh^2}$$

where

> P = the breaking (maximum) load (N)
> L = the distance between supports (span) (m)
> b = the width of the beam (m)
> h = the depth of the beam (m)

This derived equation is valid for rectangular section beams, freely supported at both ends, and loaded at the center of the span.

The flexure formula ($\sigma = Mc/I$) can also be used to estimate the stress (σ), below the proportional limit, that develops on the top and bottom surfaces of a beam as a result of any given load. For instance, if P is a load below the proportional limit, the solution of the equation will give the compression and tension stress at the top and bottom surface of the beam. Note: units of pounds and inches can be substituted for newtons and meters to solve for stress in units of psi. Consider the following examples illustrating the use of the flexural stress equation.

Example 9.4

The same sample of red oak from Example 9.1 is loaded flexurally to failure in a testing machine. The breaking load is found to be 9400 N. The MOR can then be calculated as follows:

$$\text{MOR} = \frac{1.5PL}{bh^2} = \frac{(1.5 \times 9400\,\text{N} \times 0.711\,\text{m})}{\left[0.050\,\text{m}(0.050\,\text{m})^2\right]} = 80.2\,\text{MPa}.$$

Example 9.5

Suppose it is of interest to determine the load at which the 75×75 mm sample in Example 9.2 would probably fail (break). This could be estimated as follows:

$$P = \frac{(MOR)bb^2}{1.5L} = \frac{\left[(80.2 \times 10^6\,Pa)0.075\,m(0.075\,m)^2\right]}{(1.5 \times 1.27\,m)} = 17760\,N\,or\,(1810\,kg)$$

Example 9.6

If this same 75 by 75 mm sample is loaded to only $13350\,N$ ($1360\,kg$), as described in Example 9.2, what would be the stress at the center of the span on the top and bottom surfaces of the beam? This can be determined using essentially the same equation as for the MOR, i.e.

$$\text{Flexural stress} = \frac{1.5PL}{bb^2} = \frac{(1.5 \times 13350\,N \times 1.27\,m)}{\left[0.075\,m(0.075\,m)^2\right]} = 60.3\,MPa$$

These examples are intended to provide some insight into the application of mechanical properties. An array of equations and design considerations are used by wood scientists and engineers to cover the varied applications of structural members. To become proficient in wood structural design requires a knowledge of (i) the mechanical and moisture-related behavior of wood, (ii) sound engineering principles, (iii) appropriate safety and service adjustment factors, and (iv) building codes and standards.

The most common procedures for wood and wood-based materials are put forth by the American Society for Testing and Materials (ASTM). Figure 9.6 shows a laminated hardwood beam being tested, as per ASTM 5456. This flexural test is used to develop design values for structural composite lumber products. A vast array of standardized tests are compiled in the annual ASTM publications: section four, V. 4.10 – Wood.

Shear Stress and Strain

Shear stress is that which makes one part of a material slip past the material adjacent to it. Figure 9.7 shows a method of determining the *shear strength* of wood parallel to the grain. Wood is relatively low in shear strength parallel to the grain but extremely high in shear strength across the grain. In fact, if an attempt is made to shear wood across the grain using the device illustrated in Figure 9.7, the wood tends to crush and "broom" over rather than shear.

Horizontal or parallel-to-grain shear stresses develop in beams and thus shear strength is important in design. When a beam is loaded, the wood cells attempt to slip horizontally past each other as the beam bends. This can be visualized by considering a beam made of six 25 mm-thick boards or strips that are nailed only at the center of the span, as shown in Figure 9.8. As this composite beam bends, the ends of the boards slip with respect to each other. In essence, this beam behaves like six small independent beams. If these six boards are glued together to form a laminated beam, the slipping will

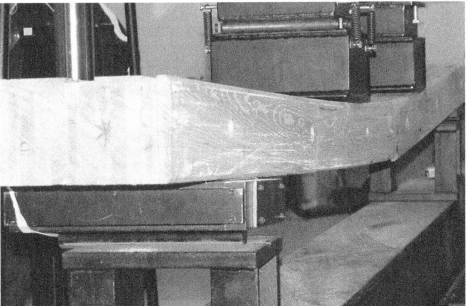

FIGURE 9.6.   Testing the bending strength of 3.7 m (12 ft) long laminated beams from low-grade oak lumber. The top photo illustrates the initial setup; the bottom photo illustrates the beam at failure.

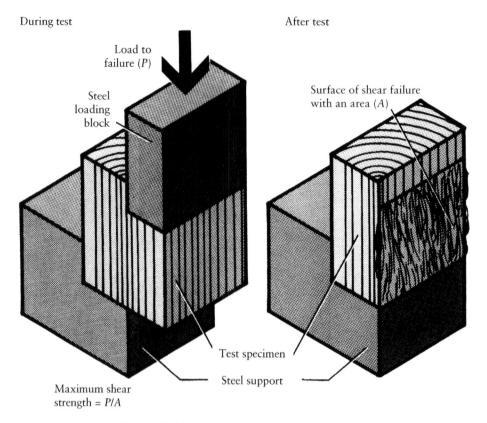

During test

Load to
failure (P)

Steel
loading
block

After test

Surface of shear failure
with an area (A)

Maximum shear
strength = P/A

Test specimen

Steel support

FIGURE 9.7.   Standard method of determining the horizontal shear strength in wood. Horizontal shear failure can also be induced by testing beams that are relatively close depth to span ratios, often on the order of 1:5 through 1:8.

be resisted and the beam will be much stiffer. In that case, greater horizontal shear stresses will develop in the beam. Such shear stresses are usually not high enough to be a limiting factor in the engineering of light-frame wood beams such as rafters and floor joists. In large solid timber beams and laminated beams, however, the shear strength may be the limiting factor regarding safe design load. In contrast to bending compression and tension stresses, which are greatest at the top and bottom surfaces of a beam, shear stresses are greatest in the central horizontal plane of the beam, that is, at the neutral axis.

Another type of shear stress, sometimes termed *rolling shear*, is important in plywood and OSB components where veneers and strands of wood are glued with the grain directions in adjacent pieces perpendicular to one another. Rolling shear is a stress that acts in the plane of the panel. It is important in the engineering design of panelized-wood components such as stress-skin panels (also called structural insulated panels, or SIPs) and box beams. In such products, rolling shear develops at the plane where the solid wood member is glued to the plywood or OSB, or where the face panels are glued to the core material. In solid wood products, shear strength parallel-to-grain is also important when designing the metal connections between structural elements and in relatively short span, deep beams.

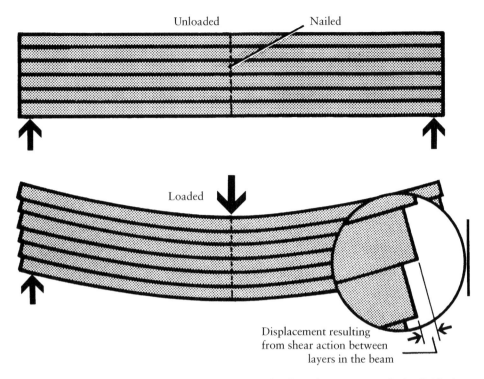

FIGURE 9.8. Illustration of horizontal shear as visualized in a beam composed of individual boards or strips. This shear behavior can readily be demonstrated with a stack of 4 or 5 m (or yard) sticks.

Figure 9.9 illustrates the action of rolling shear in a plywood floor or subfloor panel. The wood cells at the glueline tend to roll (or so it can be visualized) when the stress is applied. This stress must be considered when engineering the components, because the rolling shear strength of wood is less than shear strength parallel-to-grain.

Anisotropic Nature of Wood

Materials that have the same mechanical properties in each direction, such as many metals, plastics, and cement, are termed *isotropic*. In contrast, wood has drastically different properties parallel to the grain versus when it is perpendicular to the grain, and thus it is termed *anisotropic* (meaning not isotropic). More specifically, wood can be considered as an *orthotropic* material, i.e. one that exhibits different properties in the three mutually perpendicular directions or axes which correspond to the three primary axes of a tree. The strength and elastic properties of wood are different in the longitudinal, tangential, and radial directions. The properties in the radial and tangential directions, however, usually do not differ greatly. Because it is not always possible to predict the radial–tangential grain orientation of lumber (i.e. whether flat-sawn or edge-grain lumber will be used), a common strength value for the tangential and radial direction is standard for engineering purposes. Commonly for each species, the strength properties along (parallel to) the grain

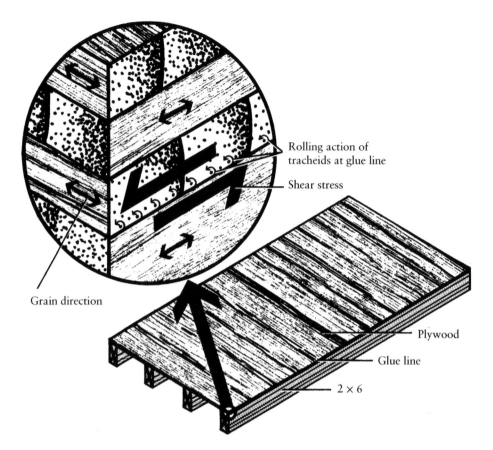

FIGURE 9.9. Development of rolling shear as it can occur in a plywood floor panel.

are 10–20 times higher than the same properties as measured across (perpendicular to) the grain.

Figure 9.10 illustrates the maximum load a Douglas fir 235×235 mm (10×10 in.) post could safely support if designed and loaded in the normal way (loaded *parallel* to the grain) and the amount it would be compressed under this load. This is compared with a post produced by stacking pieces sideways (loaded *perpendicular* to the grain). The MOE of Douglas fir parallel to the grain is 20 times as large as the MOE perpendicular to the grain. This behavior results in a greater change in length of the unusual post despite the fact that it can be loaded to only one-eighth the capacity of a normal post.

Products that are manufactured from cross-laminated veneers, particles, or fibers tend to be more isotropic than solid wood because the reconstitution process randomizes the effect of grain direction. Panel products such as plywood, OSB, and cross-laminated timber have an advantage over solid wood panels because they possess more or less uniform strength in both directions of the plane of the panel. However, the trade off to this uniformity is that it tends to reduce the maximum or design strength and stiffness as compared to the parallel-to-grain direction for solid wood. Because of the orthotropic nature of wood, it is generally not possible to subdivide wood into small, randomly oriented elements and reconstitute it to a product as strong in one direction as the original solid wood at the same density. It is difficult to beat the elegance and efficiency of Mother

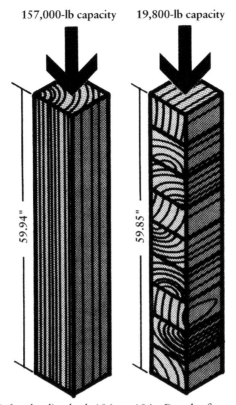

157,000-lb capacity   19,800-lb capacity

59.94"   59.85"

(Before loading both 10 in. × 10 in. Douglas fir posts
exactly 60 in. long)

FIGURE 9.10.   Comparison of strength and stiffness in the parallel versus perpendicular to grain directions.

nature. When wood is taken apart and reconstituted as some type of composite, however its uniformity is often improved which can, in turn, improve its design properties.

Products laminated parallel-to-grain such as parallel-laminated veneer lumber (LVL) and glue-laminated beams (glulam) are, on average, typically as strong as the highest grades of lumber. Their design values however are often even higher than those of comparably sized solid lumber. This difference occurs largely because in the laminated product the knots and other defects are randomized and do not occur together, while in the single solid member the defects may proceed through the entire piece. As such the strength variation in laminated lumber is less than that for solid; this is one of the reasons that the design values can be higher. Also, the presence of distinct layers discourages crack or fracture initiation and propagation.

Relationship of Strength to Specific Gravity

As discussed in Chapter 8, the strength of wood is closely correlated and directly with specific gravity. It is possible to make reasonably good estimates of strength based only upon specific gravity without knowing the species. In some countries, where grading

TABLE 9.2.   Relationship between mechanical properties and specific gravity of clear, straight-grained wood for mixed softwoods.

Property	Green wood	Wood at 12% MC
Static bending		
MOR (kPa)	$109\,600\,G^{1.01}$	$170\,700\,G^{1.01}$
MOE (MPa)	$16\,100\,G^{0.76}$	$20\,500\,G^{0.84}$
WML (kJ m^{-3})	$147\,G^{1.21}$	$179\,G^{1.54}$
Impact bending (N)	$353\,G^{1.35}$	$346\,G^{1.39}$
Compression parallel (kPa)	$49\,700\,G^{0.94}$	$93\,700\,G^{0.97}$
Compression perpendicular (kPa)	$8800\,G^{1.53}$	$16\,500\,G^{1.57}$
Shear parallel (kPa)	$11\,000\,G^{0.73}$	$16\,600\,G^{0.85}$
Tension perpendicular (kPa)	$3800\,G^{0.78}$	$6000\,G^{1.11}$
Surface hardness (N)	$6230\,G^{1.41}$	$8590\,G^{1.49}$

G represents specific gravity of oven dry wood, based on the volume at the moisture condition indicated. Compression parallel to grain is maximum crushing strength; compression perpendicular to grain is fiber stress at proportional limit. MOR is modulus of rupture; MOE is modulus of elasticity; and WML is work to maximum load. For green wood, use specific gravity based on ovendry weight and green volume; for dry wood, use specific gravity based on ovendry weight and volume at 12% moisture content.
Source: US Forest Products Laboratory (USFPL) (2010).

rules for lumber are not highly developed or where many species are sawn and used interchangeably, the grading of structural lumber is based largely upon specific gravity.

Mechanical properties are not all affected to the same degree by changes in specific gravity. Table 9.2 shows some strength-specific gravity relationships established at the US Forest Products Laboratory. Note that properties such as MOR and MOE in bending and maximum crushing strength parallel to grain increase nearly linearly with specific gravity. The relationships for many other properties are exponential functions. Thus some properties increase with specific gravity much more rapidly than others, such as surface hardness as compared with MOE.

It is sometimes desirable or necessary to compare test results from a single specimen with the published average strength values for that species. These comparisons can be done even if the specific gravity of the specimen varies from the published average. Thus, it is possible to estimate the strength of a specimen at one specific gravity if the strength property at another specific gravity is known.

Comparative Strengths of Important Species

Although it is possible to estimate the strength of wood knowing only the specific gravity, more precise information can be obtained by referring to data collected for that particular species. One of the most widely used sources of such information is the *Wood Handbook* (USFPL 2010). Mechanical property information from that text is reproduced in Online Appendix Tables A.5, A.6, and A.7. Another source of information on properties of United States and Canadian species is ASTM D2555. In addition to the mean values, this ASTM standard also indicates the standard deviation (variability) of each property mean.

The data presented in Online Appendix Tables A.5, A.6, and A.7 show strength values for small, clear, straight-grained, defect-free wood at both the green and 12% MC conditions. These tables can be very helpful in making comparisons among species. Note that they contain information on the MOE and MOR in bending, shear strength parallel to grain, and maximum crushing strength parallel to grain. Other strength properties included in Online Appendix Table A.5 are *surface hardness,* which is the force required to half-way embed a 11.3 mm (0.444 in.) diameter ball into the side of a piece of wood, *impact bending,* which is the height of drop at which a 22.7 kg (50 lb) hammer will break a 50.8×50.8 mm (2×2 in.) beam supported on a 711 mm (28 in.) span, and work to maximum load, which is also a measure of the maximum energy developed when a beam is broken in bending. In this latter case, the load is applied slowly in contrast to the impact load used in the 22.7 kg hammer drop test. One should become familiar with the type of information available in these tables, as this will provide the ability to answer many questions about the comparative properties of alternative species that may be used in various situations. For example, impact bending can be useful in detecting early decay while surface hardness is helpful in evaluating wood flooring. Keep in mind that these values are for small, clear, straight-grained wood and that larger sizes containing defects such as knots and slope of grain have higher variability, greater probability of containing strength-reducing characteristics, and lesser design strength properties.

Following are some examples that illustrate the use of the clear wood strength values in Online Appendix Tables A.5–A.7.

Example 9.7

A manufacturing plant has been producing interior (hidden) furniture parts from American sycamore but is considering changing to sweetgum. The most important property for these parts is bending and screw-holding strength. Will sweetgum have adequate strength? For wood at 12% MC, the information needed from Online Appendix Table A.5 is:

Species	Specific gravity	MOR (kPa)	MOE (psi)
Sweetgum	0.52	86 000	12 500
Sycamore	0.49	69 000	10 000

Sweetgum should be satisfactory in bending strength because it is 25% stronger than sycamore, but remember that this comparison is valid only for clear wood. Grade differences between the species could affect performance. Screw-holding strength is directly related to specific gravity, so sweetgum should be about 6% stronger than sycamore in this regard. Therefore, it appears that sweetgum would perform acceptably with respect to these two properties.

Example 9.8

Inexpensive skateboards are being successfully manufactured from hackberry, but the manufacturer is considering the use of yellow poplar instead. The important properties are impact bending and work to maximum load. Will yellow poplar be an adequate substitute for hackberry? The information needed from Online Appendix Table A.5 is:

Species	Specific gravity (green)	Impact bending mm (in)	Work to maximum load kJ m⁻³ (in lbf in⁻³)
Hackberry	0.53	1090 (43)	88 (12.8)
Yellow poplar	0.42	610 (24)	61 (8.8)

Because yellow poplar has significantly lower values in both properties, it appears that this substitution would not likely be satisfactory. If employed, more frequent breakage will almost certainly occur.

Example 9.9

Two imported species, jarrah from Australia and keruing from Southeast Asia, are being considered for use as industrial flooring. The most important property for this use is surface hardness. Which of these appears to be better suited to this application? Online Appendix A.7 indicates that:

	Specific gravity (green)	Surface hardness @ 12% MC N (lbf)
Jarrah	0.67	8500 (1910)
Keruing	0.69	5600 (1270)
Black Maple	0.52	5200 (1180)

Jarrah is clearly more resistant to denting than keruing, as indicated by the surface hardness value. However, keruing has about the same surface hardness as black maple, which is a common industrial flooring material. Therefore, either of the species should perform satisfactorily. When considering imported species, it is wise to obtain data on wood properties from the region where the material is produced because much variability exists over the range of many of these species.

These examples are much simpler than most situations faced by wood scientists, specifiers, and engineers. In actual cases, many other factors, such as dimensional stability, machinability, weight, availability, alternatives, and economics, must also be considered. Wood scientists must also apply their judgment regarding the expected variability in the material to be delivered.

Allowable Stresses

Allowable design stresses are the maximum stresses to which wood can be exposed with reasonable assurance that the wood will not fail. This concept contrasts greatly with average strength values because 50% of the pieces will fail at or below the average value. Knowledge of the average properties of clear, defect-free wood may be sufficient to answer many questions. Engineers, foresters, and wood scientists, however, are frequently involved in the use of structural lumber, laminated lumber products, and other structural wood-based composites. These forest products contain knots, slope of grain, and other characteristics that reduce strength. These strength-reducing characteristics are accounted for in the allowable design values of lumber and engineered composites that are used in buildings and other wood structures. In the case of lumber and plywood, each commercial and structural species group has a series of grades assigned to it. Each grade and species combination has allowable stresses assigned. These grades allow stronger species, and clearer pieces with fewer or smaller strength-reducing characteristics to be used in higher-stress situations. In addition, the strength "bonus" of short-term loading, the detrimental effect of continuous loading (long-term loading), natural variability within the species, effect of temporary overloading, and other uncertainties involved in manufacture and application must also be included for assurance of safety in the structural design.

Procedures used in the United States for deriving allowable stresses for lumber are detailed in ASTM Standards D245, D2555, and D5456. These procedures are complex when a group of species is sold under one trade name, e.g. southern pine. In general, the clear strength of wood is calculated and then reduced for statistical variation, moisture content, safety, size, and other factors. Alternatively, in-grade testing values based on full-size tests of lumber of various grades are used. The following equation explains the parametric derivation of the basic *allowable bending stress* (F_b). Often, however, nonparametric derivation is used because the nature of the statistical distribution of MOR may not be known.

$$F_b = \frac{(\text{average MOR} - 1.645s)}{\left[2.1(\text{combined load duration and safety factors})\right]}$$

where
   s = the standard deviation of the MOR (ASTM D2555) (average MOR − 1.645s) determines the lower 95% exclusion limit for infinitely large populations. For smaller sample sizes, values higher than 1.645 are required and are based on the number of samples)

Once developed, this basic allowable stress can be adjusted for specific loading situation as shown below:

$$F_{b\,\text{adjusted}} = F_b \times C_{MC} \times C_{SR} \times C_S$$

where

   C_{MC} = moisture content correction factor (ASTM D2555 provides green-strength values)
   C_{SR} = strength ratio to account for strength-reducing defects, i.e. knots, slope-of-grain, etc.
   C_S = correction factor for beam depth

Alternatively, F_b can also be derived nonparametrically, which can be useful in cases where the nature of the population distribution is not known or not readily explained. Average strength values are given in Online Appendix Tables A.5–A.7. Allowable stress values for softwood lumber are published by the American Wood Council, by regional lumber associations, and by trade organizations. The American Wood Council also

TABLE 9.3.   Comparison of average clear wood at 12% MC (US Forest Products Laboratory (USFPL) 2010) strength properties with allowable stresses, KD 19% (American Wood Council 2014) for structural design.

Species	Property	Clear wood	No. 1 grade (38×235 mm)	No. 2 grade (38×235 mm)
Douglas fir, south	Strength, MOR	82 000	7070	6450
	kPa (psi)	(11 900)	(1025)	(935)
	Stiffness, MOE	10 300	8960	8270
	MPa (psi)	(1.5×10^6)	(1.3×10^6)	$1.2 \times 10^6)$
Eastern hemlock	Strength, MOR	61 000	5860	4340
	kPa (psi)	(8900)	(850)	(630)
	Stiffness, MOE	8300	7580	7580
	MPa (psi)	(1.2×10^6)	(1.1×10^6)	(1.1×10^6)
Southern pine	Strength, MOR	88 000[a]	7240	5520
	kPa (psi)	(12 800)	(1050)	(800)
	Stiffness, MOE	12 300[a]	11 000	9650
	MPa (psi)	(1.8×10^6)	(1.6×10^6)	(1.4×10^6)

[a] Loblolly pine.

publishes the National Design Specification, which details procedures for the design of wood structures based on allowable stress design.

In some cases the allowable stresses for lumber are only a fraction of clear wood strength. Table 9.3 compares clear wood strength with the allowable stresses for 38×235 mm (2×10 in.) dimension lumber for three important commercial species. Note that allowable MOE values are not reduced for variability and use conditions as are strength values because in general, no threat to safety would result from overestimating stiffness. The allowable design MOE values are higher than average green values primarily because of the effect of drying.

Over the past century, structural lumber has been visually graded by skilled and certified graders who make rapid judgments as to the strength-reducing characteristics present in each piece. Once a grade is assigned, each piece thus has a corresponding suite of allowable design properties. Historically, most structural lumber has been graded in this manner. Increasingly however, at larger mills lumber grades are determined by machine. Machines in general do one of two things. The first is optically scan and measure each piece, compile all of the digital characteristics, and assign a visual grade (typically select structural, or number 1, 2, 3 or 4). Alternatively machines may use nondestructive testing to evaluate each piece. These tests, in general, measure the stiffness of each piece and correlate the predicted strength properties to the measured MOE. Such lumber is termed MSR (machine stress-rated lumber) or MEL (machine evaluated lumber). MSR and MEL require expensive machinery and have thus been somewhat limited to mills with high-quality timber resources and target highly engineered or high-value applications such as roof and floor trusses, ladder stock, glulam, and wood I-joists. More recently, however, this type of technology has become increasingly adopted in softwood lumber production. It is important for timber managers to keep this fact in mind as MSR and MEL machines have the ability to differentiate mechanical value, and thus economic value, in lumber. Timber tracts that produce higher quality lumber are worth more and vice versa.

9.1 Variability of Clear Wood Strength

Strength varies widely within and among species. There is a wide overlapping range of mechanical properties between hardwoods and softwoods used in the United States. Strengths of selected woods are shown in Table 9.4 to illustrate the range of properties that exist in some commercially important species.

Within any species there is considerable variation in clear wood strength, which corresponds to the variation in density and to the density–strength relationship for that property. The coefficients of variation for selected strength properties are illustrated in Table 9.5. These coefficients were derived from tests conducted at the United States Forest Products Laboratory on green specimens of 50 species.

Factors Affecting the Strength of Clear Wood

Moisture Content

As wood dries below the fiber saturation point, most strength and elastic properties increase. The increase in strength begins as the moisture level approaches and drops below the fiber saturation point, usually around 28–30% moisture content. This relationship is important given the moisture content range to which wood equilibrates in service. The relationship between the moisture content and strength properties of white ash are shown in Figure 9.11.

An exponential relationship can be used to estimate the strength at any given moisture content by using the properties of green wood and wood at 12% MC given in Online Appendix Tables A.5–A.7. This relationship is:

$$P = P_{12} \times \left(\frac{P_{12}}{P_g} \right)^{\left[(12-M)/(M_p - 12) \right]}$$

where

P = the property at M percentage moisture content
P_{12} = the property at 12% MC (available from the Online Appendix)
P_g = the property green MC (available from the Online Appendix)
M_p is usually taken at 25%; the Wood Handbook (USFPL 2010) provides other values for a few selected species.

As an example, suppose the MOR of northern red oak at 6% MC is desired. Online Appendix Table A.5 gives the property at 12% MC and green as 99 000 and 57 000 kPa (14 300 and 8300 psi) respectively. Therefore, the MOR at 6% MC would be approximately:

$$P = 99\,000\,\text{kPa} \times \left(\frac{99\,000\,\text{kPa}}{57\,000\,\text{kPa}} \right)^{\left[(12-6)/(25-12) \right]} = 99\,000\,\text{kPa} \times 1.74^{0.462}$$
$$= 128\,000\,\text{kPa}\,(18\,600\,\text{psi})$$

TABLE 9.4. Average strength and stiffness values at 12% MC for defect-free straight-grained wood of selected species.

Species	Bending MOR		Bending MOE		Maximum crushing parallel to grain		Compression perpendicular to grain at proportional limit	
	Average		Average		Average		Average	
	kPa	Ratio[a]	MPa	Ratio	kPa	Ratio	kPa	Ratio[a]
Hardwoods								
Quaking aspen	58 000	1.00	8100	1.00	29 300	1.00	2600	1.00
Red alder	68 000	1.17	9500	1.17	40 100	1.37	3000	1.19
Yellow poplar	70 000	1.20	10 900	1.34	38 200	1.30	3400	1.35
Southern red oak	75 000	1.30	10 300	1.26	42 000	1.43	6000	2.35
White ash	103 000	1.83	12 000	1.47	51 100	1.74	8000	3.14
Sugar maple	109 000	1.88	12 600	1.55	54 000	1.84	10 100	3.67
Softwoods								
Eastern white pine	59 000	1.00	8500	1.00	33 100	1.07	3000	1.07
Engelmann spruce	64 000	1.08	8900	1.05	30 900	1.00	2800	1.00
Ponderosa pine	65 000	1.09	8900	1.04	36 700	1.19	4000	1.41
White fir	68 000	1.14	10 300	1.20	40 000	1.30	3700	1.29
Radiata pine	80 700	1.37	10 200	1.20	41 900	1.36	4200	1.50
Coastal Douglas fir	85 000	1.44	13 400	1.57	49 900	1.62	5500	1.95
Scots pine	89 000	1.51	10 000	1.18	47 400	1.53	4800	1.71
Longleaf pine	100 000	1.69	13 700	1.60	58 400	1.89	6600	2.34

[a] Ratio of strength to weakest species in the group.

TABLE 9.5. Average coefficient of variation for selected properties of 50 species of clear wood.

Property	Coefficient of variation (%)
Static bending	
MOR	16
MOE	22
Compression parallel to grain	18
Compression perpendicular to grain	28
Shear parallel to grain	14
Specific gravity	10

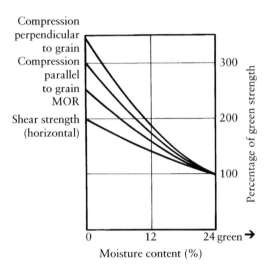

FIGURE 9.11. Relation of strength properties to moisture content for white ash. Source: US Forest Products Laboratory (USFPL) (2010) data.

The moisture–strength relationship varies considerably among species, so one should not rely heavily upon estimates of the type shown above or any others that generalize for all species. Also, moisture adjustment models are most accurate as relatively small changes in moisture content, on the order of ±5%. Beyond that that, their reliability becomes increasingly variable.

It should also be noted that some strength properties appear to peak between 4 and 7% moisture content. Thus, overdrying wood below these levels can cause detrimental effects on strength and make wood appear brittle. Low wood strength at very low moisture contents can lead to an increased incidence of torn grain during planing of dry lumber in the winter when, in a nonhumidified workshop, lumber can become extremely dry, commonly below 5%. The tension strength perpendicular to grain decreases at these moisture levels and one result may be chipped or torn grain.

Time

The strength of wood does not decrease greatly over time unless the product is subjected to continuous high levels of loading, the deleterious effects of microorganisms, high temperature, drastic moisture fluctuation, strong chemicals, or some other external detractor.

After centuries changes do occur, but these are usually the result of environmental factors and not aging per se.

Some strength loss occurs if aging is accompanied by continuous loading of the member. Most mechanical properties of wood are affected by the length of time a load is imposed. In short-duration or impact-loading situations, wood performs the best. The longer a load is supported, the lower the load magnitude is that can be safely carried. The relationship between stress level and time to failure for a variety of forest products is shown in Figure 9.12. Even if a load is small enough that there is no danger of ultimate failure, the member may continue to deflect or deform very gradually under constant stress. A common example is the gradual sagging of a shelf heavily loaded with books. This phenomenon is known as *creep*. For design with lumber, load-duration factors are applied to base design values (Figure 9.13). In general, wood exhibits higher maximum strength with loads of shorter duration.

It is believed that creep occurs at the molecular level from slippage in the relative position of the long-chain molecules in the cell wall. The bonding sites that hold water or are mutually occupied in dry wood may slowly shift or slide with respect to adjacent molecules when the whole matrix is placed under stress. The more water is present within the wall, the more easily this slippage can occur. Alternate addition and removal of water from the bonding sites greatly accelerates this creep deflection (Erickson 1999).

Creep most commonly shows itself as deflection in a wood beam after many years of loading. It can be particularly troublesome in long wood beams such as headers over large doors, long floor joists, or beams that are used in high-moisture applications. It has been shown that creep is accelerated under cyclic humidity conditions as compared with wood subjected to a constant environment.

Creep can occur in any grain direction but it occurs at much lower stress levels when the stress occurs in the transverse rather than the longitudinal direction. In particle and fiber products where the stresses act in the transverse grain direction within some of the small elements, creep is more evident than in solid wood materials. Figure 9.14 shows the results of an experiment to compare the rate of creep in particleboard and solid wood stressed parallel and perpendicular to the grain. The behavior of particleboard shows the effect of transverse stress that results from the random orientation of particles and glue bonds.

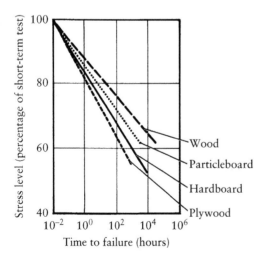

FIGURE 9.12. Relation between load level and time to failure for solid wood and composite forest products. Source: From Gerhards (1977).

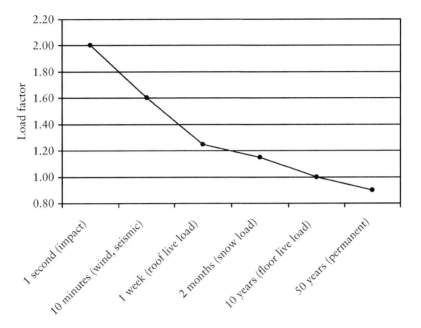

FIGURE 9.13. Load duration factors for solid lumber. Baseline for light-frame design is 10 years at full anticipated live load. For short-duration loads, wood's ultimate strength increases. During the development of design values, the tested strength values (acquired over ~5–10 min) are reduced, in part, for anticipated load duration.

FIGURE 9.14. Relative flexural creep for three materials at 80% relative humidity (relative creep is the ratio of creep deflection to instantaneous deflection). Source: From Gnanaharan and Haygreen (1979).

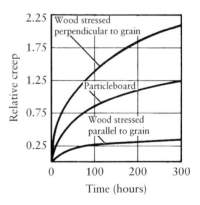

In the derivation of allowable stresses, the strength of wood determined by standard short-term test methods is reduced about 38% to adjust it to a 10-year loading period. Ten years is considered the cumulative time that wood in a building is normally subjected to maximum design loading. In situations where the load will be imposed permanently, the stress is reduced another 10% beyond the reduction for a 10-year load. When designing a structure for loads of shorter duration, such as snow or wind, engineers may increase the allowable design stresses (see Figure 9.13).

Temperature

Most mechanical properties decrease when wood is heated and increase when it is cooled. This temperature effect is shown in Figure 9.14 for wood at 12% MC and when ovendry. As long as temperatures do not exceed 100 °C, there is little permanent loss of strength in the wood. Exposure to high temperatures for long periods can cause permanent strength loss. This point should be considered when extremely high dryer or kiln temperatures are used to dry critical structural members. As long as the pieces of wood are in a heated dryer, the evaporative cooling associated with the drying process keeps the temperature of the wood cooler than that of the air. Overdrying, however, and its resultant overheating, should be avoided.

The effect that high temperature has on wood tends to be cumulative, i.e. the sum of short exposure times at a high temperature can often be as great as a single exposure of equal duration. There is one factor, however, that may result in short exposures being less severe, namely, the interior of the wood may not have adequate time to heat to the equilibrium temperature. In that case, strength loss only occurs on the surface. Nevertheless, any time wood is to be used where it will continually be subjected to temperatures in excess of 100 °C, consideration should be given to the loss of strength that may result.

Fatigue

The *fatigue* strength of a material is its ability to retain its strength when subjected to cyclic loading. According to the *Wood Handbook* (US Forest Products Laboratory (USFPL) 2010), clear, straight-grained wood subjected to 60% of its predicted strength will sustain approximately two million cycles of bending (note that this statement is based on tests in which loading was applied and removed at 30 Hz). As compared with other materials such as steel and aluminum, wood exhibits excellent fatigue performance.

Cyclic stress may have a more severe effect when defects such as knots are present or when the slope of grain is involved. Most types of structural materials are subject to fatigue if repetitively stressed at a high level of loading. Only rarely, however, is the fatigue characteristic of wood an important factor in structural design because most structural wood members are not frequently stressed above the safe levels assumed in the design of a structure. The beams in railroad bridges are an example of an application in which fatigue strength is important. There, beams are cyclically loaded each time the wheels of a rail car pass. This can occur millions of times in the life of a bridge. Also boat hulls and masts, performance composite automobile floor boards, marine piling, and wood sign posts are examples wherein long-term stress cycling can cause fatigue failure.

Reaction Wood

The erratic effects of reaction wood in structural lumber (primarily compression wood) on mechanical properties significantly limits its utility value. While the density of this aberrant wood is often higher than normal wood, its tension strength and bending strength are often unpredictably low. Brash failure is common among compression wood specimens primarily owing to its relatively high lignin and low cellulose content. Grading rules often reduce grade and thus deduct strength in compression wood pieces and it is prudent to cull compression wood pieces from high-liability, non-redundant-member products such as ladders and scaffold planking.

Exposure to Chemicals

The strength of wood may be reduced by exposure to severe acidic or alkaline environments; yet wood is more resistant than steel to acidic conditions. Chemical fertilizer and highway-salt storage buildings are often built from wood because of its ability to resist corrosion and deterioration when in contact with these chemicals.

Chemical deterioration of the cell wall can result in loss of strength from hydrolysis of cellulose, oxidation, or delignification by alkalis. Long-term exposure can ultimately reduce wood to pulp. Softwoods tend to be more resistant to chemical deterioration than hardwoods. Woods that are less permeable to moisture movement tend to be more resistant because the chemicals penetrate more slowly. For this reason, under conditions where chemical deterioration might occur, white oak is preferable to red oak, Douglas fir heartwood is preferable to sapwood, and old-growth cypress is preferable to second-growth. Cypress, Douglas fir, southern pine, and redwood are often used where contact with acidic materials or conditions is expected.

A type of chemical deterioration can also occur around iron or steel fasteners such as nails or bolts. When moisture is present in acidic woods or those that contain high extractive levels, particularly tannins, a chemical reaction occurs that produces iron salts. With high moisture, these salts can act as galvanic cells and promote localized weakening of the wood around the fastener, reducing the natural resistance of the wood to decay. This action may continue until the nail or other connector is corroded and becomes loose in the wood. This effect is sometimes seen in old unpainted buildings where it is easy to pull nails from the lumber. It is also a nontrivial issue with some types of metallic salt-type preservatives. When using nails in an exterior application with redwood, cedar, oak, or cypress, or when fabricating structures made from lumber or plywood treated with

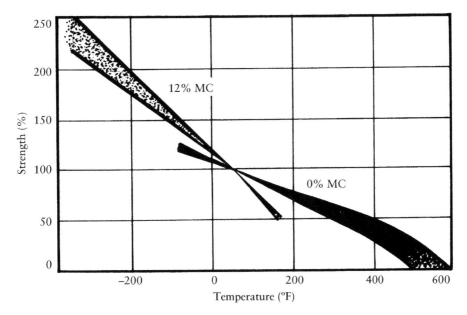

FIGURE 9.15.   The immediate effect of temperature on strength properties (expressed as a percentage of the value at 20 °C). Source: US Forest Products Laboratory (USFPL) (2010).

waterborne salt-based treatments (such as ACQ), it is always advisable to use specially coated or stainless steel fasteners and hardware to avoid such problems.

Factors Affecting the Strength of Lumber Products

The strength of wood varies with moisture content changes below about 30%. It would be very difficult, however, for structural engineers to consider the variable moisture content and induced variation in the strength of wood products in the design of a wood structure. To avoid this complication, the allowable stresses are established at a fixed moisture content. The allowable stresses of lumber, glue-laminated beams, and plywood are given fixed MC values: 19% for softwood lumber and 16% for plywood and glue-laminated beams. If these moisture content limits are to be exceeded when the product is in use, the allowable stresses must be reduced with a moisture correction factor. Reductions range from 3 to 40%. In addition to the factors that affect the strength of the clear wood, tree growth characteristics are also important to the strength of lumber. The effect of these characteristics is considered when deriving allowable stresses.

Knots

Knots are the most common wood characteristic that reduces the strength of lumber. The effect of a knot is often considered equivalent to that of a hole. Thus, tree pruning schemes generally produce logs that yield stronger lumber. In some cases, the knot may have a greater effect than a drilled hole because of the distortion in the grain that accompanies it. The extent to which a knot reduces strength depends not only upon the size of the knot but also upon its location in the piece. A knot on the top or bottom edge of a beam, at its extreme face, is much more critical than the same knot located near the centerline or neutral axis. Recall that the maximum bending stresses occur on the top and bottom edges of a beam. Knots on the bottom edge of a floor joist or rafter are more serious than on the top edge because they have a more deleterious effect in tension than compression. Skilled carpenters often inspect and install floor joists such that the largest knots are on the top edge, not the bottom.

The amount of strength loss from knots of various sizes is outlined in ASTM Standard D245. Table 9.6 shows the strength loss that results from knots in the center and on the edge of beams of several widths. Note the strength-loss difference between boards with knots located in the center as compared with the edge. A 50 mm (2 in.) diameter knot in a 38×185 (2×8 in. nominal) beam will reduce the strength by 24% if it is in the center and by 43% if it is on one edge.

Decay

Decay is generally prohibited in grades of lumber used for structural purposes (some localized types such as "white specks" are sometimes permitted), because it is often impossible to estimate by visual inspection the extent to which the decay has weakened the piece. By the time decay is visually apparent, the loss in strength may be severe. Impact strength declines as a result of decay much faster than static strength. Figure 9.16 shows the zone of fracture from two toughness (small impact-bending) test specimens of

Table 9.6.   Strength reduction in lumber resulting from knots and slope of grain as shown in ASTM Standard D245.

Strength reduction (%) from knots in the center of the wide face of a beam

Knot diameter: mm (in.)	Beam depth: mm (in.)			
	90 (3.5)	140 (5.5)	185 (7.25)	235 (9.25)
25 (1)	25	16	12	10
51 (2)	51	33	24	20
76 (3)	n.a.	50	37	30

Strength reduction (percent) from knots on one edge of the wide face of a beam

Knot diameter: mm (in.)	Beam depth: mm (in.)			
	90 (3.5)	140 (5.5)	185 (7.25)	235 (9.25)
25 (1)	43	30	23	18
51 (2)	81	55	43	35
76 (3)	n.a.	79	63	50

Strength reduction (percent) in bending and compression parallel to the grain resulting from slope of grain

Slope of grain ratio	Bending	Compression
1 in 6	60	44
1 in 8	47	34
1 in 10	39	26
1 in 15	24	0
1 in 20	0	0

n.a. = Not applicable.

FIGURE 9.16.   Normal (splintering tension) and brash failure in Douglas fir toughness specimens. The characteristic brash failure resulted from decay.

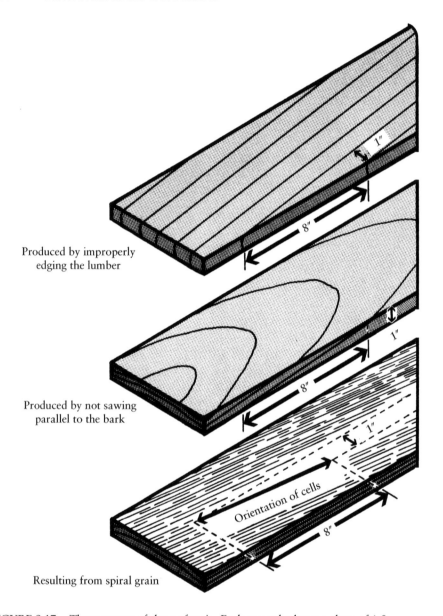

Produced by improperly
edging the lumber

Produced by not sawing
parallel to the bark

Resulting from spiral grain

FIGURE 9.17.   Three sources of slope of grain. Each example shows a slope of 1:8.

Douglas fir. The sapwood specimen on the left is normal, and the heartwood specimen on the right has a small amount of decay. The splinter-free fracture on the right (*brash failure*) is typical of decayed or partially decayed wood. In this case the scaffold plank from which the specimen was cut had its toughness reduced by 85% by decay but appeared normal to the user until failure occurred. Mills must be alert for decay when producing lumber from large, overly mature logs, where decay from the standing tree may be encountered.

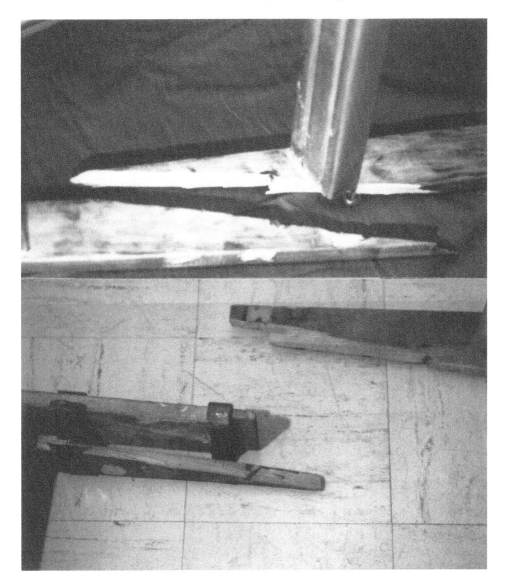

FIGURE 9.18.   Slope of grain contributed to the failure in the side rails of these two wood ladders.

Some grades of dimension lumber permit blue stain because these fungi do not cause a weakening of the wood. This is because they live upon food materials in the cell lumina and not upon the cell wall substance.

In high-moisture situations, stain and decay can occur in buildings within wood that was previously kiln dried. Water leakage from roofs, walls, and pipes, condensation, poor ventilation, and foundation seepage are common sources of moisture in wood structures. Proper design, engineering, construction, and maintenance with regard to moisture are paramount for the life and health of a building and for its occupants. If undetected and or unremediated, the results range from wood stain and incubation of minor forms of mold allergens to complete structural strength loss owing to decay.

Slope of Grain

Slope of grain in lumber is expressed as the length through which a deviation in the grain occurs. Figure 9.16 shows three examples of how a slope of 1-in-8 could develop in a board. The first two examples show how slope of grain can result from the way the lumber is cut from the log. Slope of grain of this type is fairly easy to detect in a species that has distinct growth rings. Slope can also result from trees that naturally contain spiral grain. In this case, even though the growth rings appear parallel to the edges of the piece, the slope of grain may be quite significant. The best way to visually detect this type of grain deviation is to look at resin streaks, surface checks, mineral stains, or other minor defects that tend to be oriented with the cells. New automated technologies can use electronic permittivity, or the nature of wood's ability to store and discharge energy, to detect grain deviation. These technologies are working their way into automated lumber grading technologies.

The strength of wood is affected whenever there is slope of grain greater than about 1-in-20. The percentage of strength loss in bending and compression resulting from grain deviation is shown in Table 9.6. All grades of structural lumber carry limitations as to the slope of grain. These are especially strict for critical nonredundant structural members such as ladder rails. Figure 9.18 shows broken ladder rails that failed as a result of excessive slope of grain.

References and Supplemental Reading

American Society for Testing and Materials (ASTM). 2011. Standard methods for establishing structural grades and related allowable properties for visually graded lumber. ASTM D245-06.

American Society for Testing and Materials (ASTM). 2015. Standard methods of testing small clear specimens of timber. ASTM D143-14.

American Society for Testing and Materials (ASTM) (2016a). *Annual Book of ASTM Standards, Section 4, V. 4.10, Wood.* West Conshohocken, PA: ASTM.

American Society for Testing and Materials (ASTM) (2016b). *Annual Book of ASTM Standards, Section 15, V. 15.06, Adhesives.* West Conshohocken, PA: ASTM.

American Society for Testing and Materials (ASTM). 2016c. Standard methods for evaluating allowable properties for grades of structural lumber. ASTM D2915-10.

American Society for Testing and Materials (ASTM). 2016d. Evaluating the properties of wood-base fiber and particle panel materials. ASTM D1037-12.

American Society for Testing and Materials (ASTM). 2016e. Standard method for establishing clear wood strength values. ASTM D2555-15.

American Wood Council. 2014. National Design Specification (NDS) Supplement for Wood Construction 2015 Edition. Leesburg, VA.

Bodig, J. and Jayne, B.A. (1982). *Mechanics of Wood and Wood Composites.* New York: Van Nostrand Reinhold.

Breyer, D.E., Fridley, K.J., and Cobeen, K.E. (1999). *Design of Wood Structures,* 4e. New York: McGraw-Hill, Inc.

Burgert, I., Keckes, J., and Fratzl, P. (2006). Mechanics of the wood cell wall. In: *Characterization of the Cellulosic Cell Wall* (ed. D. Stokke and L. Groom), 30–37. Ames, IA: Blackwell Publishing.

Erickson, R.W. 1999. Summary of mechano-sorptive research at the University of Minnesota, USA. IUFRO Wood Drying Conference. Capetown, South Africa.

Faherty, K.F. and Williamson, T.G. (1995). *Wood Engineering and Construction Handbook*, 2e. New York: McGraw-Hill.

Gerhards, C.C. 1977. Effect of duration and rate of loading on strength. USDA For. Ser. Res. Pap. FPL-283.

Gerhards, C.C. (1982). Effect of moisture content and temperature on the mechanical properties of wood. *Wood Fiber Sci.* 14 (1): 9–36.

Gerhards, C.C. (2000). Bending creep and load duration of Douglas-fir 2 by 4s under constant load for up to 12-plus years. *Wood Fiber Sci.* 32 (4): 489–501.

Gnanaharan, R. and Haygreen, J.G. (1979). Comparison of creep behavior of waferboard and that of solid wood. *Wood Fiber Sci.* 11 (3): 155–170.

Green, D. and Evans, J. (2003). Effect of low relative humidity on properties of structural lumber products. *Wood Fiber Sci.* 35 (2): 247–265.

Gurfinkel, G. (1973). *Wood Engineering*. New Orleans: Southern Forest Products Association.

Hoyle, R.J. Jr. and Woeste, F.E. (1989). *Wood Technology in the Design of Structures*. Ames, IA: Iowa State University Press.

Karacabeyli, E. and Barrett, J.D. (1993). Rate of loading effects on the strength of lumber. *For. Prod. J.* 43 (5): 28–36.

Koch, P. (1985). *Utilization of Hardwoods Growing on Southern Pine Sites*, USDA For. Serv. Handb, vol. 605, 506–514.

Kollman, F.P. and Côté, W.A. Jr. (1968). *Principles of Wood Science and Technology*. New York: Springer.

Madsen, B. (1992). *Structural Behavior of Timber*. North Vancouver, BC: Timber Engineering Ltd.

US Forest Products Laboratory (USFPL). 1944. Design of wood aircraft. ANC Bull. 18.

US Forest Products Laboratory (1974). *Wood Handbook*, USDA For. Serv. Agric. Handb, vol. 72.

US Forest Products Laboratory (USFPL). 2010. Wood Handbook: Wood as an Engineering Material. USDA For. Serv. FPL-GTR-190.

10

Durability and Protection

By definition, sustainability involves making things last longer or extending their benefits, in perpetuity. One of the primary ways to enhance forest sustainability is to extend the service life of wood products. If a product or structure's service life can be doubled, then in general, only half of the trees, or timberland area, are needed to provide that material. Wood protection and preservation enhances forest and environmental sustainability in this manner. In many cases, service lives are extended 5- to 10-fold. As such, the existing timberland base can provide a continually increasing amount of goods and services for a growing population.

Key to the satisfactory employment of wood and wood-based products as building materials is an understanding of the agents and conditions that can lead to decay or other forms of deterioration. Wood structures, when properly designed and constructed, can serve satisfactorily for hundreds of years. Figures 10.1 and 10.2 show examples of buildings that are still in excellent condition after a century or more of service. As with many naturally formed organic materials, however, wood may be subject to decay, fungal stains, insect infestation, fire, and surface weathering, all of which can greatly reduce the useful life of buildings and products.

There is no reason for wood deterioration to occur within a building where exposure to water can be controlled. Generally, wood that is used out of doors, is subjected to rain, or has contact with the ground or to seawater will eventually decay or be attacked by insects or marine borers. The service life of wood products and structures is best enhanced when two factors are considered. First, building design and construction should be of a nature that minimizes direct wood and water contact and seeks to shed or drain water while avoiding water traps. Second, service life can be greatly extended by proper treatment or species selection. To avoid deterioration in buildings or to extend the life of wood materials used under severe exposure, users of wood products and structures must understand the conditions under which deterioration develops and take appropriate preventive measures.

Forest Products and Wood Science: An Introduction, Seventh Edition. Rubin Shmulsky and P. David Jones.
© 2019 John Wiley & Sons Ltd. Published 2019 by John Wiley & Sons Ltd.
Companion website: www.wiley.com/go/shmulsky

FIGURE 10.1.   Dean-Page Hall, built in about 1850 in Eufala, Alabama. An example of fine antebellum wood homes found throughout the Southern USA region.

Biological agents are the major causes of wood deterioration. Deterioration can result from a variety of organisms: fungi that cause staining, softening, and decay; marine borers, mainly small mollusks and crustaceans; insects, including termites, carpenter ants, carpenter bees, and a variety of wood-boring beetles; and bacteria that cause deterioration in water-stored logs and foundation piles. The greatest financial losses from biodeterioration result from decay fungi. These agents of biological deterioration are omnipresent. They are most active in tropical climates but also develop in temperate and colder regions.

Fire is the most catastrophic and economically important form of nonbiological deterioration. There are several less-important nonbiological agents of wood deterioration: ultraviolet light and mechanical abrasion, which contribute to surface weathering; chemical agents, such as strong acids or alkalies; and products of metallic corrosion that can cause wood to deteriorate around nails or other metal fasteners.

Following catastrophic storms that blow down timber, salvage logging operations must mobilize quickly to recover downed trees. Figure 10.3 illustrates broken mature yellow pine timber following a catastrophic hurricane along the United States gulf coast. While much of this timber will be of high utility, value recovery is costly, blue stain may be common, and lumber may contain incipient decay and ring shake. If timber is to be recovered following a fire or wind storm, recovery efforts should generally be accomplished as soon as possible, and ideally within three to six months. In dry climates, such as in arid regions of the western United States and Canada, it may be possible to successfully recover timber as much as several years following the death of trees.

FIGURE 10.2.   Swedish home. This structure, built about 360 years ago in Vimmerby, Sweden, was originally a log structure but was later covered with board-and-batten siding.

FIGURE 10.3.   Mature plantation pine sawtimber damage following Hurricane Katrina in 2005. Damaged timber that can be salvage logged within about six months can be sawn immediately or stored for future conversion. Timber may contain shake and may quickly develop bluestain, both of which reduce its economic value. Source: Photo courtesy Donald L. Buckner.

Hardwood trees that grow in soft soils will often uproot during severe storms. Because their root ball remains attached, the quality of wood remains higher for a longer period, often for a year or more following blowdown. If the market quantity of timber is sufficiently high, then logs may be kept under sprinkler storage to prolong their life. Under such storage, logs can be kept for six to nine months with little or no degrade. In some cases, water-stored logs are processed after two to three years of storage; however prolonged storage encourages bacteria, excessive moisture content, and other issues that can cause processing difficulties.

Fungi

Nature of Decay Fungus

A fungus grows in or on a suitable host by germination of a spore followed by the growth of segments of *hyphae*. Under a light microscope, the hyphae often appear as threadlike structures in the wood cells. Hyphae then go from cell to cell through pit pairs or through holes (bore holes) created in the cell wall. Many hyphae growing together are called *mycelium*. Figure 10.4 illustrates the spread of hyphae through the cells of wood. The mycelium grows on and within the wood, producing bore holes and progressively degrading cell walls. As fungal growth continues, wood material is consumed, the wood loses strength, and it eventually loses weight. The fungus digests the carbon-rich wood and carbon dioxide is emitted as the primary byproduct. There is no practical way of isolating

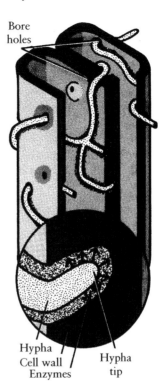

FIGURE 10.4.   Growth of fungal hypha through wood cells.

wood products from contact with spores. To prevent infection, it is necessary to prevent the conditions that allow the spores to germinate and grow. Like many plants, fungi need four factors to grow: food, water, a favorable temperature, and sufficient oxygen.

The food requirements of decay fungi are provided by the wood. To use cellulose, hemicellulose, or lignin as food, fungi break down these cell constituents into simple molecules that can be metabolized. These biochemical changes are accomplished by the catalytic action of metabolites produced by the hyphae. The metabolites are produced at the tips of the hyphae in the process of creating bore holes and also are produced along the sides of these vegetative elements.

For enzymes and other metabolites to diffuse into the cell walls and for the by-products to enter the hyphae, it is necessary for water to be present. Water is also needed for the breakdown process catalyzed by enzymes. Therefore, one of the best ways of preventing wood decay is to keep wood dry. Generally, there is very little danger of decay if wood is below the fiber saturation point. Below 20% moisture content, decay fungi are not able to advance. Some fungi are able to actively transport water from a moisture source to otherwise dry wood. These fungi grow or build a tubular structure called a *rhizomorph* that is able to actively deliver water, like a bucket brigade, to the wood, thereby enhancing decay. In most cases, buildings can be designed to avoid conditions where the wood moisture content will exceed these levels. This waterproofing is possible by proper application of roofing, roof overhangs, installation of gutters, ventilation, vapor barriers, and caulking to ensure that liquid water does not contact wood.

Besides moisture, there are other physiological requirements for the growth of fungi but there is often less chance to control them. Fungi grow most rapidly at temperatures in the range 20–32 °C (68–90 °F) and are generally inhibited at temperatures below 0 °C (32 °F) and above 38 °C (100 °F). Therefore, in tropical climates and within heated buildings in temperate climates, temperatures are typically ideal for the development of decay. High temperatures are sometimes used in drying and treating processes to sterilize wood and kill fungi that may be present. International *phytosanitary* shipping regulations for wood pallets require heat treatment or sterilization.

Fungi also require oxygen and there is no practical and effective way of sealing air out of a finished wood product in order to eliminate decay. Logs in storage may be sprinkled with water or ponded to maintain high moisture contents that retard fungal development. Pieces of wood completely buried in the ground or submerged in water do not have enough oxygen for decay to occur. Wood that is buried in the soil but protrudes above the water line, such as the exposed tops of piles that support buildings in Venice, Italy or many of the buildings on the shoreline of the Great Lakes, provides a favorable location for decay to flourish. Also, logs that have sunk in rivers, lakes, or ponds may be affected by anaerobic bacteria, which can destroy pit structures, greatly increase the permeability, and cause staining of the wood. This permeability can result in drying, treatment, and gluing problems during subsequent wood processing and also decrease aesthetic value. When buried for sufficient time, on the order of 60–80 years or more, wood such as that of foundation pilings may become weakened by anaerobic bacteria, even if the wood is treated with preservative. Bacteria, however, work much more slowly than fungi.

Fungi tend to prefer a somewhat acidic environment. The pH range of 4–6 is best for growth. Some fungi have the ability to change the acidity of the wood dramatically, and in their favor, as they develop. As with other physiological factors, species of fungi vary considerably in regard to their preferred pH.

Fungal Types

Fungi that cause deterioration of wood and other cellulosic materials are simple organisms that contain no chlorophyll. These organisms are classified in the biological Kingdom Mycetae. Unable to produce or *photosynthesize* their own food, fungi derive their energy from other organic materials. The carbohydrate and lignin components of wood provide food for a wide range of fungi. The fungi that attack living trees are *parasitic*, that is they flourish at the expense of their living host. The forest pathology discipline is concerned with such parasitic fungi. Fungi that attack milled wood products are *saprobic*. Saprophytes derive their energy from dead organic matter.

The fungi that degrade wood are classified as decay, soft-rot, stain, or mold fungi according to the type of degradation they cause. *Decay fungi* cause significant softening or weakening of wood, often to the point that its physical and mechanical characteristics are completely destroyed. Wood so affected is referred to as rotten or decayed. *Soft-rot fungi* also weaken wood and most often attack wood that is very wet and usually penetrate rather slowly. Soft rots gradually degrade wood from the surface inward. *Staining fungi* often create a bluish or blackish color when they inhabit wood and as such are detrimental to its appearance and value; yet they do not have a serious effect on the strength or the physical integrity of the wood. Molds and mildews occur only on exposed surfaces and may discolor products in use, such as house siding, but do not affect the strength. As early colonizers, molds are often an indicator that conditions are favorable to decay.

Most decay fungi belong to the biological class Basidiomycetes, named for the spore-bearing structure, the basidium. A few are Ascomycetes and Deuteromycetes (Fungi Imperfecti). Several hundred species of fungi decay wood products in North America. The most common genera include *Gloeophyllum*, *Polyporus*, *Lentinus*, *Trametes*, and *Coniophora*. The species in these genera vary widely as to the species of wood, moisture content, and temperature most conducive to their growth. The species of decay fungi found in living trees rarely infect wood products. Occasionally partially decayed trees and logs are converted into products, which can lead to misdiagnosis as active decay in a finished product or structure.

Decay fungi may be further classified as brown rots or white rots. The *brown rots* preferentially attack and consume the cellulose and hemicellulose. Brown rots are most commonly noted in softwoods. Wood seriously degraded by these fungi will have an abnormally brownish or reddish color owing to the high concentration of residual modified lignin. Brown-rotted wood develops checks perpendicular to the grain that when dried or handled break into cube-shaped pieces. Figure 10.5 shows a section from a large timber badly deteriorated by a brown-rot fungus. Brown rots attack all layers of the cell wall but the cellulose-rich S-2 layer is often the first to be degraded. Brown rots cause dramatic strength loss in early decay stages. Under a light microscope, early decay colonization appears as intermittent hyphal strands in and around the cell walls. Advanced decay often appears as missing and broken cell walls throughout. If the decay is well

FIGURE 10.5. Douglas fir timber destroyed by brown rot at the section where it was in contact with a concrete wall.

advanced, it may be difficult or impossible to slice a wood section for making a micro-scope slide.

White-rot fungi have the ability to degrade both the lignin and cellulosic compo-nents of the cell, although the lignin is usually utilized at a somewhat faster rate. *White rots* may change the color of wood only slightly but more often give it a bleached or whitish color inherent to de-lignified cellulose. These fungi are most commonly noted in hardwoods. They typically erode the cell from the lumen outward by decomposing successive layers of the cell wall. Thus the cell wall becomes progressively thinner but the wood does not tend to shrink, check, or collapse as is often the case with brown rots. White-rotted wood usually retains its shape but may eventually become a fibrous spongy mass.

Although most fungi attack only wood in which excess moisture is already present, a few species, known as *water-conducting* or *dry-rot fungi*, have the ability to transport water to the wood material affected. The two species *Meruliporia incrassata* and *Serpula lacrymans* are the best known of the water-conducting fungi. *S. lacrymans* is found in buildings in central and northern Europe and occasionally in the northeastern United States, while *M. incrassata* is found principally in the southeast and the west coast of the United States. Water-conducting fungi have the ability to transport water in mycelial tubes to otherwise dry wood, thus putting it into a condition where it is subject to bio-logical deterioration. In such a condition, wood may be mistakenly classified as "dry-rotted." Similarly, wood that has been previously deteriorated in a high-moisture environment, that has been redried, is also sometimes mistakenly classified as dry-rotted.

Stain and mold fungi do not usually deteriorate the cell walls of wood but grow either on the surface or within the cell lumina of ray parenchyma. The pigmented hyphae of *stain fungi* give the wood a bluish or blackish tinge that can greatly reduce its value. The hyphae of *mold fungi* are colorless, so they do not discolor wood, but the spores produced by molds discolor wood surfaces. Stain and mold fungi belong to the classes Ascomycetes or Deuteromycetes. Although molds do not alter the physical properties of wood, their spores lower the aesthetic value and can contribute to human respiratory issues.

Because stain fungi and molds do not degrade the cell wall, they must live on food found within the lumen, consisting mainly of sugars and starch in the ray parenchyma of sapwood. Molds and stains are particularly troublesome in freshly felled logs and recently sawn lumber where the surface is relatively wet and the sugar-starch content is high. Figure 10.6 shows the development of blue stain in a log killed by insects. Note that the stain is limited to the sapwood. In addition to color damage, stain fungi often decrease wood toughness and increase permeability. This change in permeability can increase pre-servative uptake in treated products, adhesive penetration in glued products, and finish absorbtion during the final stages of wood product manufacture.

Blue staining of freshly cut lumber can often be avoided by quickly drying the sur-face. In warm regions and with highly susceptible species such as southern pine, white pine, yellow poplar, and sweetgum, a mild fungicidal dip is often used to treat lumber immediately after sawing. Where clean bright lumber is desireable, dipping is generally a good practice with all species of lumber that cannot be immediately kiln dried.

Molds and mildew can develop under high humidity on products in storage, in use conditions where the surface remains above 20% MC for a long period, and sometimes on painted surfaces. Fungicides may be incorporated into paint formulations to reduce this problem.

FIGURE 10.6.   Bluestain in a southern pine tree that was killed by the southern pine beetle.
Source: Courtesy of S. Sinclair.

Soft-rot fungi also belong to the Ascomycetes and Deuteromycetes classes. These organisms cause progressive degradation from the surfaces of wood inward. The effects are slower to develop and less apparent than decay or staining fungi. However, in some wet service applications such as wood slats in cooling towers, transmission and distribution poles, and foundation and freshwater piling, structural failure by soft rotters has been found. In the early stages, soft rot differs from other decay in that the effected surface(s) can be removed by scraping. The decomposition of the cell walls by some types of soft rot fungi is characterized by long cavities in the longitudinal direction entirely within the secondary wall.

Preventing Wood Decay

There are three general approaches to preventing wood decay. Each is appropriate under certain situations. First, it is necessary to follow Suddarth's three rules (see Chapter 7), or simply stated: keep the wood product dry. This is usually the most effective and least expensive approach. If wood buildings in temperate regions of the world are properly designed to avoid condensation and leakage problems, the wood can last indefinitely. Building design and maintenance are key components of increasing the longevity of wood components. The use of prudent roof designs that minimize valleys and other potential water traps, incorporate liberal overhangs and utilize gutters, and keeping wood siding and trim 45–75 cm (18–30 in.) off the ground, can dramatically reduce maintenance costs and add years to building life. Inside buildings in temperate climates, wood EMC is never high enough to maintain wood at 20% moisture content unless condensation, water leaks, crawlspace moisture or some other unusual factor is present.

The EMC Table A.1 in the Online Appendix shows the high humidity required to produce such a moisture content. In tropical countries, where rainfall and humidity are high, it may not be possible to maintain wood buildings in an adequately dry condition to prevent decay. The avoidance of deterioration by maintaining a low MC content is not possible for non-pressure-treated wood in contact with rain, the ground, or other materials that may remain damp, such as concrete or stone.

Second, where it is not possible to keep wood dry, it should be treated with chemicals that inhibit fungi. If the moisture content cannot be maintained below 20%, the use of preservative-treated material is the most practical way to avoid decay. Various chemical treatments are available. The level of treatment can be varied to suit the application and expected severity of the decay hazard. The American Wood-Protection Association has developed standards for pressure-treated wood to be used in a variety of situations.

Third, in some cases it is possible to use naturally decay-resistant species rather than using preservative-treated wood. It is important to note that only the heartwood of these species is actually resistant to decay and that the current durabilty performance associated with them is not as good as it was historically, mainly owing to the increasingly young age at which the trees are harvested. Because of the limited availability and high cost of naturally decay-resistant species as lumber, this approach is most suitable for applications where both durability and appearance are important.

Historically, woods such as apitong, chestnut, cypress, greenheart, cedar, redwood, teak, and white oak were often used where a combination of decay resistance and natural appearance was desired. Examples of such applications are the use of western redcedar, redwood, or resistant tropical species for the construction of patio decks and privacy fences, as shown in Figure 10.7. Online Appendix Table A.4 lists the decay resistance of

FIGURE 10.7.   Weathered unfinished redwood privacy fence.

some woods grown in the United States. The commonly used construction woods Douglas fir, southern pine, fir, larch, hemlock, and spruce have moderate to low resistance to decay. Thus, these species must be treated to obtain a suitable degree of decay resistance.

The degree of natural resistance to decay exhibited by heartwood depends largely upon the nature and amount of fungitoxic extractives in the species (see the section "Properties of Heartwood" in Chapter 2). The amount of heartwood varies considerably within a species according to its location in the tree and its age and rate of growth. Hydrophobic extractives and other cell-occluding materials (tyloses) can improve decay resistance by decreasing permeability, thereby keeping wood moisture content low. Rapidly grown plantation timber, which is readily available, often has a lower extractive content and thus lower natural decay resistance than older slow-growth material. Also, there are kiln-drying and steaming techniques available which can darken many species, such as sweetgum, thereby making the sapwood appear like heartwood. Thus extreme caution and care should be exercised when specifying "naturally durable" species.

Bacteria

Bacterial degradation of wood products is generally limited in scope owing to the extensive time required for bacterial degradation to occur. Unlike fungi, bacteria can survive and thrive in relatively anoxic or completely anaerobic environments, thus high moisture levels even up to saturation are not an adequate defense. Bacterially infected wood can occur in the forest, commonly in bottomland trees; in the log yard amongst water-stored logs; in salvage logs from rivers or ponds; or in products such as utility poles and foundation, freshwater, and marine piling. Fortunately bacterial degradation is generally slow.

Logs that are bacterially infected from the forest can usually be carefully processed into lumber. Special care should be taken during drying as excessive splitting can occur and bacteria may be slow to give up moisture, thereby causing wet pockets. Logs that are ponded or stored under water sprinklers can become bacterially infected if storage time is prolonged beyond about six months. After that time, the bacteria begin to significantly degrade the pit and parenchyma structures, thereby causing increased permeability and creating a problem for treatment, gluing, and finishing. Bacterial infection and damage in service are usually very slow, requiring decades. Bacteria can, however, cause costly damage to buried piles, poles, and other structural members. Wood preservatives slow down the deterioration but are not completely effective because strains of bacteria often adapt to the point at which they utilize the wood preservative as part of their metabolic processes.

Insects

Wood products in use throughout the world are subject to infestation by a wide variety of termites and beetles and by a few species of ants and bees. Some insects, particularly termites and wood-destroying beetles, cause great financial loss and are of concern across wide areas of North America. A great number of other insects are important in specific regions of the world because of their damage to standing timber, logs, and wood products in use.

Termites (biological order Isoptera) have by far the greatest economic impact. Over five hundred thousand dwelling units in the United States receive annual treatment for termite control. *Subterranean termites* have been reported in every state in the United States except Alaska and are relatively rare in the tier adjacent to Canada. They are most common and represent the greatest economic problem in California, the states bordering the Gulf of Mexico, and along the Eastern Seaboard from Florida to Virginia. Also common are dry-wood termites, which are limited to an area within about 160 km (100 miles) from the coast of California and from the Gulf and South Atlantic coasts, including all of Florida.

Subterranean termites live similarly to ants – in colonies in the ground. Most species have three social castes: workers, reproductives, and soldiers. Termites differ from ants in that they do not have the narrow waist between the abdomen and thorax, and their two pairs of equal-size wings are connected to distinct wing stubs, which are apparent when the wings are lost. Also, termites are white or nonpigmented, as they spend their lives in the dark, whereas ants may be black, maroon, red, or other color or combination of colors.

Subterranean termites usually enter a wood structure from the ground or through *shelter tubes* that they construct as a means to reach the wood. They use wood both as a shelter and to obtain cellulose, their source of food. To avoid exposure to outside environments, they live entirely within the wood once a colony is established. Figure 10.8 illustrates a wood post destroyed by termites. The exterior showed no sign of its condition until failure occurred. For termites to continue their infestation, they must maintain a reasonably high level of moisture. This moisture is often brought from the ground to their sealed colonies.

Formosan termites were inadvertently introduced into the United States in the 1940s. They are significantly larger and more aggressive than their native subterranean relatives. Their zone of severe destruction radiates outward and slowly expands from the Gulf Coast area of the United States (Shupe and Dunn 2000).

FIGURE 10.8.   Wood post that failed at the ground line principally from termites, although decay was also present. Source: Courtesy of T.L. Amburgey.

TABLE 10.1. Wood-destroying beetles (does not included beetles that attack or damage standing trees).

Family	Common name	Material attacked[a]
Anobiidae	Furniture beetles	Old hardwood, old softwood
Lyctidae	True powder-post beetles	New hardwood
Bostrychidae	False powder-post beetles	New hardwood, new softwood (rare)
Cerambycidae	Long-horned beetles	New softwood
Curculionidae	Weevils	Old softwood, old hardwood

Source: Williams (1973).

[a] Lumber less than 10 years old is considered to be new.

Drywood termites, pose a different risk than the subterranean variety because they can enter wood above the ground directly from the air. Once they have gained entry, they can live in wood with moisture contents as low as 5 or 6%. Wood structures in regions subject to drywood termite attack should be inspected regularly, all cracks and exposed wood should be caulked or painted, and vents should be screened to prevent termites from coming into contact with unpainted wood.

Wood-destroying beetles in North America include lyctid beetles (true powder-post beetles), anobiid beetles (furniture beetles), false powder-post beetles, and long-horned borers. Table 10.1 lists common wood-destroying beetles and the types of material they typically attack, both hardwood and softwood, in the green or dry condition. These beetles are in addition to other beetles, such as bark beetles and various pine beetles, that can attack and kill standing trees. Evidence of *powder-post beetles* is the presence of small exit holes which vary in diameter from about 1.5 to 5 mm depending on the species and a fine flourlike powder (frass) dropped from the holes. Most in-service beetle damage is aesthetic only, caused by mature beetles exiting the wood in which they developed from eggs. Figure 10.9 shows powder-post exit holes in American basswood. By kiln drying or sterilizing wood products such that the core wood temperature exceeds ~66°C (150°F) for two to four hours, most wood-destroying beetles can be killed. Sterilization and phytosanitization processes that kill such organisms as nematodes and insects are often required of wood products, including pallets, that will be shipped internationally. It is often possible to determine if beetle damage occurred before final milling by the shape of the holes. If it occurred after final milling the exit holes are normally round. If the damage occurred before final milling then the holes may be round or oval, or the milling may have exposed elongated tunnels in the wood.

Carpenter ants and bees do not derive nutrition from wood. Rather, these insects excavate wood for nesting purposes. As such, it is possible for carpenter ants to excavate preservative-treated lumber. Carpenter ants favor warm, damp locations such as wood framing in poorly maintained (leaky) buildings, often coexisting with some level of decay fungi as wet and partially decayed wood is relatively easy for them to excavate. Carpenter ant damage is largely controlled by keeping wood structures dry. Carpenter bees sporadically bore holes in buildings (often roof trim) ~10 mm (0.4 in.) in diameter. These bees cause little more than occasional aesthetic damage.

Materials selection, building design, planning of construction details, and construction practices should be carried out with consideration for the potential severity of insect damage in the region. Protection against subterranean termites in areas of low or

FIGURE 10.9.   Powder-post beetle attack on basswood lumber; scale is shown in centimeters.

moderate attack can be accomplished by proper soil treatment and the construction of foundations, sills, and floor slabs so that termites cannot reach the wood without being visually detected. A combination of soil treatment and proper construction with effective termite barriers provides adequate protection in most temperate regions. Where extreme termite hazard exists, such as tropical regions, it may be desirable to use chemically treated lumber, plywood, and particleboard throughout the structure. Many chemicals that provide protection against decay are also effective against termite attack.

Care should be taken to avoid bringing lumber or firewood infested with powder-post beetles into a wood structure or a wood storage yard, as these can cause reinfestation. In regions where powder-post beetles are a serious problem, freshly cut lumber may be heat sterilized or treated with insecticides to prevent attack.

It is not uncommon for reclaimed lumber, such as historic barnboards or framing that is infested with powderpost beetle eggs, to be remanufactured (stripped, sawn, planed, and/or sanded) for uses such as paneling or flooring. The lumber provides a desirable, character-marked appearance for use in homes, restaurants, and retail shops. It is not so pleasant, however, when the beetles later emerge and create their exit holes, often many years after installation.

Marine Borers

Wood used in contact with sea or brackish water is subject to damage by marine borers. These small animals can cause extensive damage to posts, pilings, and wooden boats. There are three types of marine borers. Two, the *shipworms* (genera *Teredo* and *Bankia*) and the *pholads* (principally *Martesia*), are mollusks related to clams and oysters. The third type are crustaceans or "gribbles," principally species of *Limnoria*, *Chelura*, and *Sphaeroma*, which are related to crayfish and prawns.

The larvae of shipworms attach to the surface of wood and make only a small entry hole there. Once inside, they form smooth hollow galleries along the grain. As these organisms grow, the galleries become larger until the wood is completely honeycombed. The exterior of the wood remains intact. Figure 10.10 shows the damage caused by pholads and shipworms. Shipworms often concentrate near the mud line on posts or pilings and leave little exterior evidence of their presence until damage is severe.

The hard, calcified shells on the heads of shipworms grind the wood to form galleries. The rear portion of the body remains near the entry hole where it obtains water and gets rid of wastes. As the shipworm elongates and burrows deeper, it may reach a length of over 1 m.

The damage from pholads is similar to that from shipworms except that their borings tend to be shorter. They reach a length of ~6.5 cm (2.5 in.). Pholads retain their clam-like appearance as they grow. Shipworms, however, have a shell only on the head, with a long worm-shaped body behind.

Limnoria attack on pilings is less catastrophic than that of shipworms or pholads because it occurs on the wood surface and is thus more easily detected. The principal area of attack is wood exposed between high and low tides. Wave action can erode this infested portion of the pilings so that the effective diameter is continuously decreased, and this eventually results in an hourglass shape. The borings of *Limnoria* are small cylindrical channels less than 3 mm (0.12 in.) in diameter, which honeycomb the wood. The depth of penetration is apparently limited by oxygen requirements for respiration.

Preventing attack from marine borers by chemical treatment is possible but difficult. A few species of tropical hardwoods have natural resistance to marine borers (e.g. angelique, greenheart, and jarrah), but these woods are not available worldwide. No commercial timbers native to North America possess such natural durability. Whenever wood is used where it may be subject to marine borer attack, adequate and appropriate chemical treatment with durable preservatives such as creosote or chromated copper arsenate is necesary. Without such treatment, destruction is often rapid and complete.

In warm coastal waters, the presence of *Limnoria* and pholads presents a difficult preservation problem. Even heavy creosote treatments are sometimes inadequate to provide complete protection. Dual treatments, first with a waterborne preservative and then with creosote, have proved reasonably effective under such severe marine environments. Environmental concerns are important in these cases because the preservatives may leach out into the water. The rate at which preservative leaching occurs depends on the water conditions and the temperature. Loss in stagnant water is less than in the currents in tidal basins. Loss also increases with increases in temperature. Thus, retention levels that prove satisfactory in cool climates may not be adequate in the tropics.

Heat and Fire

When wood is heated above about 100 °C (212 °F) molecular decomposition occurs. If the temperature is between 100 and 200 °C (212 and 392 °F), decomposition occurs relatively slowly. Water vapor is driven off along with carbon dioxide and carbon monoxide. Flammable gases volatilize at temperatures as low as 120 °C (250 °F). In this range, wood

(a)

(b)

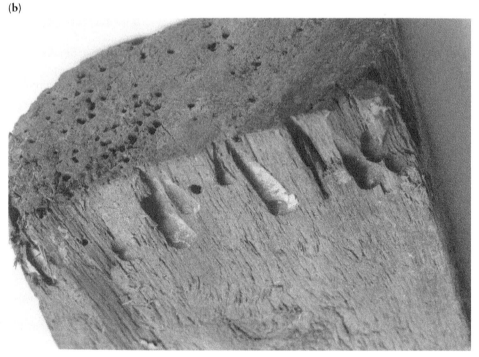

FIGURE 10.10. The attack of shipworms and pholads on material exposed to marine environments. (a) Piling almost completely consumed by shipworms. Despite the destruction, the exterior shell still appears relatively sound. (b) The end and exposed surface of a timber attacked by pholads. This piece was stored in a salt water bay in the Caribbean for only a short time prior to shipment to the United States.

gradually degrades or pyrolyzes. *Pyrolysis* is chemical breakdown caused by heat. If the temperature is raised above 200°C (392°F), rapid pyrolysis occurs. Pyrolysis accelerates between 260 and 350°C (500 and 680°F). At these temperatures, flammable gases are evolved quickly. When heated in the presence of oxygen, these gases can either ignite from an ignition source such as a flame or they can self-ignite if the temperature becomes high enough. At higher temperatures, the time for ignition to occur is relatively short, on the order of seconds. At lesser temperatures, more time, on the order of several minutes, is required to support ignition. Above about 270°C (520°F) the rate of heat produced will be greater than the heat required to generate the gas and the fire will support itself.

In a burning piece of wood the carbon char left after the volatile gases are driven off tends to increase in thickness with time but at a decreasing rate. Because wood and char are good thermal insulators, the rate of heat transfer to the interior wood decreases as the char thickens. Eventually, the heat transferred inward may not be sufficient to produce and drive off flammable gases. If this occurs, the flame will die and unless there is an outside source of heat, the piece will gradually cool. Because of this self-insulating behavior, heavy members such as glue-laminated beams, crosslaminated timber, and solid-sawn timbers provide excellent fire resistance. Improved fire resistance ratings for light-frame buildings can be obtained by the use of gypsum board on both sides of walls and on ceilings. Where a greater level of fire resistance is needed, wood members can be treated with *fire retardants*.

The impregnation of wood with fire retardants is used commercially to reduce surface flammability by reducing the production of flammable gases. Wood so treated tends to char and decompose but does not support combustion. Commonly used fire retardants incorporate combinations of organophosphates, boron, polyphosphates, and urea. Impregnation levels on the order of $35–80\,kg\,m^{-3}$ ($2.2–5.0\,lb\,ft^{-3}$) are required for proper protection. Many fire retardants can be leached from the wood by water, so their use must be restricted to interior applications. Some nonleaching fire-retardant treatments are, however, available for exterior applications such as wood shingles. Early fire retardants such as zinc chloride and ammonium sulfate often resulted in treated materials being more hygroscopic (i.e. more prone to absorbing moisture) than normal wood at low humidities. The newer systems have a lower impact on hygroscopicity and are less corrosive.

The major model building codes used in the United States permit fire retardant-treated lumber to be used in many building systems, which can then be rated as noncombustible. Some common applications for fire-retardant lumber include interior nonbearing walls in non-combustible-type buildings, exposed wood roof systems, party walls in multifamily houses, and interior wall and ceiling paneling in buildings requiring a relatively low flame-spread rating.

Weathering

A surface of wood exposed to outdoor conditions without the protection of paint or other finishes is subject to surface roughening and degradation broadly termed *weathering*. The color of the wood turns to gray or silver (depending on the species), the cells at the surface are slowly broken down, and the surface is gradually eroded. Figure 10.11 shows the surface of a redwood board weathered for eight years. This photograph illustrates the difference in the weathering between latewood and earlywood. Weathering is usually not a serious problem unless the effect on the appearance is considered detrimental. In some regions, wood siding is often left unpainted or treated with a bleach to accelerate

FIGURE 10.11.   Surface of an unfinished redwood board showing the difference in the rate of weathering of earlywood and latewood after eight years of exposure. The less dense earlywood erodes more quickly, thereby leading to the scalloped surface. A flat head screw-type fastener is shown in the center of the frame.

weathering and obtain a natural rough-gray architectural effect. In damp climates, blackish fungal stains may develop, which degrade the desired appearance. In areas where blue-stain or surface molds commonly develop, a water-repellent treatment containing a fungicide can be used to inhibit the development of such discoloration.

Weathering results from a combination of factors. Light, particularly the ultraviolet wavelengths, slowly degrades portions of the cell wall. Soft-rot fungi, which grow well in alternately wet and dry conditions, are also frequently involved. The alternate shrinkage and swelling of the surface layers produces stresses that can cause small fractures. Finally, the mechanical abrasion of wind-driven dust and water also gradually wears away the surface. If wood is left exposed to the elements and unfinished, it can weather at a rate exceeding 7 mm (0.25 in.) per century. Higher-density species generally weather at a slower rate, although they are more subject to surface checking and distortion.

Preservation

Preservative Treatments and Standards

The requirements for an ideal wood preservative include the following: (i) toxic to a range of wood-inhabiting organisms; (ii) a high degree of permanence (low volatility, resistance to leaching, chemical stability); (iii) an ability to penetrate wood readily;

(iv) hydrophobicity (i.e. moisture is repelled from the treated wood); (v) noncorrosive to metals and noninjurious to wood; (vi) safe to handle and use; (vii) low environmental impact; and (viii) favorable economics. Preservatives used today in commercial practice balance these factors. Each has its own combination with better performance in some areas and lesser performance in others. Of all of these factors, environmental performance and sensitivity along with cost are often the most important in development and use.

The effectiveness of a preservative treatment depends not only on the toxicity of the chemical to the anticipated deterioration organism(s) in its service environment, but also on how completely it has penetrated and how much is retained within the wood following treatment. Standards of the AWPA (2017) specify how deeply preservatives must penetrate and the amount of preservative per unit volume that must be retained in certain zones of the wood after treatment. Chemical retention is stated in treatment standards in terms of the weight of the active chemical per cubic foot of wood remained after treating. For 2015, in the United States, the value of treated wood shipments was on the order of seven billion USD (US Census Bureau 2017).

Standardized wood preservatives in current commercial use can be found in the AWPA standards (2017) and are of two general types: *oil-soluble chemicals* and *water-borne salts* or *metallic oxides*. These preservatives generally have broad-spectrum toxicity, that is, they are effective against a variety of organisms. The fundamental difference between the two types is the liquid used to carry the toxic chemicals into the wood structure. Heavy-oil soluble preservatives have some advantage in wet-use situations because the oily carrier retards liquid water movement. A drawback to some heavy-oil treatments is that the wood surface often remains oily, difficult to handle, and impossible to finish or paint. To overcome this problem, it is possible to use light organic solvents as carriers for toxic compounds so that the wood may be painted after treatment. These solvents evaporate rapidly, leaving the wood with an untreated appearance. Although not generally used in the United States, light organic solvent-carried preservatives are regularly used throughout Europe, Australia, and New Zealand for molding and millwork.

The most commonly used type of oilborne preservative is coal-tar creosote. The heavy creosote oil is its own carrier. In addition to toxicity to fungi, insects, and marine borers, this highly viscous organic oil is a natural water repellant, which adds to its effectiveness. Being more or less semisolid at room temperature, creosote must be heated to reduce its viscosity prior to treatment. A system of delivering creosote preservative to wood using a pigmented water carrier has been developed and commercialized in Australia. This system reduces some of the undesirable surface characteristics (namely dripping) of creosote oil treatment. Environmental issues and regulations associated with creosote increasingly limit its use in the United States. Also, creosote is a by-product of coke production from coal, the coke then being used for steel manufacture. As a significant level of steel production has shifted out of the United States in the last decade, coke production has declined, thereby decreasing the supply of North American creosote.

Pentachlorophenol dissolved in P9 oil (similar to number-two fuel oil) is another highly effective wood preservative. Although its use is restricted in many applications and some developed nations, it is still widely used in many parts of the world. Other commercial organic oilborne preservatives include copper naphthenate, copper 8-quinolinolate (oxine copper), and tributyl tin oxide (TBTO). A variety of biocides are in use as cobiocides, generally in adition to waterborne copper systems. These include carbamates, chlorthalonil, and azole-based compounds (thiazoles, triazoles, and substituted isothiazolones).

With all of these, the oil carrier can add to the preservative value. In the case of creosote treatment, the total weight of the creosote is used to calculate the retention; for other preservatives dissolved in a carrier, only the actual weight of the preservative per volume of wood is used to calculate the retention. For creosote-treated poles and pilings, the retention requirements generally range from about 125 kg m^{-3} (7.5 lb ft^{-3}) for building poles to over 400 kg m^{-3} (25 lb ft^{-3}) for marine pilings. In general, higher levels of retention are required as the deterioration and decay hazards increase, such as in warmer climates, and as the level of difficulty associated with replacement increases, such as foundation piling (difficult to replace) as compared with deck boards (relatively easy to replace).

Water is used as the carrier for moving inorganic preservatives into wood. As a preservative carrier water is more environmentally benign than oil or organic solvents. Figure 10.12 shows a sound barrier along an interstate highway in Minneapolis, Minnesota, constructed with southern pine lumber treated with waterborne preservatives.

There are a variety of waterborne preservatives currently in use in the United States and around the world. For the most part, these compounds become chemically fixed and stabilized in wood, although it has been demonstrated that when treated wood is in ground contact, components of the treating compounds may migrate short distances (0.15–0.30 m or 6–10 in.) into the surrounding soil (Cooper 1994; Morrell and Huffman 2004).

For decades, the most commonly used waterborne preservative in the United States has been chromated copper arsenate (CCA). Because of the higher cost of petro-based carriers and environmental concerns about oilborne preservatives, the

FIGURE 10.12.   A wood wall treated with waterborne preservatives serves as a sound barrier in an urban area adjacent to an interstate highway. There are thousands of miles of such sound barrier throughout the USA.

waterborne treating chemicals, and particularly CCA, became widely used. In 1997 about 80% of treated wood utilized waterborne preservatives, of which 98% was CCA (Micklewright 1998). In the case of waterborne preservatives, the weight of the chemical only, not the carrier, is used in the calculation of retention. Effective retention requirements for CCA-treated wood range from ~4 kg m^{-3} (0.25 lb ft^{-3}) for above-ground exposure to 10 kg m^{-3} (0.60 lb ft^{-3}) for use in wood foundations. Owing to increased environmental pressures, the use of metal-based wood preservatives is increasingly regulated. At the end of 2003, wood preservatives containing arsenic were federally regulated as only to be available for commercial and industrial use and CCA was voluntarily withdrawn from the consumer market. Since that time, copper-based systems have become the preservatives of choice. Generally, copper-based systems require a cobiocide because there are species of copper-tolerant wood decay fungi. Around the globe, a shift away from broad-spectrum metal oxides in favor of target-specific, more benign biocides is ongoing.

Wood-treating companies in the United States that conform to AWPA standards and that submit lumber to inspection by an authorized inspection agency may use the quality stamp of approval of that agency on their products. Utility companies and railroads using treated poles and railroad ties for purposes other than building construction often use additional standards to ensure product quality.

Treating Methods

Throughout history many methods of applying wood preservatives have been devised. Examples include painting wood ships with pine rosin, using the capillary action of freshly felled trees to treat stem butts, and boiling fence posts in oil. To some extent these methods are still used in various parts of the world.

Most modern treatment processes involve the use of differential pressure to force preservative into wood. Wood is commonly placed in a vacuum/pressure retort where preservative can then be added and a scheme of vacuum and/or pressure applied. With oilborne preservatives the preservative and the wood are often heated for the purpose of lowering liquid viscosity and increasing penetration. A vacuum can be pulled prior to introduction of the preservative to remove air from cell lumina, thereby facilitating greater retention of the treating chemicals. Many different vacuum pressure–time–temperature cycles (collectively termed "treatment mechanics") are used depending upon the species, the type of preservative, and the amount to be retained in the wood.

Some treating cycles are intended to leave the cell lumina filled with preservative for applications where the hazard of deterioration is high, such as bridge timbers and railroad ties. These are termed *full cell processes*. In *empty cell processes* an attempt is made to remove excess preservative once it has been forced to the desired depth within the wood. Figure 10.13 illustrates the pressure–time sequence used in the pressure-treating tank for one type of empty cell process, i.e., the Rueping process. These latter processes are typically used for lumber as well as distribution and transmission poles.

Species vary greatly in their resistance to penetration by preservatives although the sapwood of almost all species is readily penetrated. Species that are very difficult to treat (called *refractory*), such as inland- or mountain-type Douglas fir and white oak, are frequently incised to form passages for the preservative to penetrate more deeply into the

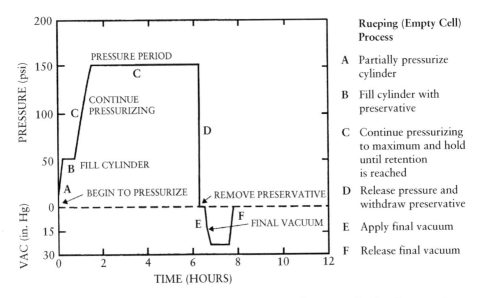

FIGURE 10.13. Treating schedule of the Reuping (empty cell) process. Similar diagrams of pressure and time are well developed for various species, products, retention levels, and preservative chemicals.

wood. Incising machines puncture the surface of wood with chisel-shaped points that expose end grain up to a depth of 2.5 cm (1.0 in.). Permeable species, such as the southern pines, are easily treated after kiln drying.

Nonpressure preservation methods are used commercially but these methods are applicable primarily for wood that will be used under conditions presenting only a mold or stain hazard or moderate decay hazard. These methods typically provide primarily an envelope treatment with minimal depth of penetration. Nonpressure methods include vacuum, brushing, dipping, and soaking. Dipping, soaking, and vacuum application of preservatives and water repellants is used extensively to protect millwork such as windows and for bright white-woods (maple, birch, cottonwood) for high-value export.

Considerations in the Use of Preservatives

The cleanliness of the preservative and the treated product is of utmost importance for uses associated with human contact such as millwork, stadium seating, or residential decks. For other uses, such as railroad ties or bridge timbers, an oily surface may be desirable and increase the resistance to weathering and water absorption and thus increases service life, but environmental issues generally no longer allow for this type of surface.

Corrosion of metal fasteners, such as nails, truss plates, and flatbed truck frames, can be a problem in treated wood where the moisture content remains high. Heavy oil-type preservatives often retard corrosion while waterborne metallics may accelerate it. The inorganic preservative chemicals in combination with moisture in wood often form galvanic cell reactions with metallic fasteners, thereby expediting fastener decomposition. In such cases, anti-corrosion chemicals may be added to the preservative and specially coated or stainless steel fasteners may be specified.

The strength of wood products may be affected by chemical treatments or treatment processes. The extent of strength loss, if any, is usually determined more by the conditions of preservative impregnation than by the chemical itself. High temperatures and pressures used to force the liquid into the wood, along with incising, can have detrimental effects upon strength. Oil-type preservatives undergo no chemical reaction with the wood that might affect strength, but waterborne salts do react and can weaken wood. Product specifications and standards provide case-specific information related to potential changes in design strength.

The *National Design Specifications of the American Wood Council* provide wood strength and design information for structural engineers (American Wood Council 2014). These specifications require that the designer check with the fire retardant-treated wood supply company for appropriate strength reduction values. Commonly, a 10% reduction in design strength is used for wood that has been pressure-impregnated with a fire retardant. High in-service temperatures can adversely affect the strength of preservative-treated wood and significantly reduce the strength of fire retardant-treated wood. This phenomenon can be critical in elevated-temperature situations such as roof framing, trusses, and sheathing.

Safety, Environment, and the Future

The act of treating with any type of preservative is, in essence, an attempt to somewhat toxify or poison the wood. The environmental impact and safety of preservatives and treated wood are critical concerns. More than ever before, environmental regulation and public scrutiny play critical roles in wood preservation. Much of this action and public involvement stems from spills and pollution incidents that have occurred. The idea of open chemical lagoons and preservatives leaching into ground water and aquifers is not acceptable in any part of the world. That said, based on life cycle analysis and the efficient cradle-to-grave costs associated with treated wood as compared with its alternatives, the production and use of treated wood is highly favorable and will continue. The best path forward is one that involves candid discussion and consideration of engineering, environmental, and economic considerations of both preserved wood and its alternatives.

Because most preservatives have some level of toxicity, care should always be exercised in the use of treated wood. Many of the most effective wood preservatives are not available to the general public and may be used only by those trained, licensed, and properly equipped for safe application. Additionally, recycling and other end-of-service-life considerations are important. Leaching and ultimate disposal issues related to wood treated with inorganic preservatives present challenges to the industry and the general public.

For the long term, research is ongoing toward developing effective preservatives with reduced environmental impacts. Much of this research focuses on examination of the molecular biochemistry of decay fungi. Rapid and accurate fungal identification by methods such as genetic markings or fatty acid analysis assists in understanding fungal development in-situ. As specific biochemical pathways of fungi are identified, methods to disrupt or halt them can be developed. Efforts to find synergistic organic preservative combinations, to mimic the effective mechanisms in extractives and heartwood (Schultz and Nicholas 2000) and to create highly organism-specific inhibitors are of keen importance. Trees that are genetically selected or enhanced for heartwood durability may also play a role in increasing wood product performance. It is likely that the use of broad-spectrum preservatives will continue to be phased out in favor of target-specific, more environmentally benign biocides. Although elusive, this is an ultimate goal that will provide a significant public benefit.

References and Supplemental Reading

AITC (2004). *Timber Construction Manual*, 5. American Institute of Timber Construction. New York: Wiley.

American Wood Council (2014). *National Design Specification (NDS) Supplement for Wood Construction 2015 Edition*. Leesburg, VA: American Wood Council.

AWPA (2017). *American Wood Protection Association Book of Standards*. Birmingham, AL: AWPA.

Barnes, H.M. and Murphy, R.J. (1995). Wood preservation – the classics and the new age. *For. Prod. J.* 45 (9): 16–26.

Boone, R.S., Winandy, J.E., and Fuller, J.J. (1995). Effects of redrying schedule on preservative fixation and strength of CCA-treated lumber. *For. Prod. J.* 45 (9): 63–73.

Cooper, P. (1994). Leaching of CCA: is it a problem? In: *Environmental Considerations in the Manufacture, Use, and Disposal of Treated Wood* (ed. Forest Products Society), 45–57. Madison, WI: Forest Products Society.

Corbett, N.H. (1975). Micro-morphological studies on the degradation of lignified cell walls by ascomycetes and fungi imperfecti. *J. Inst. Wood Sci.* 14: 18–29.

DeGroot, R.C. (1976). *Your Wood can Last for Centuries*. Washington, DC: USDA Forest Service.

Ericksson, K.L., Blanchette, R.A., and Ander, P. (1990). *Microbial and Enzymatic Degradation of Wood and Wood Components*. New York: Springer.

Feist, W.C. (1990). Outdoor wood weathering and protection. In: *Archeological Wood: Properties, Chemicstry, And Preservations*, Advances in Chemistry Series, vol. 225 (ed. R.M. Rowell and R.J. Barbour). Washington, DC: American Chemical Society Chap. 11.

Highley, T.L., Clausen, C.A., Croan S.C., Green, F III, Illman, B.L., and Micales J.A. 1994. Research on biodeterioration of wood, 1987–1992. USDA For. Ser. Res. Pap. FPL-RP-529.

Howard, J.L. 2003. US timber production, trade, consumption, and price statistics 1965 to 2002. FPL-RP_615. Madison, WI. USDA For. Serv., For. Prod. Lab.

Johnston, H.R.; Smith, V.K.; and Beal, R.H. 1972. Subterranian termites: their prevention and control in buildings. USDA Home Bull. 64.

Koch, P. 1972. Utilization of Southern Pines. USDA For. Serv. Agric. Handb. 420, Vol. 2. 650–720.

Koch, P. 1985. Utilization of hardwoods growing on southern pine sites. USDA For. Serv. Agric. Handb. 605, Vol. 2. 2268–2536.

Micklewright, J.T. (1998). *Wood Preservation Statistics – 1997*. New York: American Wood-Preservers' Association.

Milton, F.T. 1994. The preservation of wood – a self study manual for wood treaters. Pub. MI-6413-S. Univ. of Minn. Extension Ser.

Morrell, J. and Huffman, J. (2004). Copper, chromium, and arsenic levels in soils surrounding posts treated with chromated copper arsenate (CCA). *Wood Fiber Sci.* 36 (1): 119–128.

Nicholas, D.D. (1973). *Wood Deterioration and Its Prevention by Preservative Treatments*, vol. 1 and 2. Syracuse, NY: Syracuse University Press.

Richardson, B.A. (1987). *Wood Preservation*. Lancaster: Construction Press.

Scheffer, T.C. and Cowling, E.B. (1966). Natural resistance of wood to microbial deterioration. *Annu. Rev. Phytopathol.* 4: 147–170.

Schultz, T.P. and Nicholas, D.D. (2000). Naturally durable heartwood: evidence for a proposed dual defense function of the extractives. *Phytochemistry* 54 (2000): 47–52.

SFPA (2005). *Industry Statistics, Southern Pine Treated Production*. Kenner, LA: Southern Forest Products Association.

Shupe, T.F. and Dunn, M.A. (2000). The formosan subterranean termite in Louisiana: implications for the forest products industry. *For. Prod. J.* 50 (5): 10–18.

US Census Bureau. 2017. https://factfinder.census.gov/faces/tableservices/jsf/pages/productview. xhtml?src=bkmk. Accessed 25 July 2017.

Wilcox, W.W. 1968. Changes in wood microstructure through progressive stages of decay. US For. Prod. Lab. Rep. 70.

Williams, L.H. (1973). Identifying wood-destroying beetles. *Pest Control* 41 (5): 30–40.

11

Silvicultural Practices and Wood Quality

The practice of caring for and cultivating forest trees is known as *silviculture*; this activity is one of the responsibilities of the forester, whose objective is usually to accelerate growth. One method of increasing growth rate involves the reduction of competition for available sunlight, nutrients, and water, which can be achieved through such practices as control of understory vegetation, thinning, or control of the spacing between seedlings at the time of planting. Another approach is to add nutrients and water by fertilizing and irrigating. The growth rate of new forest stands can also be stimulated through genetic selection of seed or planting stock. Tremendous increases over natural growth rates are possible.

Measures of Wood Quality

Before examining how silvicultural practices affect wood quality, it is necessary to explain the concept of quality. *Quality* can be generally defined as a measure of the characteristics of wood that influence the properties of products made from it. Briggs and Smith (1986) put it differently, saying that "wood quality is a measure of the aptness of wood for a given use." A more precise definition of quality can be elusive, because characteristics important in wood that is to be used for one product are often different from those needed for wood destined for another product. In one case, quality may be measured in terms of density, uniformity of growth rings, and percentage of knot-free wood, whereas in another instance, properties such as proportion of latewood, cellulose yield, and the fiber-to-vessel ratio may be the primary quality indices. Larson (1969) provided insight into the matter of wood quality when he wrote: "During wood formation, numerous factors both inside and outside the tree lead to variation in the type, number, size, shape, physical structure, and chemical composition of the wood elements. Wood quality is the arbitrary classification of these variations in the wood elements when they are counted, measured, weighed, analyzed, or evaluated for some specific purpose. Wood quality is therefore a concept."

Forest Products and Wood Science: An Introduction, Seventh Edition. Rubin Shmulsky and P. David Jones.
© 2019 John Wiley & Sons Ltd. Published 2019 by John Wiley & Sons Ltd.
Companion website: www.wiley.com/go/shmulsky

Although the concept of quality may be difficult to pinpoint, several factors influence the suitability of wood for a variety of purposes. These factors include density, uniformity of growth rings, fiber length, percentage of clear bole, straightness of grain, proportion of heartwood, percentage of vessels (in hardwoods), and the presence of juvenile and reaction woods.

Density

As explained in Chapter 9, the density of wood is a prime determinant of its strength, and the strength–density relationship is direct. In general discussion, the term density refers to the same characteristic as specific gravity, i.e. the dry wood substance per unit of volume. (Chapter 8 discusses the technical difference.) Species that are usually high in density, such as Douglas-fir (*Pseudotsuga menzesii*), larch (*Larix* spp.), and southern yellow pine, are preferred for uses in which high strength is important. Lower-density species, such as white and sugar pine (*Pinus monticola, Pinus strobus, and Pinus lambertiana*), hemlock (*Tsuga* spp.), and true fir (*Abies* spp.), may be used for light framing or lumber products in which dimensional stability or other characteristics are of more importance than strength.

Although high-density species are preferred where high-strength lumber products are required, it must be realized that density varies considerably within a given species (Figure 11.1; see also Figure 8.4). Trees (or even zones within a tree) that form the highest-density wood will be the most valuable to the producer of structural lumber

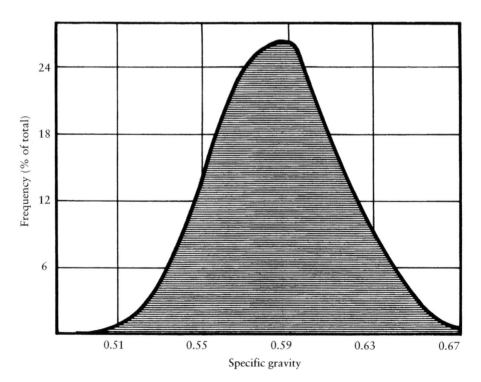

FIGURE 11.1.   Specific gravity variation in sugar maple (based on material from 12 stands in eight states). Source: From Paul (1963).

products. Therefore, any practice that might cause wood to fall in the upper part of the density range for that species can increase its value as structural lumber. This is recognized, for example, in the grading rules for southern pine. On the other hand, conditions that cause wood density to fall into the lower density range also result in lower strength. Such wood has diminished value as a structural material, although it may be easier to machine.

Species that produce relatively low- or medium-density wood are often preferred as raw material for pulp and paper manufacture over species that produce higher-density wood (see Chapter 15). The same is true with regard to the production of particleboard and fiberboard. However, the higher the density is (based on ovendry weight), the greater the yield of pulp from a given volume of wood. Thus, the forest manager may be faced with the seemingly conflicting objectives of growing low-density species, such as spruce, cottonwood, or aspen, so that both volume per acre and density of wood are maximized.

Growth rate can significantly affect wood density. This is particularly true of ring-porous and semi-ring-porous hardwoods that tend to show greater density and hardness with increases in rate of growth. The width of earlywood in ring-porous and semi-ring-porous hardwoods tends to remain relatively constant regardless of the growth rate. In these woods, a slowing of diameter growth brings the rows of large pores closer together in successive annual rings (Figure 11.2). The development of thick-walled latewood fibers and small-diameter vessels is minimized in ring-porous species when the growth rate is slow (Paul 1963). Therefore, within limits, the faster that ring-porous hardwoods are grown, the denser (and thus stronger) they are. Jane et al. (1970) pointed out that this rule holds true only within limits, because wood grown at an extremely fast rate may develop abnormally thin-walled fibers, an unusually high proportion of parenchyma cells, or both. He estimated that maximum density and strength develop in mature wood when growth rings measure 2.5–4.0 mm (6–10 per in.).

Softwoods with distinct growth rings (*distinct-ring softwoods*) produce wood characterized by bands of latewood that are clearly discernible from the bands of earlywood on either side. In such woods, latewood tracheids have distinctly thicker walls and smaller radial diameters than earlywood tracheids, and the change within a growth layer from one cell type to the next is typically abrupt. It was once commonly believed that distinct-ring softwoods decrease in density with an increase in growth rate, and numerous studies have supported this notion (Aldridge and Hudson, 1958, 1959; Ciesler and Janka 1902; Paul 1963; Wellwood 1952). The explanation advanced for this relationship is that in distinct-ring softwoods the latewood remains relatively constant in width with varying rates of growth. Despite a longstanding acceptance of a fast-growth/low-density relationship in distinct-ring softwoods, a study published in 1947 refuted this notion. In that year, Turnbull reported that density variations between plantation pines were traceable to an age effect rather than to growth rate. This result was later confirmed by Rendle and Phillips (1957), who studied samples taken from the outer portions of stems. The age effect is illustrated by a plot of density versus age for a typical forest-grown distinct-ring softwood (Figure 11.3). Here, density increases rapidly through the juvenile period, and then increases slowly and steadily through maturity. This pattern has also been observed in several diffuse-porous hardwoods. A variation of this relationship is one in which wood density declines slightly for several years following the juvenile period before tending to increase in successive growth rings.

(a)

(b)

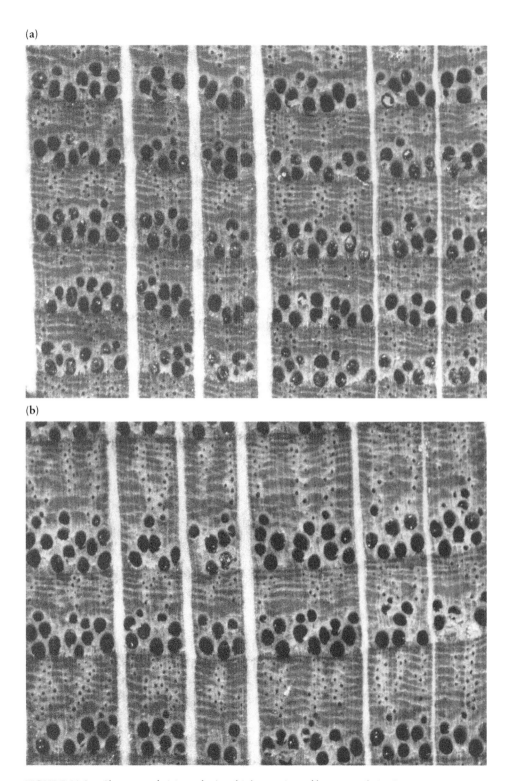

FIGURE 11.2. Slow growth (a) results in a higher potion of large vessels in ring-porous hardwoods than fast growth (b). Red oak (*Quercus* spp.) Source: Courtesy of the Department of Wood and Paper Science, North Carolina State University.

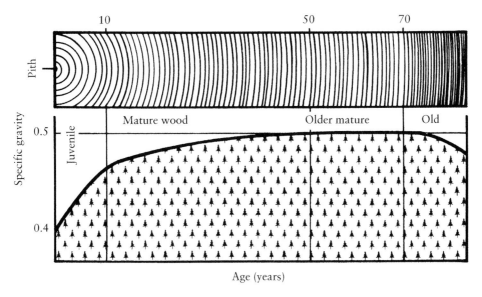

FIGURE 11.3.   Wood density versus tree age for distinct-ring softwood.

Some 30 years after Turnbull's report, Bendtsen (1978) stated that the age effect had not been sufficiently considered by a number of investigators studying the effects of growth rates on density and cited several studies that substantiated this conclusion (Dadswell 1958; Erickson and Arima 1974; Goggans 1961; Spurr and Hsiung 1954; Spurr and Hyvarinen 1954; Wardrop 1951). He summarized reservations to the notion of a growth rate–density relationship in distinct-ring softwoods by pointing out that "problems arise when growth rate–wood property relations are studied independently from other factors that affect wood. For example, if the specific gravity of fast and slow growth pine trees of *similar diameter* is measured, the slow growth trees will generally have a higher specific gravity because of the consistent pattern of increasing density from pith to bark associated with age. If the age effect is ignored, growth rate can be wrongly interpreted as the causal factor of differing specific gravity. Actually, if the specific gravity of the fast-growth trees is compared to that of the slow growth trees over the same number of annual rings, counting from the pith, they tend to be similar." A number of subsequent studies have supported this observation (Barrett and Kellogg 1984; McKimmy 1986; McKimmy and Campbell 1982; Taylor and Burton 1982). One study (Tasissa and Burkhart 1998) is particularly notable because of the extensive sample on which it was based; in this work, wood from loblolly pine trees originating from 186 plantations in 12 US southern states was analyzed. Trees harvested 12 years after thinning at two different levels of intensity – 30 and 50% of the original basal area – were evaluated and compared with wood from unthinned plots. The results conclusively indicated that thinning did not alter the proportion of the annual ring in earlywood or latewood and as a result had no effect on specific gravity.

The density of diffuse-porous hardwoods and softwoods without distinct growth rings is sometimes affected by growth rate, but results are variable. Radcliffe (1953) reported no relationship between density and growth rate in sugar maple (*Acer saccharum*); Paul (1963) found higher than normal density resulting from fast growth in the same species. Studies dealing with yellow poplar (*Liriodendron tulipifera*) also show mixed results. Paul (1957) compared the specific gravity of old-growth and second-growth yellow poplar and found the faster-grown second-growth material to have an 8% higher specific gravity. More recent studies by Thor and Core (1977) and Fukazawa (1984), however, indicated no relationship between growth rate and specific gravity. Experiments with trees of the genus *Populus* have also shown density occasionally to be higher than normal, unaffected, or lower than normal as a result of fast growth. Hernández et al. (1998) studied fast-growing hybrid poplar (*Populus ×euramericana*) and found correlations between growth rate and physical properties for various clones to be both significant and not significant. In those cases, in which significance was demonstrated, the influence of growth rate on strength was weakly negative. In Fukazawa's work with diffuse-porous hardwoods, an age effect, rather than growth rate effect, was noted in wood density differences. Similar examples could be given for *indistinct-ring softwoods* such as the soft pines, firs, and hemlocks.

Uniformity

Uniformity of growth rate has an effect upon wood structure and density variation both within and between growth rings. Larson (1967, 1969) indicated that lack of uniformity represents one of the greatest wood quality problems facing all wood-using industries. In the manufacture of pulp and paper, for instance, nonuniformity in wood density may result in losses from the overprocessing or underprocessing of some components. Thus, strength and surface qualities of paper products may be difficult to maintain precisely. The use of uniform material, on the other hand, increases the probability of producing paper of uniformly high quality.

Uniform wood is desirable not only for manufacture of fiber products but for solid wood products as well. A case in point is a product such as siding that must accept and then retain a coat of paint. Wood characterized by significant within-ring density variation (that is, by bands of very dense latewood and alternate zones of low-density earlywood) often presents a problem when painted and exposed to the elements. The very dense wood shrinks and swells more than the lower-density wood, thus causing a relative movement between the paint film and wood surface. Thus, the paint film tends to flake off after a period of weathering. Such wood is also difficult to machine to a smooth condition or to peel on a veneer lathe because of differing hardness between earlywood and latewood bands. In addition to within-ring variation, variation between growth rings also causes problems in solid wood products. Difficulties traceable to the presence of juvenile and adult woods in the same timber are discussed in Chapter 6.

Uniformity of wood structure both within and between growth rings is determined to a great extent by growth rate and conditions under which growth occurs. Silvicultural treatments should therefore be prescribed with uniformity as well as the density of wood in mind.

Proportion of Heartwood

In certain solid wood products, the proportion of heartwood is important. As explained in Chapters 2 and the 10, heartwood of some species is less susceptible to attack by fungi and insects the than sapwood and is often less penetrable by preservatives. In products designed for use where insects or decay are potential problems, resistance to attack by wood-destroying organisms significantly influences the value of wood. This is evidenced by the expensive all-heart grades of redwood and some cedars. Therefore, redwood (*Sequoia sempervirens*), cedar (*Thuja* spp.), bald cypress (*Taxodium distichum*), and other wood grown to produce naturally durable lumber for the siding, fence, or patio deck market should be managed with consideration of heartwood formation. The proportion of heartwood can also be important in products where natural color patterns are desirable. A lack of well-developed heartwood in species such as walnut (*Juglans nigra*), cherry (*Prunus serotina*), or rosewood (*Dalbergia nigra*) would negatively affect value.

As explained in Chapter 2, heartwood formation begins at various ages ranging from only a few years of age to 60 years or more. In species with naturally durable heartwood, the least durable heartwood forms first. Durability increases from the center of the tree outward until some degree of maturity is reached (Dadswell 1958). Moreover, the proportion of heartwood varies directly with tree age (Hillis and Ditchburne 1974) and inversely with growth rate (Nepveu 1999), but it is negatively impacted by very close tree spacing in young forest stands (Gominho and Pereira 2005). These relationships indicate that short *rotation ages*, very fast growth rates, or both should be avoided where maximum production of heartwood (particularly durable heartwood) is desired.

Fiber Length

Fiber length has an effect upon a number of pulp and paper properties, including tear resistance and tensile, fold, and burst strength. In most cases, long lengths are desirable.

In both hardwoods and softwoods, rotation age and growth-accelerating treatments such as fertilization and irrigation have an effect upon the average fiber length in a stem. Rotation age affects fiber length for two reasons. First, juvenile wood near the core is made up of shorter fibers than those in subsequently formed wood. Therefore, unless a stem is grown to a relatively large diameter, so as to minimize the proportion of juvenile wood, the average fiber length will be small. Second, fusiform initials of the cambium continue to grow in length for many years. The result is that with increased age, longer fibers are formed from these initials with fiber lengths at age 15–20 years of age 2–2.5 times longer than at age 1.

Fertilization and irrigation treatments affect fiber length differently in hardwoods than in softwoods. Numerous experiments with southern yellow pine have shown reduced fiber length as a result of intensive culture (Klem 1968; Posey 1965; Youngberg et al. 1963). The same relationship has been documented for jack pine (*Pinus banksiana*), balsam fir (*Abies balsamea*), black spruce (*Picea mariana*), and white spruce (*Picea glauca*) i n Canada (Fujiwara and Yang 2000). In contrast, studies involving hardwoods almost invariably show longer fiber lengths after growth stimulation (Einspahr et al. 1972; Fujiwara and Yang 2000; Kennedy 1957; Kennedy and Smith 1959; Maeglin et al. 1977; Murphey and Bowier 1975). Whether it is the increase in growth rate resulting from intensive culture or plentiful nutrients and water that more greatly affects fiber length is uncertain. For example, findings correlating fiber length and growth rate (Bissett et al. 1951) are contradicted by more recent studies that indicate little or no correlation (Zobel et al. 1961, 1969).

Although more information is needed, it is nonetheless clear that fiber length is affected by forestry practices. This factor should therefore be considered whenever product quality is significantly related to fiber length.

Occurrence of Juvenile Wood and Reaction Wood

The proportion of juvenile wood that develops in a stem is related to the growth rate at a young age. Softwood stems grown rapidly during the juvenile period will have a relatively high proportion of juvenile wood as compared to those grown more slowly early in the rotation period. Not only do wide rings near the stem center mean that a substantial amount of wood is formed during the normal juvenile period, but also steps taken to stimulate growth at an early age may even cause an extension of the juvenile period (Larson 1969; Megraw and Nearn 1972; Zobel et al. 1972). This notion was later supported by Bryant (1984) when he observed that in slower-growing trees an optimum or high average density is reached at an earlier age than in more rapidly growing trees. Recent research suggests that the period of juvenile wood may be governed by more than simply genetics and growth rate (Kučera 1994; Peszlen 1995). Kučera, for example, hypothesized that the period of juvenile wood formation is closely linked to height growth, with transition to mature wood coinciding with slowing of height growth as crown closure occurs.

Bendtsen (1978) reported that the combined effects of early accelerated growth can be quite significant. He explained that it is not unusual for improved trees to reach sawtimber size in 15–25 years, depending upon the species and locality of growth. He indicated that loblolly pine trees of that age may contain 75% or more juvenile wood. A later study produced estimates of 60% juvenile wood in loblolly pine at age 20, and as much as 80% juvenile wood in cottonwood of the same age (Bendtsen and Senft 1986). In contrast, trees of similar dimensions from a natural forest, growing more slowly, contain only a small fraction of juvenile wood when they reach merchantable size. Briggs and Smith (1986) described a theoretical 0.9 m (36 in.) diameter tree that had grown at a uniform rate of 30 rings per inch and had produced juvenile wood for the first 20 years. The wood of this 540-year-old tree would be less than 5% juvenile wood. In comparison, a 36-year-old, 0.9 m (36 in.) diameter tree grown at a uniform rate of two rings per 2.5 cm (1 in.) would contain over 60% juvenile wood. Trees like the latter example could become increasingly common in the southern United States in the coming decades as fast-growing pine plantations replace slower-growing, naturally regenerated forests, especially if forest rotations as short as 30 years are employed; in this region, it is inevitable that much more juvenile wood volume will be harvested in the future. In any event, the timing of silvicultural treatments is obviously important if the occurrence of juvenile wood is to be minimized.

Reaction wood formation is apparently also affected by growth conditions. As explained in Chapter 6, the juvenile core often contains an appreciable amount of reaction wood. Therefore, rapid growth at a young age increases the proportion not only of juvenile wood but also of reaction wood. Several investigators have reported reaction wood formation in rapidly grown wood even without lean occurring in the stems

(Crist et al. 1977; Isebrands and Bensend 1972; Isebrands and Parham 1974; Pillow and Luxford 1937). Timell (1986) stated that a connection between rapid growth and compression wood formation in conifers is not surprising and suggested a linkage to abundant supplies of auxin.

Cellular Composition

In the manufacture of pulp and paper from hardwoods, the relative proportions of vessels, fibers, and rays are important because ray cells and vessels (particularly large vessels) tend to be lost in the pulping process. Vessels that are retained sometimes cause difficulty with respect to surface quality of paper, because their shape is not conducive to development of strong intra-element bonds. "Picking," or lift-off of vessels from the paper surface during printing, is one result (Allchin 1960). Any increase in the number or volume of vessels resulting from growth manipulation is then viewed as undesirable in pulp to be used for many kinds of paper products.

Although determinations have been made for only a few species, studies have generally indicated that the proportion of vessels and ray tissue is unaffected by growth acceleration treatments such as fertilization and irrigation (Einspahr et al. 1972; Saucier and Ike 1972). However, genetics have been shown to significantly affect the mix of various cell types; for example, Crist and Dawson (1975) showed 19% vessels in one clone of *Populus tristis* as compared with 30% in another. There may also be a relationship between genetic characteristics and associated growth rate; Ganchev (1971) found that clones exhibiting high growth rates tended to have fewer but larger vessels.

Knots

From the standpoint of sawlog and veneer log production, size, and frequency of knots is perhaps the single most important aspect of quality. Knots greatly affect both appearance and strength, and because of this their occurrence is a primary factor in the determination of log and lumber grades.

Despite the reality that all trees produce branches (and thus knots), minimizing their development is possible. A number of silvicultural practices can influence knot development, including spacing at the time of planting, the timing of thinnings, treatments to accelerate the rate of growth, and pruning. The selection and breeding of less branchy trees is also a possibility.

Harris et al. (1975) noted the combined importance of density and the size and frequency of knots in structural lumber and suggested that the development of large knots could be compensated for by an increase in density. Working with *Pinus radiata*, they calculated that bending resistance can be maintained even when knot diameters increase 60% if a relatively modest 10% increase in density is also achieved. This work clearly indicates the importance of density to strength and illustrates the substantial effect of knots.

Boyce (1965) wrote that the largest percentage of top-grade lumber and veneer can be grown in the shortest possible time by concentrating cultural and genetic practices on the butt log. He also referred to comments by Krajicek (1959) and Brinkman (1955) indicating that treatment of the second log is more costly and difficult because it is higher from the ground and has more and larger branches. The concept that treatment to minimize knottiness should be concentrated on the butt log is generally accepted today.

Grain Orientation

Grain orientation that is not parallel to the long axis of a stem often results in slope of grain in manufactured products. As explained in Chapter 9, slope of grain can drastically reduce strength. This kind of grain orientation also adversely affects machining properties and the nature of moisture-induced dimensional changes. Grain orientation that is not parallel to the pith, such as spiral grain (see Chapter 2), is thus to be avoided, if possible, in wood to be used for many kinds of products.

A number of investigators have noted an apparent connection between the development of spiral grain and growth conditions. Fortunately, it appears that *intensive culture practices*, (see the section "Intensive Culture" later in this chapter) and irrigation in particular, tend to reduce the occurrence of spiral grain (Boyd 1968; Brazier 1977; Smith et al. 1972). Pruning stems at an early age is also suspected of suppressing the development of spiral grain (Noskowiak 1963). Because intensive culture apparently reduces grain deviation, this factor is a minor consideration when prescribing traditional silvicultural treatments. However, because growth conditions influence grain orientation, this relationship should still be considered when evaluating the merits of new techniques for accelerating growth.

Chemical Composition

The portion of wood weight composed of cellulose, lignin, hemicellulose, and extractable compounds significantly affects pulp yields (Chapter 15). Consumption of cooking liquor, a chemical solution containing lignin-dissolving compounds, used in the chemical pulping of wood, and the potential for bleaching are also affected. Generally, it is desirable to have the largest possible fraction of cellulose coupled with minimal fractions of lignin and extractives.

Because acceleration of growth can affect the proportion of juvenile wood in a stem, this kind of activity can also affect the chemical composition of a stem. As noted in Chapter 6, juvenile wood tends to be low in cellulose and high in lignin compared with mature wood. Acceleration of growth early in a rotation increases the proportion of juvenile wood and can thereby cause an undesirable reduction in the proportion of cellulose. But what happens if growth is accelerated after the juvenile period? Numerous investigations into this important matter have yielded conflicting results, and further work is needed to define the relationships more precisely.

There is some evidence that stand density may influence the chemical composition of wood. For example, Shupe et al. (1995, 1996) found that lower stand densities yielded greater lignin, holocellulose, and alpha cellulose content in wood and that higher stand densities resulted in higher extractive content.

Concepts of Wood Quality

A number of factors determine the suitability of wood for a specific end use: density, uniformity, proportion of heartwood, fiber length, occurrence of juvenile or reaction wood (or both), cellular composition, presence of knots, grain orientation, and chemical composition. The combined effect of these factors determines wood quality; the importance of each is dependent upon the intended use.

To the forester, the fact that quality is determined by a number of factors is matched in importance by the fact that silvicultural practices affect virtually all of them. The net effect of specific forestry practices upon wood quality is the subject of the following section.

Growth Manipulation and Wood Quality

The amount of space in which a tree grows is an extremely important determinant of growth rate and thus of wood properties. The spacing between trees and the extent of surrounding vegetation defines the degree of competition for such critical growth elements as nutrients, water, and sunlight. When competition is slight, crowns and root systems can develop fully because these critical elements are not limiting factors. On the other hand, when crowding occurs, trees must compete intensely for the elements needed for growth. For example, consider sunlight-tolerant seedlings that have taken root in a clear-cut area. They first must compete with weeds and shrubs. Young trees that develop quickly have the best chance of survival. Others may remain stunted and eventually die. Trees that do become established face a new battle – with each other. If a large number of trees survive the first stage of competition with other vegetation, crown development will quickly fill the open space between trees. As competition for sunlight, nutrients, and moisture intensifies, growth of the entire stand slows. Despite the slow growth, slight differences in the rate of height extension will eventually begin to favor a few trees, giving them a significant advantage over their neighbors. This slight advantage is magnified as success in competition for sunlight improves growth rate even more. Suppressed trees are left further and further behind, and the weakest fail. Even the trees that grow fastest lose branches through natural pruning caused by limited light from lateral crowding; such crowding may eventually diminish the capacity for wood production.

The scenario described above serves to illustrate one role of the forester: to minimize the effects of competition and thus maximize the rate of growth. For example, fertilization and irrigation can be applied so that critical elements are present even in densely stocked stands. Another approach to control of competition levels is manipulation of the stand to ensure that it is in balance with naturally occurring growth elements. By spacing trees widely at planting time and then controlling competing vegetation, early growth can be accelerated greatly. Later, timely thinnings can reduce competition so that remaining stems continue to grow at a fast pace. Such techniques have been used successfully for centuries. However, growth manipulation affects wood properties. The questions to be addressed now are what properties are affected, in what ways, and by how much.

Spacing Trees at Planting Time

Assuming that nutrient and water availability is limited and maximum sunlight is desirable for growth, widely spaced trees will grow faster than crowded ones. When this relationship is combined with the knowledge that growth rate in some tree species is related to wood density, and properties such as strength and dimensional stability are closely correlated with wood density, it is easy to see how the spacing of trees can affect wood properties.

By increasing the growth rate, a wide spacing of young ring-porous hardwoods tends to maximize the density and thus the strength of these trees. However, the effect of accelerated growth upon the density of softwoods with distinct latewood is not as clear. For these species, then, the effect of wide initial spacing is open to question. Some would argue that density is decreased. Others would predict no effect, as discussed earlier. There are conflicting findings on the question of whether spacing affects the period of juvenile wood formation. Yang (1994), when working with white spruce (*P. glauca*), found more growth rings of juvenile wood at a 3.7 m × 3.7 m (12 ft × 12 ft) spacing than when trees were spaced more closely. He did not, however, find this effect in black spruce (*P. mariana*). Clark and Saucier (1989) studied juvenile to mature wood transition in southern pine under different plantation spacings and concluded that spacing had no effect on the age of transition. They did acknowledge that wide initial spacing (4.6 × 4.6 m, or 15 × 15 ft) led to a significantly larger juvenile core (16 cm or 6.3 in.) than a narrower spacing (a 1.8 × 1.8 m or 6 × 6 ft spacing resulted in an 11 cm or 4.3 in. juvenile core). One point upon which research findings are in almost total agreement is that by maximizing ring width in the early years, the size of the juvenile core is also enlarged. For instance, one study of plantation loblolly pine, in which aggressive treatment of competing vegetation was combined with high annual rates of nitrogen fertilizer application, showed a 62% increase in the size of the juvenile core as compared with controls (Clark et al. 2004). Because juvenile wood is low in density, a large juvenile core will in itself reduce average stem density. Thus, although open-grown stands of distinct-ring softwoods will produce merchantable-size trees in a shorter time than dense stands, the average annual production of fiber per acre – or of woody biomass – will not necessarily be greater, because average density may be less.

Rapid early growth, resulting in a large core of juvenile wood, is an important consideration regardless of the species involved. Should a large juvenile core be developed so that later growth of mature wood occurs in the form of the largest possible cylinder? Or should early growth be subdued to minimize the size of the juvenile core? Although this point has been debated, no clear answer has emerged. The answer appears to vary, depending upon management objectives and product alternatives. Kellogg and Kennedy (1986) shed some light on the spacing dilemma. They agreed that planting to an initial spacing such as 2.4 m (8 ft) and then thinning later would result in a minimal juvenile core. They also noted, however, that unless there was a commercial market for the thinned material – and in many regions, there was not – the initial narrow spacing/thinning strategy made the remaining trees at least twice as expensive as planting to a wider spacing. More recent research suggests little cause for concern related to tree spacing at the time of planting. Spacing has been found to not affect the juvenile period in at least three species: Norway spruce (Kučera 1994), black spruce (Yang 1994), and Douglas-fir (Gartner et al. 2002).

Another result of wide spacing is a greater proportion of heartwood. Gominho and Pereira (2005), for example, found considerable differences in heartwood proportion and height in nine-year-old eucalyptus planted at 2 × 1 m (6.5 × 3.2 ft) versus 3 × 3 m (9.75 × 9.75 ft) spacing. Trees spaced at 2 × 1 m had 20.3% heartwood that extended to 45% of tree height at age nine years, compared with 41.1% of heartwood that extended to 62% of tree height for trees planted at a 3 × 3 m spacing.

One direct result of growth under wide spacing is that knots are larger and more prevalent than if growth occurs under crowded conditions; the availability of light reduces natural pruning and tends to cause branches to attain relatively large diameters. Thus, when harvesting and converting open-grown trees to solid wood products, more material is wasted than when using wood from more densely grown stands. One study of a situation in which a commercial market for thinnings was available (Clark et al. 1994) showed significant financial benefits from close initial spacing and late thinning. Loblolly pine, thinned at age 18 and at five-year intervals thereafter to age 38, was shown to yield more than 60% no. 2 and better lumber when planted at a 1.8×1.8 m (6×6 ft) spacing and later thinned to about 23 m^2 ha^{-1} (100 ft^2 per acre), but less than 42% no. 2 and better when planted at 3.6×3.6 m (12×12 ft) spacing and thinned to the same basal area. Another study (Zhang et al. 2002) examined the effects of tree spacing on lumber grade yield for black spruce. A comparison of stand densities at 3086, 2500, and 2066 trees ha^{-1} showed a steady increase in branch diameter with decreasing stand density, but little impact on structural lumber grade yield or lumber strength. However, at the lowest stand density (1372 trees ha^{-1}) lumber grade yield, strength, and stiffness were considerably lower than in trees from higher-density stands; at the lowest stand density only 6.2% of lumber met design values for standard visual grades as compared with 26.7% for the highest stand density. Lumber strength and stiffness were 20 and 17% lower for wood from the lowest stand density compared with the highest. Overall, the stiffness of lumber from the black spruce plantations studied was 28.9% lower than average values for lumber originating in natural black spruce stands. Similar results were reported by Chuang and Wang (2001) and Ishiguri et al. (2005) for Japanese cedar (*Crytomeria japonica*) and by Wang et al. (2005) for Taiwania (*Taiwania cryptomerioides*). In the Chuang and Wang study, the quality of standing trees in stands with different tree spacings was assessed using stress-wave and ultrasonic-wave methods; regardless of the testing method used, trees grown under what is described as medium and poor growth conditions had significantly better strength properties than trees grown under superior growth conditions (i.e. in trees grown more rapidly). Wang et al. evaluated lumber yield from trees grown in four different stand densities using visual grading and found that the highest yield of first and second grades (62.5%) came from logs originating in the highest-density stand (6940 trees ha^{-1}), with progressively lower yields as stand density decreased. The yield of lumber of first and second quality in the lowest stand density studied (1000 trees ha^{-1}) was 41.6%.

Wide spacing of trees in replanting of open areas can result in higher proportions of juvenile wood and heartwood, larger and more frequent knots, and somewhat greater stem taper than that which develops with closer spacing. When growing trees for lumber production, it may be better from a wood quality standpoint to manage for relatively slow growth for the first 5–10 years, with steps taken to improve growth rate thereafter. The practical closeness of planting is largely dependent upon a combination of planting costs, later response to thinning, markets for material removed in the thinning process, and the future impact of quality considerations on the value of sawlogs and veneer logs. Forest health issues also influence initial spacing if thinning is not done on a timely basis. For example, pine bark beetle risk increases as stand basal area increases.

Thinning

Trees remaining when surrounding trees are removed by thinning or partial cutting respond to the more open environment by stimulated crown development and the formation of wider growth rings along the bole. Effects upon growth are often dramatic.

One study of slash pine (*Pinus elliottii*) in Georgia indicated the benefits of thinning. A yield of 251 m³ ha⁻¹ (28 cords per acre) was obtained at age 23 for a plot that had been thinned at age three, as compared with a yield of less than 45 m³ ha⁻¹ (5 cords per acre) in an adjacent unthinned plot (Jones 1977). In addition to increased yield at harvest, another benefit of periodic thinning is that material normally lost through natural mortality is salvaged.

A note of caution is in order at this point with respect to the results of thinning and other silvicultural treatments. Larson (1972) indicated that it was important to distinguish between very young and mature trees when measuring the results of silvicultural practices, pointing out that young and mature trees may react quite differently to the same treatments. An example of this is provided by Phelps and Chen (1991), who in working with white oak (*Quercus alba*) observed significantly higher diameter growth rates after thinning in intermediate-size trees but little change in codominant trees. Cutter et al. (1991), also working with oak, did find that thinning affected growth rates in understory trees as well as dominants and codominants, but noted that after a period of subsequent growth, thinning primarily increased the yield of the lowest grades of lumber in understory trees, whereas thinning resulted in significant increases in upper grades of lumber in dominant and codominant trees. The impact of thinning of hardwoods has also been found to be dependent on the degree of thinning (Mitchell et al. 1988). Moderate to heavy thinnings in oak (that is, to 40–50% stand stocking levels) generally yielded the greatest increases in wood production.

Because of the impact on crown development and growth rate, thinning may adversely affect some wood properties. These effects can be minimized, however, because much of the impact on wood properties is traceable to the timing of thinnings. Done early in the life of a stand, thinning can have the same result as wide spacing in a plantation. One effect of early thinning is an increase in the size of the juvenile core through greater ring width in this zone. Early thinning can also cause the juvenile period to be extended, resulting in a broader transition zone. To the extent that the proportion of juvenile wood in stems is increased as a result of thinning, wood can be expected to have lower density and strength, shorter fibers, higher longitudinal shrinkage upon drying, and a greater proportion of lignin. Delay of thinning until after the juvenile stage would avoid these problems.

Whether or not a tree is in the juvenile stage, thinning decreases the tendency for upward crown recession and may stimulate the growth of existing branches as well as development of new ones. The results are increased knot development and stem taper. The tendency for resurgence of branch growth is most serious in very young softwood stands in which lower branches are still alive. In thick stands of pine, lower branches will usually be dead by the time pole size is attained. Because pine cannot regain lost lower branches, there is no chance to revitalize branch growth by thinning; thus, trees can never revert to an open-grown form. The same is true of many hardwoods as well, but some species tend to produce new branches (called *epicormic branches*) in the lower stem if competition within a stand is reduced.

Middleton et al. (1995) addressed the issue of stem taper and knot development in a study of lodgepole pine (*Pinus contorta*) and found that although lumber recovery increased in general with increasing tree diameter classes, lumber recovery was lowest in trees from stands containing few trees per ha; such trees were highly tapered and knotty. A significantly higher combination of lumber grade and yield was obtained from trees in medium-density stands (1000–1200 trees ha^{-1} or 400–500 trees per acre) as compared with trees from a stand density class of 700 trees ha^{-1} (300 trees per acre).

Thinning has been found to have an effect upon the development of compression wood. Working with ponderosa pine (*Pinus ponderosa*), Barger and Folliott (1976) discovered that thinning or partial harvest cutting can substantially increase the incidence of compression wood in remaining trees. They found that compression wood was more likely to form in leaning stems following thinning than before this treatment. Because of this, they suggested that guidelines for thinning should reflect a critical appraisal of stem form irregularities and lean in residual trees.

Thinning can affect the percentage of latewood in subsequent growth rings, although the degree of effect depends upon the age and prior history of the stand as well as the degree of release (Larson 1973). Larson indicated, for instance, that thinning of a very dense and stagnated stand of distinct-ring softwoods can be expected to result in an immediate and substantial increase in growth accompanied by a marked decrease in the proportion of latewood and thus density (suggesting a strong growth rate–density relationship). A more recent study by Koga et al. (2002) examined the effect of precommercial thinning of balsam fir (*A. balsamea*) on radial growth ring width and wood density. In this case, radial growth increased following thinning, with the growth response limited primarily to earlywood. Wood density of both early- and latewood was unaffected by thinning, meaning that the increase in earlywood growth rate resulted in a decrease in wood density. On the other hand, several studies of the distinct-ring softwoods larch and pine indicated either no change or an increase in the wood's specific gravity after the thinning of older stands (Choong and Fogg 1989; Lowery and Schmidt 1967; Moschler et al. 1989; Nicholls 1971). In several of these studies, it was noted that reduced competition served to reduce late-season depletion of soil moisture, thereby prolonging the latewood portion of the growing season. Moschler et al. (1989) also observed that heavy thinning tended to decrease the density of earlywood and increase the density of latewood. The implication is that although overall wood density may change little, if at all, wood quality may be nonetheless negatively impacted by thinning owing to a decrease in wood uniformity. The same effect has been noted in trees genetically selected for fast-growth characteristics (see the section "Genetic Improvement and Wood Quality," later in this chapter). An indication of the response of hardwoods to thinning is provided by a study involving ash (*Fraxinus* spp.), a ring-porous hardwood. In this case, specific gravity was found to increase only 1% as a result of thinning compared with that of wood taken from dense, unthinned stands (Paul 1963). The trees examined ranged from 50 to 65 years of age, with thinning done at age 30.

A study of Norway spruce found an increase in microfibril angle in the S_2 layer of the cell wall as a result of thinning-induced acceleration of growth (Herman et al. 1999). The Herman et al. findings are tempered by a subsequent study (Chiu et al. 2005) of *Taiwania* that found no effect of thinning on either tracheid length or microfibril angle. Whether thinning-induced growth rate impacts on microfibril angles is highly species dependent or more generally influenced by specific site and growth conditions remains to be resolved. Measurements of microfibril angles in selected growth rings in both mature and juvenile wood zones revealed significantly larger S_2 microfibril angles (29% on average) in wide growth rings of Norway spruce than in narrower growth rings formed prior to thinning (21%), suggesting the likelihood of greater warping owing to longitudinal shrinkage.

As outlined above, a strong argument can be made for delay of thinning until the juvenile period is completed if lumber or other solid products are to be produced from the resulting wood. This strategy avoids the development of an unduly large juvenile core and results in the formation of uniformly dense mature wood. Delay of thinning also minimizes knot development in the lower bole. In contrast, in at least some species there appears to be a critical period for thinning beyond which release seems to have little stimulus on crown enlargement. Furthermore, by delaying thinning and restricting diameter growth at a young age, later growth will be formed over a smaller core. It might also be added that pulpwood quality is not necessarily decreased by rapid early growth. The forester must weigh these factors against wood quality considerations in establishing a thinning schedule.

Fertilization and Irrigation

The forestry practices discussed thus far – spacing trees at planting time, thinning, and control of understory vegetation – are designed to maximize the potential for growth, given the limitations of a particular site. Another strategy that can be used to achieve increased rates of growth involves modification of the site through such cultural practices as fertilization, irrigation, and a combination of the two. The objective here is to improve site quality. Thinning, for example, reduces competition for available nutrients, water, and sunlight and concentrates growth on fewer stems. Application of fertilizer, on the other hand, stimulates growth by increasing the availability of nutrients generally, thereby enhancing crown development and the size of photosynthesizing surfaces. Both the size and the number of leaves in the crown are affected. Because the growth of all vegetation on the site is stimulated, it is usually necessary to employ thinning, understory control, or both in conjunction with a fertilization program. The use of fertilizer may serve to improve the growth potential of an already productive site or bring a nonproductive or nutrient-deficient site into productivity. Similarly, irrigation can be used on moisture-deficient sites to improve growth potential.

Effects of Fertilization on Wood Quality.   Foresters and users of wood have long been concerned about the effects of fertilizer and water application on wood quality. Larson (1973) lent some perspective to this concern, pointing out that these techniques simply improve the site. He went on to say that all too often fast-grown trees from better sites were readily accepted while the quality of similar wood produced by improved growth conditions was questioned.

Changes in wood properties resulting from fertilization or irrigation-induced fast growth are similar to those occurring as a result of increased growth rate associated with other practices. Wood density, for example, is affected according to the general rules given earlier: fast growth may cause a density increase in ring-porous hardwoods and a variable effect upon density in other woods.

The impact of fertilization upon wood density of softwoods can be substantial. Posey (1965) found that fertilization of a loblolly pine (*Pinus taeda*) stand with 179 kg N, 90 kg P, 90 kg K ha^{-1} (160 lb N, 80 lb P, 80 lb K per acre) caused wood specific gravity to decrease from 0.48 to 0.39 and the proportion of latewood to decrease from 47 to 36%. The almost 20% drop in specific gravity was sizable and represented a decrease of about

$90\,kg\,m^{-3}$ of solid wood (475 lb of wood fiber per cord of wood). However, Posey's figures also showed that rate of growth increased almost 50%, with the result that the increased rate of growth more than compensated for the decrease in density. Similar results have been found in numerous studies dealing with the fertilization of distinct-ring softwoods including Douglas-fir (Erickson and Lambert 1958; Parker et al. 1976; Siddiqui et al. 1972), slash pine (Gooding and Smith 1972; Williams and Hamilton 1961), red pine (*Pinus resinosa*) (Gray and Kyanka 1974), spruce (Klem 1964; Weetman 1971), loblolly pine (Clark et al. 2004) and radiata pine (Downes et al. 2002). The average wood density loss noted in these studies was about 6–10%. Although such studies have clearly indicated an inverse relationship between fertilization and density in distinct-ring softwoods, the magnitude of density decrease resulting from large-scale application may have been overestimated. Losses of less than 2% have been realized in commercial fertilization of spruce in Sweden (Hagner 1967). Larson (1973) indicated that density decreases of this order may be more realistic, because fertilizer application rates on experimental plots are commonly higher than those recommended for commercial practices. Moreover, when care has been taken to examine growth ring width–wood density relationships on a growth ring-by-growth ring basis, the growth rate has been found to be independent or very weakly correlated with wood density (Nyakuengama et al. 2002).

Zobel and Talbert (1984), after a review of findings by a number of investigators, found that heavy nitrogen fertilization, in particular, caused decreases in wood specific gravity. They also indicated that some species produced wood of very high specific gravity when grown on phosphorus-deficient sites; in this case, addition of phosphorus to the site resulted in a lowering of wood specific gravity to normal levels in subsequent growth. The same results were obtained with radiata pine in New Zealand (Nyakuengama et al. 2002). These observations are interesting because they raise the possibility of adjusting fertilizer formulations to achieve wood quality as well as growth rate objectives.

Klem (1974) commented on the effect of different fertilizer application rates, saying that his work had shown that varying the types and amounts of fertilizers did not influence density of wood differently as long as treatments caused the same increase in growth rates. This statement implies that although density decreases associated with commercial use of fertilizer may be relatively small as compared with experimental sites, the cause of this is that the increase in growth rate relative to the potential is small as well.

Along with a slight to significant decrease in density, fertilization has also been noted to cause greater within-ring uniformity of density, a phenomenon that Zobel and Talbert (1984) indicated was one of the most desired benefits of fertilization. Gladstone and Gray (1973) found an increase in cell wall thickness of earlywood tracheids and a simultaneous thinning of latewood tracheid walls in fertilized red pine and Douglas-fir. The same phenomenon has been observed in other studies dealing with Douglas-fir (Megraw and Nearn 1972; Parker et al. 1976), loblolly pine (Clark et al. 2004), and Japanese larch (*Larix leptolepis*) (Isebrands and Hunt 1975). Larson (1973) explained this by pointing out that there was a transition zone between the earlywood and latewood consisting of tracheids that were neither true earlywood nor true latewood but which most closely resembled earlywood cells. He indicated that the transition zone was broadened in response to accelerated growth, resulting in lower proportions of true latewood (and thus lower density) and a generally more uniform cell structure within the ring. Klem (1967) concluded that all wood properties become more uniform after fertilization.

Examination of Figure 11.4 shows that response to fertilization is greatest immediately following treatment. As time passes, within-ring density variation gradually returns to the pretreatment pattern. It appears that the effect of fertilization upon wood quality generally lasts only three to five years.

There is some disagreement as to whether fertilization of softwoods has an effect upon tracheid length; some investigators have found a 5–10% decrease (Gray and Kyanka 1974; Posey 1965), while others have noted a slight increase in the average length after fertilization (Manwiller 1972). A fertilization effect on microfibril angle, with increased microfibril angle following fertilization, has also been reported (Downes et al. 2002). Slight decreases in the *holocellulose fraction* (consisting of cellulose and the hemicelluloses) and small increases in the extractive content of rapidly grown wood have also been reported.

The impact of fertilization on wood quality in hardwoods has been less widely investigated than effects upon softwoods. However, it appears that density is little affected by rapid growth induced by fertilizer. Studies with a number of diffuse-porous woods have shown no significant density effects from fertilizer treatments. The lack of any relationship has been noted in sycamore (*Platinus occidentalis*) (Saucier and Ike 1972) and yellow poplar (*L. tulipifera*) (Thor and Core 1977). Slight decreases in wood density have been observed in fast-grown quaking aspen (*Populus tremuloides*) (Einspahr et al. 1972; Kennedy 1968). The density of ring-porous woods tends to be slightly increased by fertilization, as it is by thinning. Slightly increased specific gravity in trees from fertilized stands has been shown in tests involving red oak (*Quercus* spp.) (Szopa et al. 1977) and red oak and white ash (*Fraxinus americana*) (Mitchell 1972).

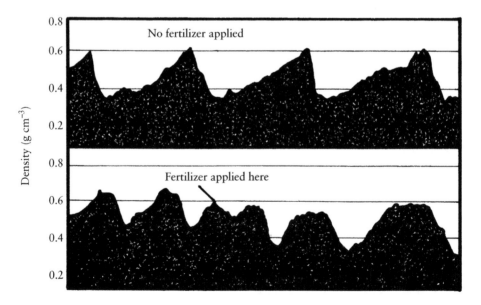

FIGURE 11.4.   Acceleration of softwood growth broadens the earlywood-to-latewood transition zone and decreases wood density. Source: Parker et al. (1976). Density profile is for radiata pine (*Pinus radiata*). Peaks in each profile correspond to latewood produced in each growth cycle. Source: From Larson (1973).

Investigations of fertilization effects upon hardwood quality have dealt with the impact upon vessel-to-fiber ratio, fiber length, and intrinsic values such as machining properties. A review of a number of such studies showed no change in vessel-to-fiber ratios and only a slight increase in the proportion of volume occupied by vessels, owing to a tendency for rapid growth to slightly increase vessel diameter. Fiber lengths have been found to either remain unchanged or increase slightly.

As with thinning, fertilizer applied to young stands results in a large juvenile core because of increased ring width and extension of the juvenile period. Additionally, such treatment often increases the size of individual branches and delays natural pruning. Timing is therefore important when fertilizing, just as it is in thinning. Carmean and Boyce (1974) pointed out that although branch size was increased on better sites, the internode distance between branch whorls also increased. Thus if fertilization treatments are combined with a pruning program, log quality can be markedly improved.

Whereas the preceding discussion refers to experiments with fertilization, most of the relationships also apply to wood formed during irrigation or under the combined effects of fertilization and irrigation; that is, irrigation results in moderate changes in cell dimensions, often results in increased uniformity of density, and can cause extension of the juvenile period. An exception is the effect of irrigation upon density levels. A number of studies have shown increases in the latewood percentage and thus density as a result of irrigation (Chalk 1951; Howe 1968; Paul and Marts 1931; Posey 1965). This increase is thought to be from reduced competition for soil moisture during late-season growth.

Effects of Fast Growth on Product Properties.   A major consideration in the use of intensively grown wood is how growth rate affects properties of products. In solid wood products, strength is important. Rapidly grown wood has been evaluated for this quality by several investigators. They found that although some aspects are diminished by accelerated growth, others are improved. For instance, in studies of wood from fertilized versus unfertilized red pine, Gray and Kyanka (1974) noted significant decreases in work to proportional limit and work to maximum load with accelerated growth (see Table 9.1 and Figure 9.3). They also found significant increases in modulus of elasticity (MOE) and only minimal changes in modulus of rupture (MOR) and *fiber stress at proportional limit* (fspl) in wood from fertilized trees. (The significance of these properties is discussed in Chapter 9.) These trees were 16 years of age at the time of treatment and 36 years old at the time of harvest.

Other experimentation (Hsu and Walters 1975) revealed complex relationships between silvicultural treatments and strength in loblolly pine treated at age 18. For example, heavy fertilization of a wet plot yielded wood having an fspl of 26,000 kPa (3790 psi) as compared with 49,000 kPa (7120 psi) for wood from heavily fertilized but drier sites. The MOE and MOR were also affected. Strength values of wood from the fertilized but dry sites tended to slightly exceed accepted average strength values for the species involved. In contrast, strength figures determined for wood from wet, fertilized sites were only about one-half the accepted values. However, no effect of treatment upon work to maximum load or work to proportional limit was found. This study further indicated that fertilization reduced the fspl, MOE, and MOR, whereas irrigation reduced only the MOE. Fertilization and irrigation together were noted to affect strength values more than either treatment individually.

Murphey and Brisbin (1970) also found that a combined fertilizer-irrigation treatment had a marked effect upon strength of red pine. The equivalent of 50 mm week^{-1} (2 in. week^{-1}) of sewage plant effluent was found to adversely affect almost all mechanical properties, whereas a 25 mm week^{-1} (1 in. week^{-1}) application showed no such effects. They concluded that adverse effects resulted from an abnormally high soil moisture content and recommended that excess irrigation be avoided with species sensitive to wet sites. Thus, the strength of mature softwoods can be affected greatly by silvicultural treatment. The results are not always bad, however, and appear controllable with the proper treatment mix.

Because current grading rules for some lumber products limit maximum growth rates by specifying a minimum number of growth rings per inch, Koch (1972) urged caution in the development of fast-growth southern pine (less than or 1.2 rings cm^{-1} or 3 rings in.$^{-1}$). He suggested that mechanical properties as well as paint retention and gluing potential would be adversely affected by accelerated growth. He indicated that strength would be inadequate in wood used for small poles or squared timbers if harvesting was done too early – that is, if rotation age was too short. Koch calculated that the strength of large timbers would probably be adequate, although reduced, but said that such material would be too low in strength for use in the outer layers of large laminated beams. Koch's concerns regarding lumber were substantiated by Fight et al. (1986). Biblis et al. (1995) and Biblis and Carino (1999) have examined the rotation age question through strength testing of wood from loblolly pine plantations of various ages. These investigators examined the lumber yield by grade from rapidly grown logs that met dimension grade requirements for no. 2 and no. 3 common grades and found it to be only one-fifth to one-half that obtained from more slowly grown material. Biblis et al. (1995) found that only 52–64% of lumber from 35-year-old trees met stiffness values required in existing lumber standards. They concluded that even dense stands of loblolly pine must be older than 35 years at harvest to ensure that a high percentage of lumber attains the desired strength properties. Biblis and Carino later concluded that a harvest age of 50 years for loblolly pine was likely to yield a high proportion of lumber (that is, 94–95%) meeting prescribed strength standards. A similar study of Canadian jack pine showed the same trend, but on a different time scale (Duchesne 2006); in this study 50-, 73-, and 90-year-old stands were evaluated. The results showed the mean annual increment to be greatest in the 50-year-old stand and least in the 90-year-old stand, but with an inverse relationship between mean annual increment and strength. The strength (MOR) of lumber cut from the 50-year-old stand was 16% lower than lumber from the older trees, and the stiffness (MOE) was 16–19% lower.

Another study of red pine examined lumber grade yield from a 57-year-old plantation and a 125-year-old natural stand in eastern Maine. In this case only 9% of lumber from plantation trees graded as select as compared with 31% of lumber from the older natural stand (Deresse et al. 2002). In addition to the problems mentioned by Koch, the presence of juvenile wood is the cause of serious downgrade losses in lumber because of warping upon drying. Longitudinal shrinking of juvenile zones causes a particularly high proportion of rejects when producing small lumber sizes.

Probable difficulties in manufacturing plywood were also cited by Koch. This conjecture was later confirmed by MacPeak and Weldon (1987), who found unusually low yields of upper veneer grades and full-sheet veneers, and marginal stiffness in finished plywood in a mill trial involving logs from small-diameter, rapidly grown trees.

Studies of machining properties of rapidly grown hardwoods have generally shown no adverse effects, nor have significant detrimental effects upon strength been noted. Much the same conclusion was reached following an evaluation of the machining properties of plantation-grown white spruce (*P. glauca*) (Hernández et al. 2001). Although turning and mortising properties were found to be poor, it was noted that higher-density spruce wood machined better than material of lower density.

In both hardwoods and softwoods, problems related to an expanded juvenile core and increased knottiness in wood grown quickly when young appear to be a serious concern when dealing with solid wood products. Other growth effects upon properties of solid products appear to be relatively minor. Regardless of the properties under consideration, it is generally acknowledged that growth rate-related changes in wood quality are smaller in hardwoods than in softwoods (Bendtsen 1978).

Pulp yields from wood produced under accelerated growth conditions are reported to be similar to yields from wood of more slowly grown trees. Effects of accelerated growth upon pulp quality have generally been found to be mixed. For example, a study of 35 botanical families of fast versus more slowly grown Canadian white spruce (*P. glauca*), with differences in growth rate attributable to site, showed significantly shorter fibers in faster-grown material for 33 of the 35 families; the result was lower tear index and pulp yields, but higher sheet densities (Duchesne and Zhang 2004). Negative impacts of fast growth on pulping and paper properties were also implied by Gladstone et al. (1970), who found a 2–7% higher pulp yield from latewood of loblolly pine as compared with earlywood. This finding led to suggestions that growth acceleration resulting in decreased quantities of latewood would cause similar reductions in yield. Subsequent pulping studies of fast-grown slash pine and Douglas-fir have shown pulp yields to range from 1% lower to 2% higher than slower-grown controls (Gooding and Smith 1972; Parker et al. 1976; Siddiqui et al. 1972). On the other hand, many pulp properties improved. Gooding and Smith found, for example, that paper produced from fertilized wood had higher density, burst factor, and tensile strength than that from unfertilized wood. This agreed with the earlier findings of Posey (1965). Posey's report of lower tear strength in paper from fertilized wood was confirmed by Gooding and Smith only for material taken from the basal portion of trees. Siddiqui and his co-workers reported no difference in paper properties between fertilized and unfertilized material except for tensile strength; again, the advantage in tensile strength favored the fertilized material. Similar results were obtained by Hussein et al. (2006), who studied the effect of precommercial thinning on kraft pulping and pulp quality of balsam fir. In this case, a stand thinned at age 12–3000 and 1500 trees ha^{-1} and harvested 35 years later was evaluated. Although tensile strength improved in comparison with unthinned material, pulping yields and fiber length were lower and chemical consumption in pulping higher. With heavy thinning (to 1500 trees ha^{-1}) pulp yield per unit volume was 8% lower than the unthinned stand (10,000 trees ha^{-1}); the overall effect of thinning on volume yield was not reported. Silvicultural treatments involved trees that had passed the juvenile stage in all of these studies. As indicated in Chapter 6, even juvenile material yields pulp with certain properties that are superior to those associated with mature wood. Juvenile pulps tend to exhibit higher burst and folding strength but lower tear strength and opacity than pulps made from adult wood. Even development of a large juvenile core, therefore, is not necessarily undesirable in material that is to be pulped. Experimentation with material from short-rotation, intensively cultured plantations has, in fact, shown that poplar trees harvested after only three to five years yielded satisfactory kraft pulp for use by itself or in a blend with softwood pulp from mature stands (Marton et al. 1968; Phelps et al. 1985; Zarges et al. 1980). The findings of Hussein et al. (2006) notwithstanding, many people, including Snook et al. (1986), have observed that yields obtainable from rapid growth tend to outweigh disadvantages and that pulpwood trees should be grown as fast as is economically feasible.

In conclusion, the impact of fertilization and irrigation upon wood product properties depends upon the products involved and the timing of silvicultural treatments. The most serious problem associated with growth acceleration is an enlarged juvenile core. This problem is particularly important when solid wood products are involved. It is sufficiently serious that delays in growth-stimulating treatments until after completion of the juvenile period may be warranted. A practice of growing trees over a relatively long rotation so that the proportion of juvenile period in the stem is minimized may also be a good idea if solid products are the objective. Where trees are being grown for pulpwood, the importance of minimizing juvenile wood depends upon specific properties desired in the paper. Advantages of increases in the rate of wood production in many cases offset disadvantages related to modified pulp properties.

Pruning

Pruning is the practice of trimming branches from chosen portions of standing trees to reduce the occurrence of knots in subsequently produced wood. In softwoods, when a branch is removed from the bole of a tree, the sheath of new growth will eventually cover the stub, producing knot-free wood thereafter. Such wood has markedly higher value than knotty wood for solid wood products and veneer because of increased strength and improved appearance. Considerable improvements in lumber grade and veneer grade yields can result from pruning (Cahill 1991). In hardwood species, pruning does not necessarily result in higher-quality or higher-value lumber (Dwyer and Lowell 1988; Lowell and Dwyer 1988).

It is perhaps obvious that branches should be trimmed as close to the bole as possible. The closer the cut, the sooner new growth will overtake and cover the stub. Calvert and Brace (1969) determined, for example, that pruning branches to produce 0.6 cm (0.25 in.) rather than 1.9 cm (0.75 in.) stubs would cut healing time in half (from 10 to 5 years). They also pointed out that 1.3 cm (0.5 in.) branch stubs were not uncommon where pruning was done more than 3.7 m (12 ft) from the ground. Advice regarding trimming branches as close to the bole as possible must be tempered with the observation that pruning should never involve cutting into the branch collar. The branch collar is a swollen area of the main stem around the base of the branch.

Ideally, the knotty core of a tree should be as small as possible. As noted by Cahill et al. (1988), the size of the knotty core can be minimized by pruning trees while they are young, a practice that also allows maximum growth of clear wood and produces small limb scars that heal quickly. However, several investigators have determined that pruning larger trees yields the most immediate payoff. Calvert and Brace (1969) recommended pruning only second-growth pine trees 0.20–0.25 m (8–10 in.) in diameter and larger, pointing out that even slower-growing large trees tended to produce a greater volume of wood than faster-growing smaller ones. Brown (1965) had earlier come to a similar conclusion. Cahill et al. did not specify a minimum diameter for

FIGURE 11.5 Thinned, pruned New Zealand radiata pine at age 13.

pruning treatment but did state that it was important from an economic viewpoint to concentrate treatments on trees that could be recognized as likely survivors until the time of harvest. This criterion in itself suggested a lower diameter limit for pruning that was above pulpwood size. Recommendations that pruning be restricted to larger trees must be balanced against the reality that many years pass before healing of pruning wounds occurs (Figure 11.5).

Genetic Improvement and Wood Quality

The field of genetics offers perhaps the greatest potential for improvement of wood yield and quality in the future. Through the last half of the twentieth century efforts in this area have concentrated on identifying trees in the natural forest that exhibit superior growth or form. Seeds have been collected from these trees and planted in nurseries to raise new generations of trees having many of the same characteristics. Vegetative propagation has also been employed in reproducing clones of certain species. A relatively recent development is the perfection of the technique of tissue culture, whereby a tiny portion of parent tissue is placed in a test tube with the proper chemical medium and grown to a small plantlet, which is then transferred to a nursery. Even more recent is the application of biotechnologies, including recombinant DNA technology, to the tree improvement field. Increasingly common in the literature are references to techniques such as gene transfer, protoplast fusion, and microinjection. These techniques are based upon the ability to identify DNA *fragments*, a single gene (or group of genes) that is responsible for certain tree or wood characteristics. Genes so identified and subsequently isolated are then multiplied and transferred to a host cell or cells, with cells thereby regenerated to form a plant. Such techniques offer opportunities for very rapid gains in tree improvement by bypassing the long developmental periods of woody plants. Spectacular gains are possible in the future.

As a follow-up to superior tree identification and production of seedlings from such trees, considerable effort has been invested in crossbreeding of selected tree progeny to develop trees that combine the best features of several superior trees. Wood quality characteristics are among the features upon which selection is based. Genetic selection of both hardwoods and softwoods has resulted in the development of fast-growing trees that produce wood of normal to higher-than-normal density. It has been shown to be possible to breed for long fibers, high proportions of fibers relative to vessels, uniform density, amount of heartwood extractives, low proportions of lignin and juvenile wood, and minimal branch development. Even the tendency to produce spiral grain and cell wall microfibril angles has been shown to be heritable. In fact, recent research has determined that the large microfibril angle in juvenile wood of softwoods, which is a major factor in reduced stiffness and high longitudinal shrinkage of such wood, is also genetically controlled (Cave and Walker 1994; Donaldson and Burdon 1995; Ying et al. 1994). Cave and Walker observed, "Opportunities to improve wood quality through selection of juvenile material with a low microfibril angle are great and promise greater benefits than selection based on density." It has also been demonstrated that genetic selection or transgenic engineering can be used to reduce bark thickness, the number of sclereids and other undesirable elements in the bark, and herbicide and insect resistance (Bauer 1997; Strauss et al. 1997).

One of the terms used to express the potential for genetics in tree and wood improvement is *heritability*. As explained by McElwee (1963), heritability, which is expressed as a numerical value between 0 and 1, indicates the degree to which variation is determined by the parents. Thus, a heritability of 1 for a given characteristic means that the variability of that characteristic is determined entirely by parental origin. A heritability of 0 indicates variability in a characteristic to be totally governed by the environment, with no parental influence. To provide an indication of the potential for genetic selection in improvement of tree and wood properties, heritability values established experimentally for various tree and wood characteristics of pine species are presented in Table 11.1. Note that two factors – wood specific gravity and tracheid length – have very high heritability values. Latewood cell wall thickness and percentage of latewood, both of which are closely related to wood specific gravity, also rank high in heritability, at least in radiata and loblolly pine.

Via et al. (2004) note that although, to date, some tree improvement programs have targeted specific gravity improvement and adopted such goals as improvement of pulp yield, lumber strength, and stiffness, most programs have focused more narrowly on functions such as growth rate (volume), stem straightness, stem form, and disease resistance.

Although Table 11.1 indicates the relationship of various properties to parental influence, it does not give an idea of the magnitude of change possible through genetic selection. Even characteristics low in heritability can represent very significant opportunities for improvement if the variability associated with that characteristic is high. As an example, consider volumetric growth rate, which has a heritability of only 0.15. Despite this low value, Zobel (1977) indicated that volumetric yield gains of at least 10–20% could be expected in young plantations of southern pine made up of trees

TABLE 11.1.   Heritability of various tree and wood properties in pine.

Property	Heritability	Remarks
Wood specific gravity	0.5–0.7	Reported by McElwee (1963) as an average of results obtained in a number of studies of pine. Individual findings ranged from 0.2 to 1.0.
	0.5–0.8	Values for southern pine as reported by Barker (1972).
	0.52	*Pinus taeda* – Stonecypher et al. (1973).
	0.5–0.7	*Pinus radiata* – Dadswell et al. (1961).
Tracheid length	0.73–0.83	*P. radiata* – Dadswell et al. (1961).
	0.31–0.48	*Pinus sylvestris* – Hannrup et al. (2000).
	0.30–0.59	*Picea glauca* – Ikovich et al. (2002)
	0.36	*Eucalyptus regnans* – Raymond et al. (1998).
Latewood percentage	0.47–0.54	*P. radiata* – Dadswell et al. (1961).
	0.80	*P. taeda* – Goggans (1962).
Latewood cell wall thickness	0.84	*P. taeda* – Goggans (1962).
Earlywood cell wall thickness	0.13	*P. taeda* – Goggans (1962).
Growth rate (expressed in volumetric gain)	0.15	*P. taeda* – Stonecypher et al. (1973)
Stem straightness	0.14	*P. taeda* – Stonecypher et al. (1973).
Oleoresin yield	0.55	Squillace and Dorman (1961).
Resistance to disease (fusiform rust)	0.22	*P. taeda* – Stonecypher et al. (1973).

from first-generation seed orchards. He further estimated that volume yield gains of 35–45% (over non-improved stock) could be reasonably expected in plantations established from second-generation orchards. Although these figures were impressive, Zobel pointed out that such estimates were conservative and referred to reports of 30–50% or greater gains in volume yield in established first-generation plantations.

Wood specific gravity is an important determinant of wood quality and dry weight yields per volume of woody material. Because of this, even large increases in volumetric growth might be viewed as undesirable if the results were a significant decrease in specific gravity. Thus, breeding programs are generally aimed at controlling specific gravity as well as improving growth rate. Using an example provided by Zobel (1977), an assessment of eight-year-old southern pine trees in North Carolina revealed an average specific gravity for trees in the plantation of 0.425 compared with 0.410 for commercial controls. This meant that the plantation trees had produced about 90 more lb dry wt 100 ft^{-3} (or about 15 kg dry wt m^{-3}) wood than the controls. Translated to an area basis for a 25-year rotation, the difference amounts to about 4300–4500 lb additional dry wood per acre (or 4820–5040 kg ha^{-1}), even assuming equal volumetric growth rates of plantation and control trees. This estimated large increase was described by Zobel as conservative, because genetic differences in specific gravity among tree families generally increase with age.

Given the impact of genetics on wood specific gravity it is not surprising that genetic factors have a major impact on strength and other properties of wood in solid form, a reality that is demonstrated in numerous studies. Investigations of the impact of genetic variation on properties of particle and fiberboards are more recent. Such studies reveal

TABLE 11.2.   Gains from mass selection of loblolly pine based on 280 open-pollinated families from random parents.

Characteristic	Gain, as percentage of mean
Height	14
Basal area	18
Straightness	7
Crown	4
Volume	25
Specific gravity	10
Dry weight	26
Fusiform rust resistance	18–42

Source: Stonecypher et al. (1973).

that tree breeding and genetic manipulation have considerable potential for positively influencing properties of products such as medium density fiberboard and structural flakeboard (Peters et al. 2002; Shi et al. 2005). On the other hand, Knudson et al. (2006) reported that white spruce genetically selected for fast growth (the trait most commonly focused on) and harvested at a short rotation age (36 years) yielded veneer of considerably lower stiffness than other Canadian species typically used in producing veneer products.

In view of current concerns about juvenile wood, a recent study of lodgepole pine (*Pi. contorta* Dougl Ex Loud var. latifolia Engelm) is particularly interesting. Wang et al. (2000) found evidence indicating that by selecting trees exhibiting fast growth and high wood density, it may be possible to increase wood density and homogeneity in juvenile wood.

Estimates of the first-generation gains possible in other characteristics are presented in Table 11.2. Although these figures refer only to pine, estimates of similar magnitude have been obtained for other softwoods and a variety of hardwood species as well.

After an extensive review of research related to hardwood quality, Cutter et al. (2004) concluded that, "One of the glaring areas where noticeable effort has been lacking has been in the area of genetic improvement of hardwood quality." They indicated that except for black walnut (*J. nigra*), on which research has been extensive, results have been particularly paltry in areas other than the use of hardwood for fiber and bioenergy. Whether the future will bring significantly greater attention to hardwood genetics remains to be seen.

In summary, by identifying trees in the natural forest with the most desirable characteristics and then selectively breeding their offspring to combine into one tree the best features of the superior parents, fast-growing new trees are being developed that will be more suitable for product conversion than their ancestors. Based upon initial experience, it appears that the benefits from this kind of activity will be enormous.

Intensive Culture

A relatively new concept in forestry is *intensive culture*. The idea here is that forestland that has the highest productive capacity can be intensively farmed to achieve high yields, thereby easing pressure on less-productive lands. In intensive-culture plantations, genetically superior trees are ideally spaced and then fertilized and irrigated over

(a)

(b)

FIGURE 11.6.   Spectacular growth in Aracruz plantation in northeastern Brazil. These eucalypts are (a) two months old (top), (b) six months old (bottom), and (c) three years old (next page). The intensive management of these plantation trees produces this dramatic growth.

(c)

FIGURE 11.6.   *Continued.*

optimally selected rotations. Under these conditions, growth rates are quite high. The main application of such plantations appears to be for production of wood for reconstituted products such as paper and flakeboard or for energy generation. In these plantations, short rotations will be selected to maximize the production of biomass per unit time. As an example of what is possible, yields as high as 6.0 dry tons per acre per year ($13.6 \, m^3 \, ha^{-1} \, year^{-1}$) have been reported for intensively cultured hybrid aspen stands in northern Wisconsin. This compares with an average yield in natural aspen stands within the same region of just 0.31 dry tons per acre per year ($0.70 \, m^3 \, ha^{-1} \, year^{-1}$) and as much as 1.0 ton acre^{-1} year^{-1} on the best sites. Attainment of even higher rates of growth is thought possible with the development of improved genotypes and cultural practices.

Tree plantation growth in regions of the world where rainfall is evenly distributed and temperatures moderate can be spectacular (Figure 11.6). The Aracruz project on the northeast coast of Brazil has achieved growth rates as high as 11.1 dry tons per acre per year ($25.8 \, m^3 \, ha \, year^{-1}$) using various species of eucalyptus.

References and Supplemental Reading

Aldridge, F. and Hudson, R.H. (1958). Growing quality softwoods. I. *Q. J. For.* 52 (2): 107–114.
Aldridge, F. and Hudson, R.H. (1959). Growing quality softwoods. II. *Q. J. For.* 53 (3): 210–219.
Allchin, P.C. (1960). The I.G.T. vessel picking test. *Appita* 13 (6): 186–190.
Amateis, R.L., Radtke, P.J., and Burkhart, H.E. (1996). Growth and yield of thinned and unthinned plantations. *J. For.* 94 (12): 19–23.

Anderson, E.A. (1951). Healing time for pruned Douglas-fir. *Timberman* 52 (12): 74–80.

Barger, R.L. and Ffolliott, P.F. (1976). Factors affecting occurrence of compression wood in individual ponderosa pine trees. *Wood Sci.* 8 (3): 201–208.

Barker, J. 1972. Location effects on heritability estimates and gain predictions for ten-year-old loblolly pine. Ph.D. diss., NC State Univ., School of Forest Resources.

Barrett, J.W. 1968. Pruning of ponderosa pine – effect on growth. USDA For. Serv. Pac. Northwest For. Range Exp. Stn Res. Pap. PNW-68.

Barrett, J.D.; and Kellogg, R.M. 1984. Strength and stiffness of second-growth Douglas-fir dimension lumber. Forintek Canada Corp. Rep. FR22.

Bauer, L.S. (1997). Fiber farming with insecticidal trees. *J. For.* 95 (3): 20–23.

Bendtsen, B.A. (1978). Properties of wood from improved and intensively managed trees. *For. Prod. J.* 28 (10): 69–72.

Bendtsen, B.A. and Senft, J. (1986). Mechanical and anatomical properties in individual growth rings of plantation-grown eastern cottonwood and loblolly pine. *Wood Fiber Sci.* 18 (1): 23–28.

Biblis, E.J. and Carino, H.F. (1999). Flexural properties of lumber from 50-year-old loblolly pine plantation. *Wood Fiber Sci.* 31 (2): 200–203.

Biblis, E.J., Carino, H.F., Brinker, R., and McKee, C.W. (1995). Effect of stand density on flexural properties of lumber from two 35-year-old loblolly pine plantations. *Wood Fiber Sci.* 27 (1): 25–33.

Bissett, I.J.W., Dadswell, H.E., and Wardrop, A.B. (1951). Factors influencing tracheid length in conifer stems. *Aust. For.* 15 (1): 17–30.

Boyce, S.G. 1965. Improved hardwoods for increased utilization. Proc. 8th Conf. Forest Tree Improvement, pp. 1–6.

Boyd, J.D. 1968. Effect of plantation conditions on wood properties and utilization. FAO World Symp. Manmade Forests and Their Industrial Importance, pp. 789–822.

Brazier, J.D. (1977). The effect of forest practices on quality of the harvested crop. *Forestry* 50 (1): 49–66.

Briggs, D.G. and Smith, W.R. (1986). Effects of silvicultural practices on wood properties of conifers – a review. In: *Douglas-fir: Stand Management for the Future* (ed. C. Oliver, D. Hanley and J. Johnson), 220–229. Seattle: University of Washington Press.

Brinkman, K.A. 1955. Epicormic branching on oaks in sprout stands. US For. Serv. Cent. States For. Exp. Stn Tech. Pap. 146.

Brown, G.S. 1965. The improvement of the quality of the timber from coniferous plantations in New Zealand by silvicultural means. Presented at IUFRO Sect. 41 (For. Prod.) Meet., Melbourne, Aust.

Bryant, P.A.V. 1984. The impact of fast growth in plantations on wood quality and utilization. Proc. Symp. Site Quality and Productivity of Fast Growing Plantations. IUFRO, Pretoria, South Africa.

Cahill, J.M. (1991). Pruning young-growth ponderosa pine: product recovery and economic evaluation. *For. Prod. J.* 41 (11/12): 67–73.

Cahill, J.M., Snellgrove, T.A., and Fahey, T.D. (1988). Lumber and veneer recovery from pruned Douglas-fir. *For. Prod. J.* 38 (9): 27–32.

Calvert, W.W.; and Brace, L.G. 1969. Pruning and sawing eastern white pine. Can. For. Serv. Dep. Fish. For. Publ. 1262.

Carmean, W.H.; and Boyce, S.G. 1974. Hardwood log quality in relation to site quality. USDA For. Serv. North Cent. For. Exp. Stn Res. Pap. NC-103.

Cave, I.D. and Walker, J.C.F. (1994). Stiffness of wood in fast-grown plantation softwoods: the influence of microfibril angle. *For. Prod. J.* 44 (5): 43–48.

Chalk, L. (1951). Water and growth of Douglas-fir. *Q. J. For.* 45 (3): 237–242.

Chiu, C.-M., Lin, C.-J., and Wang, S.-Y. (2005). Tracheid length and microfibril angle of young Taiwania grown under different thinning and pruning treatments. *Wood Fiber Sci.* 37 (3): 437–444.

Choong, E.T. and Fogg, P.J. (1989). Effect of cultural treatment and wood-type on some physical properties of longleaf and slash pine wood. *Wood Fiber Sci.* 21 (2): 193–206.

Chuang, S.-T. and Wang, S.-Y. (2001). Evaluation of standing tree quality of Japanese cedar grown with different spacing using stress-wave and ultrasonic-wave methods. *J. Wood Sci.* 47 (4): 245–253.

Ciesler, A. and Janka, G. (1902). Studien uber die Qualitat Rasch erwachsenen Fichtenholzes. *Centralbl. Gesamt. Forstwes.* 28: 337–416.

Clark, A., Borders, B., and Daniels, R. (2004). Impact of vegetation control and annual fertilization on properties of loblolly pine wood at age 12. *For. Prod. J.* 54 (12): 90–96.

Clark, A. and Saucier, J.R. (1989). Influence of initial planting density, geographic location, and species on juvenile wood formation in southern pine. *For. Prod. J.* 39 (7/8): 42–48.

Clark, A., Saucier, J.R., Baldwin, V.C., and Bower, D.R. (1994). Effect of initial spacing and thinning on lumber grade, yield, and strength of loblolly pine. *For. Prod. J.* 44 (11/12): 14–20.

Crist, J.B.; and Dawson, D.H. 1975. Anatomy and dry weight yields of two Populus clones grown under intensive culture. USDA For. Serv. North Cent. For. Exp. Stn Res. Pap. NC-113.

Crist, J.B.; Dawson, D.H.; and Nelson, J.A. 1977. Wood and bark quality of juvenile jack pine and eastern larch grown under intensive culture. Proc. Tech. Assoc. Pulp and Paper Ind. For. Biol. Wood Chem. Conf., pp. 211–216.

Cutter, B.E., Lowell, K.E., and Dwyer, J.P. (1991). Thinning effects on diameter growth in black and scarlet oak as shown by tree ring analyses. *Forest Ecology and Management* 43 (1–2): 1–13. https://doi.org/10.1016/0378-1127(91)90071-3.

Cutter, B., Coggeshall, M., Phelps, J., and Stokke, D. (2004). Impacts of forest management activities on selected hardwood wood quality attributes – a review. *Wood Fiber Sci.* 36 (1): 84–97.

Dadswell, H.E. (1958). Wood structure variations occurring during tree growth and their influence on properties. *J. Inst. Wood Sci.* 1: 11–32.

Dadswell, H.E., Fielding, J.M., Nicholls, J.W.P., and Brown, A.G. (1961). Tree to tree variations and the gross heritability of wood characteristics of *Pinus radiata*. *TAPPI* 44 (3): 174–179.

Deresse, T., Shepard, R., and Rice, R. (2002). Longitudinal shrinkage, kiln-drying defects, and lumber grade recovery of red pine (*Pinus resinosa* Ait.) from a 125-year-old natural stand and a 57-year-old plantation. *For. Prod. J.* 52 (5): 88–93.

Dimock, E.J.; and Haskell, H.H. 1962. Veneer grade yield from pruned Douglas-fir. USDA For. Serv. Pac. Northwest. For. Range Exp. Stn Res. Pap. PNW-48.

Dobie, J. and Wright, D.W. (1978). Economics of thinning and pruning – a case study. *For. Chron.* 54 (1): 34–38.

Donaldson, L. and Burdon, R. (1995). Clonal variation and repeatability of microfibril angle in *Pinus radiata*. *NZ J. For. Sci.* 25: 167–174.

Downes, G., Nyakuengama, J., Evans, R. et al. (2002). Relationship between wood density, microfibril angle and stiffness in thinned and fertilized *Pinus radiata*. *IAWA J.* 23 (3): 253–265.

Duchesne, I. (2006). Effect of rotation age on lumber grade yield, bending strength and stiffness in jack pine (*Pinus banksiana* Lamb.) natural stands. *Wood Fiber Sci.* 38 (1): 84–94.

Duchesne, I. and Zhang, S. (2004). Variation in tree growth, wood density, and pulp fiber properties of 35 white spruce (*Picea glauca* (Moench.)(Voss)) families grown in Quebec. *Wood Fiber Sci.* 36 (4): 467–475.

Dwyer, J.P. and Lowell, K.E. (1988). Long-term effects of thinning and pruning on the quality, quantity, and value of oak lumber. *Northern Journal of Applied Forestry.* 5: 258–260.

Einspahr, D.W.; Benson, M.K.; and Harder, M.L. 1972. Influence of irrigation and fertilization of growth and wood properties of quaking aspen. Proc. Symp. Effect of growth acceleration on the properties of wood. USDA For. Serv. For. Prod. Lab.

Erickson, H.D. and Arima, T. (1974). Douglas-fir wood quality studies. II. Effects of age and stimulated growth on fibril angle and chemical constituents. *Wood Sci. Technol.* 8: 255–265.

Erickson, H.D. and Lambert, G.M.G. (1958). Effects of fertilization and thinning on chemical composition, growth, and specific gravity of young Douglas-fir. *For. Sci.* 4 (4): 307–315.

Fight, R., Snellgrove, T., Curtis, R., and Debell, D. (1986). Bringing timber quality considerations into forest management decisions: a conceptual approach. In: *Douglas-fir: Stand Management for the Future* (ed. C. Oliver, D. Hanley and J. Johnson), 20–25. Seattle: University of Washington Press.

Fujiwara, S. and Yang, K.C. (2000). The relationship between cell length and ring width and circumferential growth rate in five Canadian species. *IAWA J.* 21 (3): 335–345.

Fukazawa, K. (1984). Juvenile wood of hardwoods judged by density variation. *IAWA Bull.* 5 (1): 65–73.

Ganchev, P. (1971). Studies on wood anatomical structure in several Euro-American poplar species. *Gorskostop. Nauk.* 8 (4): 15–22.

Gartner, B., North, E., Johnson, G., and Singleton, R. (2002). Effects of live crown on vertical patterns of wood density and growth in Douglas-fir. *Can. J. For. Res.* 32 (3): 439–447.

Gartner, B., Robbins, J., and Newton, M. (2005). Effects of pruning on wood density and tracheid length in young Douglas-fir. *Wood Fiber Sci.* 37 (2): 304–313.

Gazo, R., Beauregard, R., and Kimberley, M. (2000). Incidence of pruning and log height class on incidence of defects in radiata pine random-width boards. *For. Prod. J.* 50 (9): 28–31.

Gladstone, W.T., Barefoot, A.C. Jr., and Zobel, B.J. (1970). Kraft pulping of earlywood and latewood from loblolly pine. *For. Prod. J.* 20 (2): 17–24.

Gladstone, W.T.; and Gray, R.L. 1973. Effects of forest fertilization on wood quality. For. Fertil. Symp. Proc. USDA For. Serv. Gen. Tech. Rep. NE-3.

Goggans, J.F. 1961. The interplay of environment and heredity as factors controlling wood properties in conifers with special emphasis on their effects on specific gravity. NC State Coll. Sch. For. Tech. Rep. 11.

Goggans, J.F. 1962. The correlation, variation, and inheritance of wood properties in loblolly pine (*Pinus taeda* L.). NC State Coll. Sch. For. Tech. Rep. 14.

Gominho, J. and Pereira, H. (2005). The influence of tree spacing on heartwood content in *Eucalyptus globulus* Labill. *Wood Fiber Sci.* 37 (4): 582–590.

Gooding, J.W.; and Smith, W.H. 1972. Effects of fertilization on stem, wood properties, and pulping characteristics of slash pine (*Pinus elliottii* var. eliottii Engelm.). Proc. Symp. Effect of Growth Acceleration on the Properties of Wood. USDA For. Serv. For. Prod. Lab.

Gray, R.L. and Kyanka, G.H. (1974). Potassium fertilization effects on the static bending properties of red pine wood. *For. Prod. J.* 24 (9): 92–96.

Hagner, S.O. 1967. Fertilization as a production factor in industrial forestry. MacMillan Lect. Ser., Univ. BC For. Lect. 37.

Hannrup, B., Ekberg, I., and Persson, A. (2000). Genetic correlations among wood, growth capacity, and stem traits in *Pinus sylvestris*. *Scand. J. For. Res.* 15 (2): 161–170.

Harris, J.M., James, R.N., and Collins, M.J. (1975). Case for improving wood density in radiata pine. *NZ J. For. Sci.* 5 (3): 347–354.

Herman, M., Dutilleal, P., and Avella-Shaw, T. (1999). Growth rate effects on intra-ring and inter-ring trajectories of microfibril angle in Norway spruce (*Picea abies*). *IAWA J.* 20 (1): 3–21.

Hernández, R.E., Bustos, C., Fortin, Y., and Beaulieu, J. (2001). Wood machining properties of white spruce from plantation forests. *For. Prod. J.* 51 (6): 82–88.

Hernández, R.E., Kaubaa, A., Beaudoin, M., and Fortin, Y. (1998). Selected mechanical properties of fast-growing poplar hybrid clones. *Wood Fiber Sci.* 30 (2): 138–147.

Hillis, W.E. and Ditchburne, N. (1974). The prediction of heartwood diameter in radiata pine trees. *Can. J. For. Res.* 4: 524–529.

Horton, K.W. (1966). Profitability of pruning white pine. *For. Chron.* 42 (3): 294–305.

Howe, J.P. (1968). Influence of irrigation on ponderosa pine. *For. Prod. J.* 18 (1): 84–92.

Hsu, J.K. and Walters, C.S. (1975). Effect of irrigation and fertilization on selected physical and mechanical properties of loblolly pine (*Pinus taeda*). *Wood Fiber Sci.* 7 (3): 192–206.

Hussein, A., Gee, W., Watson, P., and Zhang, S. (2006). Effect of precommercial thinning on residual sawmill chip kraft pulping and pulp quality in balsam fir. *Wood Fiber Sci.* 38 (1): 179–186.

Ikovich, M., Namkoong, G., and Koshy, M. (2002). Genetic variation in wood properties of interior spruce II. Tracheid characteristics. *Can. J. For. Res.* 32 (9): 2128–2139.

Isebrands, J. and Bensend, D. (1972). Incidence and structure of gelatinous fibers within rapid-growing eastern cottonwood. *Wood Fiber Sci.* 4 (2): 61–71.

Isebrands, J. and Hunt, C.M. (1975). Growth and wood properties of rapid-grown Japanese larch. *Wood Fiber Sci.* 7 (2): 119–128.

Isebrands, J. and Parham, R.A. (1974). Tension wood anatomy of short rotation *Populus* spp. before and after kraft pulping. *Wood Sci.* 6 (3): 256–265.

Ishiguri, F., Kasai, S., Yokota, S. et al. (2005). Wood quality of sugi (*Cryptomeria japonica*) grown at four initial spacings. *IAWA J.* 26 (3): 375–386.

Jane, F.W., Wilson, K., and White, D.J.B. (1970). *The Structure of Wood*. London: Adam & Charles Black.

Jones, E.P. 1977. Precommercial thinning of naturally seeded slash pine increases volume and monetary returns. USDA For. Serv. Southeast. For. Exp. Stn Res. Pap. SE-164.

Jozsa, L. (1995). An overview of forest pruning and wood quality in British Columbia. In: *Forest Pruning and Wood Quality of Western North American Conifers* (ed. D. Hanley, C. Oliver, D. Maguire, et al.), 36–64. Seattle: College of Forest Resources, University of Washington.

Kellison, R.C.; Lea, R.; and Frederick, D.J. 1982. Effect of silvicultural practices on wood quality of southern hardwoods. Proc. TAPPI R&D Conf., pp. 99–103.

Kellogg, R.M. and Kennedy, R.W. (1986). Implications of Douglas-fir wood quality relative to practical end use. In: *Proceedings of a Symposium: Douglas-fir: Stand Management for the Future, 18–20 June 1985, Seattle, WA* (ed. C.D. Oliver, D.P. Hanley and J.A. Johnson), 103–107. Seattle: College of Forest Resources, University of Washington Press.

Kennedy, R.W. (1957). Fiber length of fast- and slow-grown black cottonwood. *For. Chron.* 33 (1): 46–50.

Kennedy, R.W. (1968). Anatomy and fundamental wood properties of poplar. Growth and utilization of poplars in Canada. *Can. Dep. For. Rural Dev. Dep. Publ* 1205: 149–168.

Kennedy, R.W. and Smith, J.H.G. (1959). The effects of some genetic and environmental factors on wood quality in poplar. *Pulp Pap. Mag. Can.* 60: T35–T36.

Klem, G.S. 1964. The effect of fertilization on three quality properties of Norway spruce. For. Commun. Lond. 235. Transl. (1965) by C. Clayre from Nor. Skogbr. 10(18):491–494.

Klem, G.S. 1967. Some aspects of the qualities of wood from fertilized forests. Proc. 4th Tech. Assoc. Pulp Paper Ind. For. Biol. Conf., Pointe Claire, Quebec, pp.120–130.

Klem, G.S. (1968). Quality of wood from fertilized forests. *TAPPI* 51 (11): 99A–103A.

Klem, G.S. 1974. Properties of wood from fertilized pine and spruce forests. Nor. Treteknisk Inst. 51.

Knudson, R., Wang, B., and Zhang, S. (2006). Properties of veneer and veneer-based products from genetically improved white spruce plantations. *Wood Fiber Sci.* 38 (1): 17–27.

Koch, P. 1972. The three-rings-per-inch dense southern pine – should it be developed? Proc. Symp. Effect of Growth Acceleration on the Properties of Wood. USDA For. Serv. For. Prod. Lab.

Koga, S., Zhang, S., and Bégin, J. (2002). Effects of precommercial thinning on annual radial growth and wood density in balsam fir (*Abies balsamea*). *Wood Fiber Sci.* 34 (4): 625–642.

Krajicek, J.E. (1959). Epicormic branching in even-aged, undisturbed white oak stands. *J. For.* 57: 372–373.

Krinard, R.M.; and Johnson, R.L. 1975. Ten-year results in a cottonwood plantation spacing study. USDA For. Serv. South. For. Exp. Stn Res. Pap. SO-106.

Kučera, B. (1994). A hypothesis relating current annual height increment to juvenile wood formation in Norway spruce. *Wood Fiber Sci.* 26 (1): 152–167.

Larson, P.R. 1967. Silvicultural control of the characteristics of wood used for furnish. Proc. 4th TAPPI For. Biol. Conf., Pointe Claire, Quebec.

Larson, P.R. 1969. Wood formation and the concept of wood quality. Yale Univ. Sch. For. Bull. 74.

Larson, P.R. 1972. Evaluating the quality of fast-grown coniferous wood. Proc. 1972 Ann. Meet. West. Stand. Mgmt. Comm., Seattle, WA.

Larson, P.R. 1973. Evaluating the quality of fast grown coniferous wood. Proc. 63rd West. For. Conf., pp. 146–152.

Lowell, K.E. and Dwyer, J.P. (1988). Stem form and lumber yield: long-term effects of thinning and pruning. *North. J. Appl. For.* 5 (1): 56–58.

Lowery, D.P.; and Schmidt, W.C. 1967. Effect of thinning on the specific gravity of western larch crop trees. USDA For. Serv. Int. For. Range Exp. Stn Res. Note INT-70.

MacPeak, M.D., Burkhart, L.E., and Weldon, D. (1987). A mill study of the quality, yield, and mechanical properties of plywood produced from fast grown loblolly pine. *For. Prod. J.* 37 (2): 51–56.

Maeglin, R.R.; Hallock, H.; Freese, F.; and McDonald, K.A. 1977. Effect of nitrogen fertilization on black walnut – growth, log quality, and wood anatomy. USDA For. Serv. Res. Pap. FPL-294.

Manwiller, F.G. 1972. Volumes, wood properties, and fiber dimensions of fast- and slow-grown spruce pine. Proc. Symp. Effect of growth acceleration on the properties of wood. USDA For. Serv. For. Prod. Lab.

Marton, R., Stairs, G., and Schreiner, E. (1968). Influence of growth rate and clonal effect on wood anatomy and pulping properties of hybrid poplars. *TAPPI* 51 (5): 230–235.

McElwee, R.L. 1963. Genetics in wood quality improvement. Proc. 7th Conf. Forest Tree Improvement, pp. 21–24.

McKimmy, M.D. (1986). The genetic potential for improving wood. In: *Douglas-fir: Stand Management for the Future* (ed. C. Oliver, D. Manley and J. Johnson), 118–120. Seattle: University of Washington Press.

McKimmy, M.D. and Campbell, R.K. (1982). Genetic variation in the wood density and ring width trend in coastal Douglas-fir. *Silvae Genet.* 31 (2–3): 43–51.

McQuilkin, R.A. 1975. Pruning pin oak in southeastern Missouri. USDA For. Serv. North Cent. For. Exp. Stn Res. Pap. NC-121.

Megraw, R.A. 1986. Efffect of silvicultural practices on wood quality. In Proc., TAPPI Res. Devel. Conf., Raleigh, NC, pp. 27–34.

Megraw, R.A.; and Nearn, W.T. 1972. Detailed DBH density profiles of several trees from Douglas-fir fertilizer thinning plots. Proc. Symp. Effect of Growth Acceleration on the Properties of Wood. USDA For. Serv. For. Prod. Lab.

Middleton, G.R., Jozsa, L.A., Munro, B.D. et al. (1995). Lodgepole pine product yield related to differences in stand density. Abstract, XX IUFRO World Congress, Finland. *IAWA J.* 16 (1): 13.

Mitchell, H.L. 1972. Effect of nitrogen fertilizer on the growth rate and certain wood quality characteristics of sawlog size red oak, yellow poplar, and white ash. Proc. Symp. Effect of Growth Acceleration on the Properties of Wood. USDA For. Serv. For. Prod. Lab.

Mitchell, R.J., Dwyer, J.P., Musbach, R.A. et al. Crop tree release of a scarlet-black oak stand. *Northern Journal of Applied Forestry* 5 (2): 96–99.

Møller, C.M. (1960). The effect of pruning on the growth of conifers. *Forestry* 33 (1): 37–53.

Moschler, W.W., Dougal, E.F., and McRae, D.D. (1989). Density and growth ring characteristics of *Pinus taeda* L. following thinning. *Wood Fiber Sci.* 21 (3): 313–319.

Murphey, W.K. and Bowier, J.J. (1975). The response of aspen to irrigation by municipal waste water. *TAPPI* 58 (5): 128–129.

Murphey, W.K.; and Brisbin, R.L. 1970. Influence of sewage plant effluent irrigation on crown wood and stem wood of red pine (*Pinus resinosa* Ait.). Pa. Agri. Exp. Stn Bull. 772.

Nepveu, G. (1999). Possible effects on wood quality to expect from accelerating tree growth in Europe: tentative answers and questions to accommodate. In: *Causes and Consequences of Accelerating Tree Growth in Europe* (ed. T. Karjalainen, H. Spiecker and D. Laroussinie), 197–206. Joensuu, Finland: European Forest Institute.

Nicholls, J.W.P. (1971). The effect of environmental factors on wood characteristics. II. The effect of thinning and fertilizer treatment on the wood of *Pinus pinaster*. *Silvae Genet*. 20 (3): 67–73.

Noskowiak, A.F. (1963). Spiral grain in trees – a review. *For. Prod. J.* 13 (7): 266–277.

Nyakuengama, J., Downes, G., and Ng, J. (2002). Growth and wood density responses to later-age fertilizer application in *Pinus radiata*. *IAWA J.* 23 (4): 431–448.

Parker, M.L.; Hunt, K.; Warren, W.G.; and Kennedy, R.W. 1976. Effect of thinning and fertilization on intra-ring characteristics and kraft pulp yield of Douglas-fir. Proc. 28th Symp. Applied Polymer, pp. 1075–1086.

Paul, B.H. (1957). Second growth is good. *South. Lumberman* 195 (2432): 29–30.

Paul, B.H.1963. The application of silviculture in controlling the specific gravity of wood. USDA For. Serv. Tech. Bull. 1288.

Paul, B.H. and Marts, R.O. (1931). Controlling the proportion of summerwood in longleaf pine. *J. For.* 29: 784–796.

Peszlen, I. (1995). Juvenile wood characteristics of plantation wood species. Abstract XX IUFRO Wood Congress, Finland. *IAWA J.* 16 (1): 14.

Peters, J., Bender, D., Wolcott, M., and Johnson, J. (2002). Selected properties of hybrid poplar clear wood and composite panels. *For. Prod. J.* 52 (5): 45–54.

Phelps, J.E. and Chen, P.Y.S. (1991). Wood and drying properties of white oak from thinned and unthinned plantations. *For. Prod. J.* 41 (6): 34–38.

Phelps, J., Isebrands, J., Einspahr, D. et al. (1985). Wood and paper properties of vacuum airlift segregated juvenile poplar whole-tree chips. *Wood Fiber Sci.* l7 (4): 529–539.

Pillow, M.Y.; and Luxford, R.F. 1937. Structure, occurrence, and properties of compression wood. USDA For. Serv. Tech. Bull. 546.

Posey, C.E. 1965. Effects of fertilization upon wood properties of loblolly pine (Pinus taeda L.). Proc. 8th Conf. Forest Tree Improvement, pp. 126–130.

Radcliffe, B.M. 1953. The influence of specific gravity and rate of growth upon the mechanical properties of sugar maple in flexure. Purdue Univ. Agric. Exp. Stn Bull. 597.

Raymond, C., Banham, P., and Macdonald, A. (1998). Within tree variation and genetic control of basic density, fiber length, and coarseness in *Eucalyptus regnana* in Tasmania. *Appita J.* 51 (4): 299–305.

Rendle, F.J.; and Phillips, E.W.J. 1957. The effect of rate of growth (ring width) on the density of softwoods. Proc. 7th Conf. Br. Commonw. For.

Saucier, J.R.; and Ike, A.F. 1972. Response in growth and wood properties of American sycamore to fertilization and thinning. Proc. Symp. Effect of Growth Acceleration on the Properties of Wood. USDA For. Serv. For. Prod. Lab.

Senft, J.F., Bendtsen, B.A., and Galligan, W.L. (1985). Weak wood: fast grown trees make problem lumber. *J. For.* 83 (8): 476–484.

Shi, J., Zhang, S., Riedl, B., and Brunette, G. (2005). Flexural properties, internal bond strength, and dimensional stability of medium density fiber panels made from hybrid poplar clones. *Wood Fiber Sci.* 39 (4): 629–637.

Shupe, T.F., Choong, E.T., and Yang, C.H. (1995). The effect of cultural treatments on the chemical composition of plantation-grown loblolly pine wood. Abstract XX IUFRO World Congress – Finland. *IAWA J.* 16 (1): 18.

Shupe, T.F., Choong, E.T., and Yang, C.H. (1996). The effects of silviculture treatments on the chemical composition of plantation-grown loblolly pine wood. *Wood Fiber Sci.* 28 (3): 286–294.

Siddiqui, K.M.; Gladstone, W.; and Marton, R. 1972. Influence of fertilization on wood and pulp properties of Douglas-fir. Proc. Symp. Effect of Growth Acceleration on the Properties of Wood. USDA For. Serv. For. Prod. Lab.

Smith, D.; Wahlgren, H.; and Bengtson, G.W. 1972. Effect of irrigation and fertilization on wood quality of young slash pine. Proc. Symp. Effect of Growth Acceleration on the Properties of Wood. USDA For. Serv. For. Prod. Lab.

Snook, S., Labosky, P., Bowersox, T., and Blankenhorn, R. (1986). Pulp and paper-making properties of a hybrid poplar clone grown under four management strategies and two soil sites. *Wood Fiber Sci.* 18 (1): 157–167.

Spurr, S.H. and Hsiung, W.Y. (1954). Growth rate and specific gravity in conifers. *J. For.* 52 (3): 191–200.

Spurr, S.H. and Hyvarinen, M.J. (1954). Wood fiber length as related to position in the tree and growth. *Bot. Rev.* 20: 561–575.

Squillace, A.E. and Dorman, K.W. (1961). Selective breeding of slash pine for high oleoresin yield and other characters. *Proc. Int. Bot. Congr.* 2 (14): 1616–1621.

Staebler, C.R. (1963). Growth along the stems of full-crowned Douglas-fir trees after pruning to specified heights. *J. For.* 61 (2): 124–130.

Stonecypher, R.; Zobel, B.; and Blair, R. 1973. Inheritance patterns of loblolly pines from a non selected natural population. NC State Univ. Agric. Exp. Stn Tech. Bull. 220.

Strauss, S.H., Knowe, S.A., and Jenkins, J. (1997). Benefits and risks of transgenic Roundup Ready® cottonwoods. *J. For.* 95 (5): 12–19.

Szopa, P.S., Tennyson, L.C., and McGinnes, E.A. Jr. (1977). A note on effects of sewage effluent irrigation on specific gravity and growth rate of white and red oaks. *Wood Fiber Sci.* 8 (4): 253–256.

Tasissa, G. and Burkhart, H.E. (1998). Modelling thinning effects on ring specific gravity of loblolly pine (*Pinus taeda* L.). *For. Sci.* 44 (2): 212–223.

Taylor, F.W. (1977). Variation in specific gravity and fiber length in selected hardwoods throughout the mid-South. *For. Sci.* 23 (2): 190–194.

Taylor, F. and Burton, J. (1982). Growth ring characteristics, specific gravity, and fiber length of rapidly grown loblolly pine. *Wood Fiber Sci.* 14 (3): 204–210.

Thor, E. and Core, H.A. (1977). Fertilization, irrigation, and site factor relationships with growth and wood properties of yellow poplar. *Wood Sci.* 9 (3): 130–135.

Timell, T.E. (1986). *Compression Wood in Gymnosperms*, vol. III, 1703. Berlin: Springer.

Turnbull, J.M. 1947. Some factors affecting wood density in pine stems. Union, South Africa, British Empire For. Conf.

USDA Forest Service. 1976. Intensive plantation culture. North Cent. For. Exp. Stn Gen. Tech. Rep. NC-21.

Via, B., Stine, M., Shupe, T. et al. (2004). Genetic improvement of fiber length and coarseness based on paper product performance and material variability – a review. *IAWA J.* 25 (4): 401–414.

Wang, S.-Y., Chen, J.-H., and Hsu, K.-P. (2005). Effects of planting density on visually graded lumber and mechanical properties of Taiwania. *Wood Fiber Sci.* 37 (4): 574–581.

Wang, T., Aitken, S.N., Rozenberg, P., and Millie, F. (2000). Selection for improved growth and wood density in lodgepole pine: effects on radial patterns of wood variation. *Wood Fiber Sci.* 32 (4): 391–403.

Wardrop, A.B. (1951). Cell wall organization and the properties of the xylem. I. Cell wall organization and the variation of breaking load in tension of the xylem in conifer stems. *Aust. J. Sci. Res. Ser. B* 4: 391–414.

Weetman, G.F. 1971. Effects of thinning and fertilization on the nutrient uptake, growth, and wood quality of upland black spruce. Woodland Pap. Pulp Pap. Res. Inst. Can. No. 28.

Wellwood, R.W. (1952). The effect of several variables on the specific gravity of second-growth Douglas-fir. *For. Chron.* 28 (3): 35–42.

Williams, R.E. and Hamilton, J.R. (1961). The effect of fertilization on four wood properties of slash pine. *J. For.* 59 (9): 662–665.

Yang, K.C. (1994). Impact of spacing on width and basal area of juvenile and mature wood in *Picea mariana* and *Picea glauca*. *Wood Fiber Sci.* 26 (4): 479–488.

Ying, L., Kretschmann, D.E., and Bendtsen, B.A. (1994). Longitudinal shrinkage in fast-grown loblolly pine plantation wood. *For. Prod. J.* 44 (1): 58–62.

Youngberg, C.T.; Walker, L.C.; Hamilton, J.R.; and Williams, R.F. 1963. Fertilization of slash pine. Ga. For. Res. Counc. Res. Pap. 17.

Zarges, R., Neuman, R., and Crist, J. (1980). Kraft pulp and paper properties of *Populus clones* grown under short rotation intensive culture. *TAPPI* 63 (7): 91–94.

Zhang, S., Chauret, G., Ren, H., and Desjardins, R. (2002). Impact of initial spacing on plantation black spruce lumber grade yield, bending properties, and MSR yield. *Wood Fiber Sci.* 34 (3): 460–475.

Zobel, B. (1977). Increasing southern pine timber production through tree improvement. *South. J. Appl. For.* 1 (1): 3–10.

Zobel, B.J.; Kellison, R.C.; and Kirk, D.G. 1972. Wood properties of young loblolly and slash pines. Proc. Symp. Effect of Growth Acceleration on the Properties of Wood. USDA For. Serv. For. Prod. Lab.

Zobel, B.J.; Kellison, R.; and Matthias, M. 1969. Genetic improvement in forest trees – growth rate and wood characteristics in young loblolly pine. Proc. 10th South. Conf. Forest Tree Improvement, pp. 59–75.

Zobel, B.J.; McElwee, R.L.; and Browne, C. 1961. Interrelationship of wood properties of loblolly pine. Proc. 6th South. Cont. Forest Tree Improvement, pp. 142–163.

Zobel, B. and Talbert, J. (1984). *Applied Forest Tree Improvement*, 403–404. New York: Wiley.

12

Lumber

The conversion of logs to lumber in its simplest form consists of sawing boards from logs, squaring the edges, and cutting to length. This process yields one of the world's most cost-efficient and environmentally sensitive building materials. Contemporary sawmills, particularly softwood mills, are highly technical and automated operations that use electronic scanners and computers to control important steps in the process. Economics dictate that as much lumber as possible be obtained from logs using methods capable of high production rates. Historic sawmills in the western United States designed to handle large logs could cut over 2400 m³ (about one million board feet) of lumber per day. Contemporary large softwood sawmills generally produce between 450 and 1900 m³ (200 000 and 800 000 board feet) of lumber per mill each day. To achieve these high production numbers from small-diameter trees, computer-controlled optimization of the conversion process and high stem-per-hour feed rates are utilized. With 8 ft long logs, production rates on the order of 5–10 seconds per log through the primary breakdown machine center are common. Lumber yields in modern sawmills are greater than ever, in some cases approaching 70% for trees in the 200–250 mm (8–10 in.) small-end-diameter class.

In 2013 the United States's sawmills produced about 88.1 × 10⁶ m³ (37.3 billion board feet) of lumber products, with softwoods accounting for 80% of the total (Howard and Jones 2016). This production level is ~24% higher than that in 2009 – evidence of both the strong economic rebound and the close tie between economic vitality and lumber production. Demand swings of this magnitude present great challenges both for mill operators and for timberland owners and managers. Virtually all hardwood lumber is produced in the eastern United States, with production nearly equal in the North and South regions. Softwood lumber production was again dominated by the US South, with 36 × 10⁶ m³ (15.3 × 10⁹ board feet) in the South in 2013, compared with 31.8 × 10⁶ m³ (13.5 × 10⁹ board feet) in the West (Howard and Jones 2016). The North accounted for only 2.82 × 10⁶ m³ (1.2 × 10⁹ board feet) of softwood lumber production in 2013. These numbers indicate great differences and variation in the lumber industry as one looks at different regions of the country. In addition to domestic production, the United States is a net importer of softwood lumber, most of which comes from Canada. In 2013, estimated

Forest Products and Wood Science: An Introduction, Seventh Edition. Rubin Shmulsky and P. David Jones.
© 2019 John Wiley & Sons Ltd. Published 2019 by John Wiley & Sons Ltd.
Companion website: www.wiley.com/go/shmulsky

net softwood lumber imports, or foreign trade, to the United States totaled 21.5×10^6 m³ (9.1×10^9 board feet), about 20% of the volume of domestic consumption (Howard and Jones 2016).

The trend toward mill consolidation, upgrades, and production increases has continued for the past two decades. In the past decade, hundreds of millions of dollars have gone into mill capital improvements. At the time of this writing, several major new mills are being constructed in the US South. This investment is an indicator of the long-term strength of the timber producing industry. More mills mean more competition and better prices for timber and timberland.

Hardwood lumber in the United States is often classified by the geographic area in which it grows. There are three separate regions in which most US hardwoods grow, as follows: Southern, Appalachian, and Northern. Presently, the most valuable hardwood species (black cherry, walnut, hard maple, and oaks) are from the Appalachian region. Production of hardwood lumber is somewhat evenly distributed among these three Eastern regions. Only about 3% of US hardwood lumber is produced in the West, with alder being a primary species.

One of the challenges of the lumber industry or the industries that use lumber is learning the terminology. *Lumber* is often defined as a solid product sawn from a log. In North America softwood lumber is classified as boards, dimension, or timbers. These terms apply to thickness categories. When dealing with softwood lumber, the term *boards* indicates lumber less than 51 mm (2 in.) in nominal thickness. *Dimension* is material 50–125 mm (2–5 in.) in nominal thickness, and *timbers* are pieces nominally thicker than 125 mm (5 in.) or thicker. The terminology applied to hardwood lumber is often quite different from that used to describe softwoods. For example, the term *dimension* when applied to hardwood refers to material that has been cut to size for furniture or pallet manufacture, not at all similar to its meaning for softwood lumber. Entire texts such as *Terms of the Trade* (Evans 1993) are devoted to the vocabulary of lumber and the forest products industry.

Differences Between Softwood and Hardwood Lumber

The main uses for the higher grades of hardwood lumber are furniture, flooring, millwork, cabinets, case goods, and export. Lower grades are utilized for pallet and container manufacture and miscellaneous industrial applications. The largest single use of hardwood lumber in the United States is for pallets. Nearly 7.3×10^6 m³ (3.1×10^9 board feet) of hardwood lumber are used each year for pallets and crating, accounting for about 27% of the total hardwood lumber production (Barrett 2004). A lesser volume of pallets and crates is produced from softwood lumber, mostly in West Coast plants.

Railroad ties are another important product of hardwood sawmills. In 2017,estimated railroad cross-, switch-, and bridge-tie consumption accounted for about 2.42×10^6 m³ (1.03×10^9 board feet) of lumber produced (HRM 2017). Sawmills that produce ties usually also produce lumber. This diversity allows mills to produce materials that have the highest market value at any given time. The dense hardwoods such as red and white oak, maple, and hickory are the preferred species and are used for 80% of ties produced. Only a small percentage of ties in the United States are produced from softwoods.

White oak, red oak, the gums, yellow poplar, the maples, and ash are among the most commonly harvested hardwoods in the United States. Economically, red oak is the single most important hardwood species domestically. Hardwoods most commonly sought and used for high-quality furniture are: red oak, ash, black walnut, black cherry, hard maple, and yellow birch. Ash, gum, elm, and yellow poplar are also extensively used for solid wood products such as furniture, stair parts, cabinets, and trim. However, practically every hardwood found in suitable quantity and size is manufactured into lumber for some purpose.

Softwood lumber is used principally as a building construction material. Softwoods are used as structural members (see Chapter 9) and for decorative or finishing purposes such as paneling, siding, decking, exterior trim material, windows and doors, and export. The single largest use of softwoods in the United States is as dimension lumber used in light-frame buildings. About 85% of all housing units built are of this type. New housing construction utilizes more than two-thirds of the total softwood lumber production (Howard and Jones 2016).

Douglas fir, southern yellow pine, ponderosa pine, western white and sugar pines, western hemlock, and the true firs are the most economically important US softwoods cut for lumber. Douglas fir and southern yellow pine provide about three-fourths of the softwood sawtimber harvested. Another 15% of softwood lumber is obtained from western hemlock, ponderosa pine, and the true firs.

The basis for the grading and *scaling* (measuring the volume) of hardwood lumber is very different from that for softwoods. Hardwood sawmills generally produce rough lumber of random widths, while softwood mills typically surface their lumber to specific sizes, e.g. 38×89 mm (2×4 in.), 19×140 mm (1×6 in.), etc. Furthermore, grade recovery is a paramount consideration in hardwood sawmilling. This situation occurs because of the vast price difference between the higher and lower grades. As a result, there are significant differences in the mill layouts and the manufacturing processes for softwood versus hardwood lumber. Hardwood lumber grades were developed with the assumption that each board will be cut into smaller pieces during secondary manufacturing. These usable pieces are termed *cuttings*. For instance, the highest grade of hardwood lumber, *firsts and seconds* (FAS), contains 83% usable cuttings that exceed a certain minimum size. Some species and lower grades such as "2B and 3B" allow "sound" defects in the cuttings. The standard hardwood lumber grades, promulgated by the National Hardwood Lumber Association (2015), are not necessarily adaptable to applications where an entire board will be used as a single piece.

Softwood lumber grades, in contrast, are based principally on either structural uses (strength and stiffness) or uses where the board appearance, as a whole, is important. Therefore, softwood lumber is primarily graded by considering the entire piece, with a deduction for its single most detrimental characteristic. As explained in the Chapter 9, knots and slope of the grain are the main characteristics that reduce strength and stiffness. Softwood grading rules contain specifications intended to assure minimum strength performance levels while maintaining a reasonably good appearance.

Additionally, there are a variety of specialty grades that apply to both hardwood and softwood. Various types of millwork, export, trim, flooring, paneling, pallet, tie, etc. grades exist and where appropriate, lumber is graded and sold as such.

Lumber Designation by Species Group

To simplify marketing and distribution, lumber is sometimes sold by species groups rather than by individual tree species. Online Appendix Tables A.8 and A.10 list the standard commercial names for hardwood and softwood lumber produced from North

American tree species. In most cases, hardwood species are grouped by woods that are difficult or impossible to separate once they are cut from the tree (for example the red oak group). Foresters may pay a premium when purchasing logs of a certain species because of their knowledge of the grade yield that can be expected. New technologies allow foresters to assess the strength and stiffness properties of standing trees. This process allows foresters to base their bids for tracts of timber not only on species, stand density, and form, but also on mechanical properties which in large part determine the value of the end products.

Inconsistencies between hardwood tree names and lumber names can cause confusion. Some hickories are classed as pecan when cut to lumber. Sweetgum is simply called gum or is classed as red gum if it is heartwood. Lumber from yellow poplar is called poplar, which could be confused with the true poplars, i.e. aspen and cottonwood.

Softwood species are often grouped by their similar strength properties to simplify marketing. Examples of such groupings are *Doug fir–Larch*, which is Douglas fir and western larch; *Hem–Fir*, which is West Coast hemlock and various true firs; *SPF*, which is spruce, mixed pines, and true fir; and *Northern Pine*, which is jack pine and Norway pine. *Southern pine* lumber is produced principally from four tree species: longleaf, slash, shortleaf, and loblolly pine. In some instances slash and longleaf pine are exported under the name pitch pine.

Lumber Size and Measurement Standards

Worldwide, lumber is produced, bought, and sold on a cubic meter basis. This simplifies volume specifications from locale to locale. In the United States, lumber is sold on the basis of board feet. A *board foot* is a volume 1 ft² by 1 in. thick (144 in.³). The number of board feet in a piece of lumber is therefore determined by multiplying the nominal thickness in inches times the nominal width in feet times the nominal length in feet. For example, a 1×6 in. piece 10 ft long contains 1×6/12×10 = 5 board feet. Geometrically, one cubic meter contains 424 board feet; however, this number varies in practice owing to factors such as nomenclature, shrinkage, surfacing, and actual versus nominal dimensions.

Complicating the measurement of lumber when using board feet is the fact that the actual size of a piece often differs from its *nominal size*. A dry, surfaced softwood 2×4 in. board, for example, measures 1.5×3.5 in. There is a larger standard size for softwoods in the rough green condition to allow for sawing variation, drying shrinkage, and surfacing but this size is also less than the nominal size. It finishes, however, at 1.5×3.5 in. thickness and width. The US Department of Commerce (2015) puts forth the "American Softwood Lumber Standard" which describes and stipulates these factors in detail. Standard sizes of hardwood lumber account for shrinkage and surfacing but in this case the nominal and the rough dry sizes are nearly the same. A nominal 25 mm (1 in.) thick hardwood board has a standard surfaced thickness of 20.6 mm (13/16 in.) and a minimum standard rough, kiln-dried thickness of 23.8 mm (15/16 in.). If sold green (undried), hardwood lumber should be thick enough to allow for shrinkage when dried to the specified rough size.

The board footage of softwood lumber is based upon the nominal size rather than the actual size. Therefore, an 8 ft long 2×4 in. board, which actually measures

1.5 in. × 3.5 in. × 8 ft, contains 2 × 4/12 × 8 = 5.33 board feet. Online Appendix Table A.9 lists nominal and actual sizes for softwood lumber surfaced on all four sides (S4S) when dry or green. Softwood lumber is graded either by inspectors after it is planed and dried or by automated technology of one type or another.

The board footage of hardwood lumber is based upon nominal thickness and the actual width. For example, a piece of 51 mm (2 in.) thick lumber that measures 235 mm (9.25 in.) in width by 3.05 m (10 ft) in length contains 2 × 9.25/12 × 10 = 15 board feet (15.4 to the nearest board foot). Hardwood inspectors simultaneously measure and grade hardwood lumber as it moves along the green chain. If hardwoods are cut to specified widths, as in the case of pallet stock, the board footage is measured as with softwoods. The thickness of hardwood lumber is often termed in "quarters of an inch." For example, 1 in. thick lumber is referred to as four-quarter (4/4) and 2 in. thick lumber as eight-quarter (8/4).

In Europe and most countries outside North America, lumber is specified by its actual size. The volume in cubic meters (the unit used for measurement) is based on the actual volume of the material when dry. Therefore, a dry piece measuring 148 × 240 × 4 m contains 0.148 × 0.24 × 4 = 0.142 m³. If the lumber is measured when green, a shrinkage factor must be taken into account. Although, geometrically, a cubic meter of dry lumber contains 424 board feet as determined from the actual (not the nominal) lumber size, a conversion factor of 450 board feet per cubic meter for green lumber is often used in international trade. The United States has been slow to adopt metrication into lumber manufacturing. The metric system is used to some degree as needed in the case of international trade; however, the board foot, or thousand board foot, unit of measurement remains deeply entrenched.

Sawlog Scaling and Measurement

The units of measure used in buying and selling sawlogs vary widely depending upon the size of the logs, whether the logs are sold tree length or as individual bolts, and upon regional practices. The same basic methods are used to purchase logs for a plywood plant, although some specifications such as the minimum diameter and the amount of acceptable heart rot may differ from that for sawlogs.

The most commonly used methods of measuring logs are:

1. *Log scaling*, in which the diameter and length of each log is measured. These measurements are converted, by use of a log rule, to cubic meters, cubic feet, or board feet. Manual scaling is relatively slow and is therefore limited to mainly to hardwood and low-production softwood mills.

2. *Weight scaling*, in which a load of logs is weighed. Weight can then be converted to an equivalent cubic meter, board foot, cubic foot, or cord volume, or payment can be made directly on a weight basis. Table 12.1 shows the average weight of 1000 board feet of logs for the major southern yellow pine species as measured using three different log rules. Most large mills purchase logs based on weight because it is not possible to maintain high production rates when logs are scaled individually. In this case, timber buyers for the mill develop general specifications for acceptable logs and pay for the logs by weight at the mill gate. The log

TABLE 12.1.   The average weight of 1000 board feet of logs as scaled by three log rules.

Species of southern pine	Doyle	Scribner Decimal C	International ¼ in.
Loblolly	8.9	6.4	5.5
Shortleaf	9.0	6.3	5.4
Longleaf	12.1	7.2	6.1
Slash	11.9	7.5	6.4

Log rules (short tons per 1000 board feet).
Note: in contemporary production, 4–5 short tons of southern pine logs are required to produce 1000 board feet of finished lumber.

TABLE 12.2.   Wood and bark volume for stacked standard cords of 1.22 m long bolts of longleaf pine pulpwood.

Average diameter	Volume per stacked m³		Green weight
inside bark (cm)	Wood (m³)	Bark (m³)	(kg)
13	0.15	0.060	607
18	0.18	0.058	681
23	0.19	0.055	701
28	0.19	0.053	722
33	0.20	0.053	739
38	0.53	0.053	765

Source: Williams and Hopkins (1968).

conversion rates, in terms of tons of logs per thousand board feet of finished lumber, may also vary with season of the year as logs may contain more moisture at different times of the year.

3. *Volume scaling* is sometimes used in the sale of sawbolts but is more commonly applied to pulpwood. A *standard cord* is defined as the volume of stacked wood $1.22 \times 1.22 \times 2.44$ m ($4 \times 4 \times 8$ ft), i.e. 3.63 m³ (128 ft³). This volumetric unit has historically been used in the United States. There are many variations from this standard. In some regions of the Lake States a pulpstick length of 2.54 m (100 in.) is used instead of 2.44 m (8 ft). In some areas of the South, a length of 1.60 m (63 in.) is used. The actual volume of wood in a cord, regardless of its definition, varies with the length, diameter, and straightness of the bolts and with the care used in stacking. Various studies have found the solid wood content of a standard cord to vary from 1.64 to 2.66 m³ (58–94 ft³), and typically 2.26–2.41 m³ (80–85 ft³). A more standardized unit of volume is the cubic meter. Table 12.2 illustrates the relationship between bolt diameter and the actual volume of wood and bark per cubic meter of 1.22 m (4 ft) long bolts.

Log rules provide a conversion that relates the diameter and length of a log to an estimated lumber yield. Many factors determine how many board feet of lumber will be obtained per cubic foot of round logs. These include the width of saw kerf, log diameter (which affects the proportion lost as slabs), taper, sweep, and processing technology. Three commonly used log rules are briefly described:

1. *Doyle rule.* This rule is computed from the equation $V = (D-4)^2(L/16)$, where V = the board foot volume, D = the scaling diameter in inches, and L = the length in feet. This rule underestimates the volume of lumber that can be obtained from small logs. Mills that purchase small logs on Doyle may obtain over twice the volume of lumber estimated by the log rule. A mill producing twice as much lumber as the log scale would have a 100% overrun.
2. *Scribner rule.* This rule was devised using the premise of sawing 25 mm (1 in.) thick boards with 6.4 mm (1/4 in.) kerf from perfectly circular logs, based on small-end diameter, and with no taper taken into account. This rule underestimates lumber yield on small logs and on long logs with taper. If the volumes are rounded to the nearest 10 board feet, this rule is called the Scribner Decimal C.
3. *The International 1/4 in. rule.* This rule also assumes sawing 25 mm (1 in.) thick lumber with 6.4 mm (1/4 in.) kerf and allows for 12.7 mm (1/2 in.) of taper for each 1.22 m (4 ft) of length and 3.2 mm (1/16 in.) of shrinkage in board thickness. The International rule generally provides the closest estimate to actual lumber production of any of these commonly used rules, that is, it allows less overrun than the Doyle or Scribner rules for all but the largest logs.

For any log rule, log scaling involves measuring the length and diameter of a log as a first step. The scaling diameter is the average diameter of the small end inside the bark. Charts can be used to convert the diameter and length to the scale volume. Electronic scanners can also be used for scaling. When done manually, scaling sticks that read directly in cubic meters or board feet are used.

When selling or purchasing logs, the basis for measurement is not nearly as important as understanding the relationship between the different rules and the actual log volume. It is equally important that all parties agree on which log rule will be used. Table 12.3 compares the content of 4.88 m (16 ft) long logs as scaled with different log rules.

TABLE 12.3. Comparison of board foot volumes as determined by three board foot log rules for 16 ft long logs.

Scaling diameter (in.)	Cubic volume[a] (ft³)	Doyle	Scribner	International ¼ in.
6	4.3	4	12	19
8	7.1	16	31	39
10	10.6	36	55	65
12	14.8	64	86	97
14	19.7	100	123	136
16	25.3	144	166	181
18	31.5	169	216	232
20	38.5	256	272	290
22	46.2	324	334	354
24	54.6	400	403	424
26	63.6	484	478	501

[a] Assumes diameter taper of 2 in. per log.

When both the volume of logs and the lumber produced are measured in cubic units (i.e. cubic feet or cubic meters) then conversion efficiency is expressed as a simple ratio or percentage. For example, if 1000 m³ of logs yields 624 m³ of lumber, then the yield is simply 0.624 or 62.4%.

When log volume is measured in cubic units and lumber volume is measured in board feet, conversion efficiency is expressed in terms of the lumber recovery factor or LRF, which is calculated by the following formula:

$$LRF = \frac{\text{mill tally (board feet)}}{\text{cubic volume log scale (cubic feet)}}$$

For example, if 1000 board feet of lumber were recovered from 135 ft³ of logs, the LRF would be:

$$LRF = \frac{1000}{135} = 7.41$$

The LRF varies with the average diameter, the taper, the straightness of the logs being sawn, the thickness of lumber being produced, and the type of manufacturing equipment. The best mills in the United States today are achieving LRF values of 9–10, or even higher. A weighted average for North American softwood sawmills is reported at 7.4 (Spelter et al. 2007). To put these values in perspective, consider the theoretical maximum LRF that might be obtained. Using the earlier definition of a board foot as a piece of wood measuring 1 × 1 × 1 in. thick (or one of equivalent volume), the maximum possible LRF would be 12 (assuming no waste in manufacturing). However, using the minimum green thickness for nominal 1 in. softwood lumber of 25/32 in. (1.984 cm) raises the theoretical maximum LRF to 15.4. Sawing efficiency is also sometimes expressed in terms of overrun or underrun, where:

$$\text{percentage overrun (underrun)} = \frac{(\text{volume lumber tally} - \text{volume log scale})}{\text{volume log scale}}$$

It should be apparent that mill overrun varies greatly depending on the log rule used. Table 12.4 compares typical overrun expectations in the southern pine region. It is also evident that actual sawn lumber size impacts overrun. Progressively thicker and wider lumber (e.g. boards, to dimension, to timbers) has progressively fewer sawlines, less kerf loss, and better recovery.

Another means of measuring conversion efficiency is tons of logs per thousand board feet. In the case of southern pine, often 4–5 short tons of logs are required to produce 1000 board feet of finished lumber. The finished lumber weighs ~1000 kg (2200 pounds) per 1000 board feet at ~15% MC.

Measurement of the actual log volume input by use of electronic scanners has become common in the softwood lumber industry. This practice allows monitoring of the true log volume input to a mill, which is not possible when using log rules or weight scaling. This information makes possible significant improvements in process control. True shape-scanning of logs allows for highly accurate volume assessment. With that information, a given mill can compare their volume output LRF versus volume input

TABLE 12.4.   Predicted percentage overrun in southern pine sawmillls using three log rules.

Scaling diameter (in.)	Log rule (percentage overrun)		
	Doyle	Scribner Decimal C	International ¼ in.
6	172	21	23
8	95	19	24
10	59	16	25
12	39	14	25
14	25	11	26
16	16	9	26
18	8	6	27
20	2	4	28

Source: Williams and Hopkins (1968). These predictions are for one-inch thick lumber. Overrun values are greater for 2 in. thick dimension lumber.

for any given size of log. The mill can then determine exactly which log size class produces the highest LRF and design their log procurement program accordingly. Digital camera-based technology is developed, and in many cases implemented at the mill level, that can scale (measure volume and form) an entire load of logs while they are on the log truck.

Another means of measuring conversion efficiency is by comparing the weight of logs required to produce a given volume of lumber. The average requirement for softwood sawmills as reported by Spelter et al. (2007) is 1.74 metric tons per cubic meter (4.5 short tons per thousand board feet).

In many European mills, logs are sorted by size prior to entering the mill. This practice requires extra yard space for inventory and time. It is claimed, however, that mill throughput and production are increased because the sawing machine centers require fewer movements and adjustments during lumber manufacturing. Also, for these reasons, machine maintenance and downtime are reduced. In the United States, some mills have separate large and small log primary breakdown lines. Some mills with a single primary breakdown line sort logs by diameter after they are bucked. Then the line will run the smaller tops for several minutes at a higher rate and then run the larger butt logs for several minutes at a slower rate. In this manner, the primary breakdown line production is maximized.

Lumber Manufacturing

The general sequence of processing steps in lumber manufacture is illustrated in Figure 12.1. All lumber, however, does not necessarily go through all the steps shown. Additionally, there is usually a series of shunts in the sawmill through which any board can be sent back to previous machine centers for reprocessing. Hardwood lumber is usually graded and sold while rough and green. In that case, drying, planning, and remanufacturing occur at subsequent offsite operations. Most modern softwood sawmills have equipment to accomplish all of the steps shown in the block diagram.

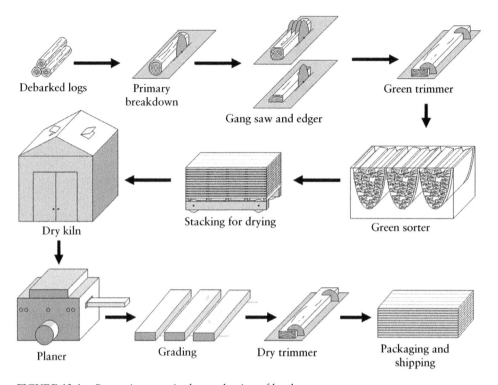

Debarked logs

Primary
breakdown

Gang saw and edger

Green trimmer

Dry kiln

Stacking for drying

Green sorter

Planer

Grading

Dry trimmer

Packaging and
shipping

FIGURE 12.1. Processing steps in the production of lumber.

Debarking

Sawmills debark logs prior to sawing. This has several advantages, but the main reason for debarking is so that bark-free chips can be produced for pulping. Many pulp mills will not purchase chips containing bark at any price. This situation will change if methods to separate bark from chips become economically feasible and are adopted by the pulp industry.

There are other important advantages of debarking. Removal of sand and grit along with bark greatly decreases the rate at which saws are dulled which, in turn, reduces mill downtime. True shape scanning for optimized sawing is more accurate with debarked logs. Bark is generally best suited to use for energy generation at the mill or as a landscape mulch.

The most widely used type of debarker in medium to large sawmills is the *ring debarker* (Figure 12.2). In this machine the log passes through a rotating ring that holds a number of pressure bars. These press against the log and tear off the bark. Large units of this type can debark logs at speeds of up to $1.0\,\mathrm{m\,s^{-1}}$ (200 lineal ft min^{-1}). Another type, the *cambio debarker*, looks similar to a ring debarker but centers the log between three rotating cylindrical heads and rotates the log as it passes through a revolving ring that holds knife-faced pressure arms.

A *Rosserhead debarker* is sometimes used in hardwood sawmills that process relatively large, rough logs and where high production rates are less critical. The rossing head is a rotating cutterhead, similar to the head on a lumber planer, that rides along the log

FIGURE 12.2.   Knife aperture of a ring debarker. As the log passes through the rotating circle of knives, they stop-down on the surface and separate the bark from the log at the cambium layer.

and cuts off the bark as the log is rotated. This debarker is also suited to situations where crooked or stubby logs must be debarked. With this type of equipment care must be taken to not remove too much wood with the bark. The high value of lumber and pulp chips makes excessive wood fiber removal an expensive mistake in any mill.

Drum debarkers can be used for processing small-diameter pre-bucked logs to feed small-log machines such as the chipping canter. These debarkers consist of a large-diameter steel cylinder or drum (about 4 m in diameter and 25 m long) through which bucked logs are passed. The drum turns at about two to four rotations per minute. There is a slight downward pitch to the cylinder that causes the logs to tumble from one end of the drum to the other. As the logs tumble inside the cylinder, paddles on the walls and the other logs abrade the bark away.

Primary Breakdown

The primary breakdown of the log is accomplished by one of two methods. One method is to place the log on a carriage that travels on rails and conveys the log back and forth through the saw. This equipment is referred to as a *carriage rig*. One piece is cut from the log with each pass by the saw. Lumber thickness is determined by placement of the log on the carriage. The mechanism that moves the log forward on the carriage is called the *setworks*. Until the 1960s almost all lumber was produced on carriage rigs. Since that time, single-pass headrigs specifically designed for small logs have come into common use. Single pass headrigs typically have either a chain-type or an overhead-dog feed system. There the log is fed through the headrig in one direction only. Chipping heads often produce a cant that is subsequently sawn into smaller boards. In new softwood sawmills, carriage rigs are usually only used for larger logs [over about 0.4 m (16 in.) in diameter], in mills that require a great deal of flexibility with regard to log and lumber sizes, or high-grade nondimension lumber mills. Most hardwood mills use carriage rigs because of their flexibility and ability to use a variety of sawing patterns for grade recovery.

Two types of saws, the *circular saw* and the *bandsaw*, have been used in conjunction with carriage rigs since the mid-1800s. Figures 12.3 and 12.4 illustrate a classic small carriage rig sawmill utilizing circular headsaws. In such mills there are three operators, the head sawyer, the edger, and the trimmer, who are responsible for the decisions that determine the quantity and grade of lumber obtained from each log. Circular headsaws vary in diameter from about 0.9 to 1.5 m (36–60 in.) and can handle logs up to about 0.9 m (36 in.) in diameter. If larger logs are to be cut, a second circular saw, located above the main saw, can be used. These types of mills are still used throughout North America. They are rugged and require minimal capital investment. Their production volumes are relatively low and their overall contribution to the national lumber picture is relatively small.

Single-band headsaws have the advantage of being able to cut almost any diameter of log that can be delivered to the saw. They are more expensive to maintain than circular saws, and their original capital investment is higher. Band mills are used in a wide variety of applications from large-scale mills as the primary single- or multi-pass headrig, to small portable sawmills, to mills that saw large-diameter logs. Figure 12.5 shows a modern bandsaw headrig, primarily for hardwoods. The hydraulic feed system used is faster, safer, and more efficient than old-style cable-driven systems. A 17° tilt of the headrig reduces carriage wear and derailment, improves log stability, and assists with the collection and routing of boards at the outfeed.

FIGURE 12.3. Photograph of an early circle sawmill. Note that the sawyer rides back and forth on the carriage. Source: Photo courtesy of Corley Manufacturing, Chattanooga TN.

Both circular- and band-saw headrigs remove a considerable amount of wood with the cut. The width of this cut, called the *kerf*, varies from about 4.8 to 9.5 mm (3/16–3/8 in.) in circular saws. Bandsaw kerf is somewhat narrower, typically about 3.2–4.8 mm (1/8–3/16 in.). Improvements in the machinery and blade, and in filing technologies now permit bandsaws to cut nearly as fast as circular mills. As such, band mills are often favored as means of primary breakdown as they combine high speed and precision cutting with low kerf loss. In either case (band or circle), nontrivial volumes of sawdust are produced, particularly if boards and dimension are cut on the headsaw. For this reason, and also to increase the production rate through a mill, headsaws often do not cut the lumber to its final thickness but instead saw logs into thick cants. A *cant* is a log that has been sawn on between one and four sides, but not yet reduced to boards. Cants are then cut to their final size by multiple-blade gang resaws or edgers that have higher feed speeds and less kerf loss than the headsaw.

Sash or frame saws can be used as head rigs and as resaws. These saws contain an assemblage of like blades mounted in a reciprocating frame through which each incoming log passes. Sash saws are not designed for grade recovery and are somewhat slower than band- or circle-type saws but have other advantages. The capital and maintenance costs of sash saws are often significantly lower than those of comparable band or circle saws. Kerf losses are minimal, similar to those available from band saws. Logs or cants

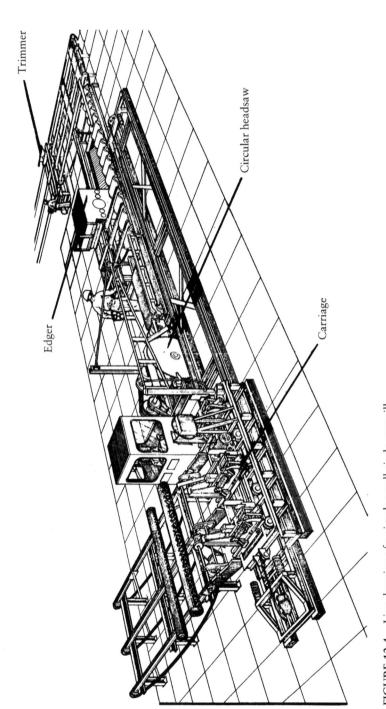

FIGURE 12.4. Line drawing of a simple small circle sawmill.

FIGURE 12.5.   Shotgun-fed 17° bandsaw headrig. Source: Photo courtesy of Corley Manufactturing – Chatanooga TN.

can also be guided through the saw to maximize recovery. These saws are employed in some parts of Europe where other mill practices such as log sorting, live sawing, and flitch drying (unedged boards) are not uncommon. There, boards from a log are reassembled for drying and edged afterward (Figure 12.6). These practices often improve production efficiency and prolong equipment life.

FIGURE 12.6.   Live-sawn logs kept together for drying and subsequent edging. Source: Photo courtesy of Koenig Sawmill Machinery, Esterer WD North America, Memphis, TN.

As mentioned above, another method of primary breakdown is by a *single-pass headrig*. In equipment of this type, the logs pass through the machine only once, in a more-or-less continuous flow. Further breakdown of the log is accomplished by secondary sawing. The three most common types of single-pass headrigs in use are chipper canters, scragg mills, and multiple-band headrigs. The increasing use of small logs for the production of lumber has stimulated the development and improvement of this type of processing equipment. Figure 12.7 shows a twin-band headrig cutting a small log.

The *chipper canter* or chipping headrig is different from carriage rig headsaws in that a squared or shaped cant is produced by the removal of chips from the sides of the log. The chips are removed with knives mounted on rotating cutterheads, so no sawdust is produced. As a coproduct of lumber manufacturing, chips (for pulp production) are typically worth more than sawdust (for fuel). Typically, these machines operate at from 1.00 to 1.5 m s^{-1} (200–300 lineal ft min^{-1}). Logs are fed in one after the other with very little space end to end. Such a mill can cut hundreds of small logs per eight-hour shift.

Chipping headrigs, developed in the 1960s, are now used throughout North America for small softwoods. Although some models handle logs up to 0.61 m (24 in.) in small-end diameter, most mills using these machines are cutting logs from 0.15 to 0.41 m (6–16 in.) in diameter. There are several manufacturers of chipper canters of various types. These differ in the way the log is held and transported through the machine, the type of cutterhead used, and the shape of the chipped profile. Some models incorporate a sawing section that cuts the cant into lumber immediately after it has been chipped to shape. Such a machine is shown in Figure 12.8. Some of the possible configurations of lumber that can be produced on chipping headrigs are shown in Figure 12.9. Chipper canters that do not contain a sawing section cut rectangular cants. These cants are then

FIGURE 12.7. Twin-band, single-pass headrig cutting a small softwood log. A 100 mm wide cant and two slabs are produced. Often the twin bands are preceded by twin chippers thus pulp chips, instead of slabs, are produced.

FIGURE 12.8. Chipping headrig (Chip-N-Saw) showing the log transport and chipping and sawing sections.

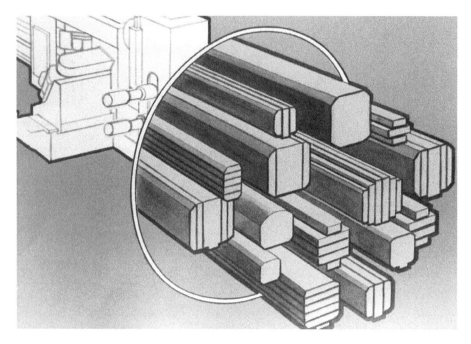

FIGURE 12.9. Illustration of various configurations of cants and lumber that can be produced in pass-through machines of the general type shown in Figure 12.8.

broken down into lumber by a *gang saw*, such as the rotary gang shown in Figure 12.10. Such gang saws can now be mounted on rollers and be positioned, in real time, via complex computer-controlled hydraulic systems for sawing curved or sweepy cants. These *curve-saw* systems can obtain higher yields and straighter-grained lumber compared with stationary saw boxes when cutting lower-quality timber.

Some sawmills utilize a chipping headrig for their small logs and a carriage rig for larger logs that have the potential to yield higher grades of lumber. Chipper canters are also used to convert cores from plywood veneer bolts into 2 × 4s. In that case, the veneer bolts are peeled down to a certain diameter, usually about 143 mm (5 5/8 in.), and then the core of the bolt is ejected from the veneer lathe and conveyed to the chipper canter.

Scragg mills consist of two or four circular saws on a common arbor (or shaft). The distance between the blades can be adjusted by the operator to produce 102, 152, or 203 mm (4, 6, or 8 in.) thick cants. Logs make one pass through this rig, producing a two-sided cant plus two slabs (on a two-saw scragg). The cant is then broken down to boards by a gang resaw.

Multiple-band headrigs usually consist of two or four opposing bandsaws of the same general type but somewhat smaller than the single-band headsaws used with carriage rigs. As with the other single-pass headrigs, multiple-band headrigs were developed to process small logs, although there is no reason why such systems could not also be used for larger logs, as long as there is no need to turn the log to recover higher lumber grades. Multiple-band headrigs are adaptable to electronic scanner and computer control applications.

Multiple-band headrigs are sometimes used in conjunction with *slab chippers* that slab two sides of the log before it reaches the bandsaws. The slab chippers have cutterheads

FIGURE 12.10. A classic double-arbor edger that can quickly break down a cant into many boards. Upper and lower arbors (shafts) carry circular saws that mesh perfectly to cut the cant from both faces.

with knives that chip flat the portion of the log that would ordinarily be removed as a slab. The cutting action of a slab chipper is similar to the cutterheads in some chipper canters, and high-value pulp chips are produced instead of slabs and sawdust. The use of two-slab chippers with a quad-band makes it possible to produce pulp chips, four boards, and one cant from each log.

In sum, there are a variety of methods of primary breakdown. Differences in raw material, desired products, mill history, and capital budget are the largest factors in determining what type of system is used.

Secondary Breakdown

After primary breakdown on the headrig, lumber may only require cutting to width (edging) and length (trimming). In most mills, however, a greater degree of secondary breakdown is required, such as sawing a cant into boards or resawing a slab to obtain a board. Many types of saws and chippers are available for secondary breakdown. Gang saws make multiple cuts simultaneously and therefore can reduce a cant to boards in a single pass. Resaws can split a plank (also called a *multiple*) into two or three boards or cut a board from the wider end of a slab. A slab chipper can also be used to salvage lumber from slabs.

Modern guided curve saws cut along the profile of two-sided cants. In simpler systems, the cant is mechanically guided through a stationary saw box. Alternatively, cants

FIGURE 12.11.   Outfeed of a sash gang resaw with feedworks designed to guide curved two-sided cants through for increased yield. Source: Photo courtesy of Koenig Sawmill Machinery, Esterer WD North America, Memphis, TN.

can be scanned electronically and the saw programmed (in real time) to follow the natural curvature in the cant or the feedworks can be programmed to guide the cant through the saw. Figure 12.11 illustrates the outfeed of a sash-type gang resaw in which the cant is guided through the saw on moveable feedworks to maximize yield. In curve saws that use circular blades, the blades are thin and flexible, which allows them to bend and flex as they cut along their curved paths. The curved but pliable boards are straightened during subsequent drying and their high grade is retained (Wagner et al. 2002). Although tighter sawing curves are possible, generally the curve is limited to a maximum deflection of 100 mm (4 in.) in a 4.88 m (16 ft) long cant because of subsequent automated handling equipment limitations. In some instances, curve sawing is not beneficial. When raw material quality is high or the logs are being sawn for grade, curve sawing is not usually justafiable.

Another popular machine for breaking down cants is the linebar resaw (Figure 12.12). There, cants are pushed against a fence or linebar and a fixed-thickness board is taken. The cant is then routed such that it passes through the saw again. This process continues until the cant is completely reduced to boards. Linebar resaws are rugged, fast, and can handle a queue of multisized cants simultaneously.

Edging and trimming are important operations requiring operators with a good knowledge of lumber grades. Electronic optimized scanners and control systems that allow automatic edging and trimming of boards for maximum volume yield are now available. These are driven primarily by size, value, and allowable wane parameters. Development is still needed for systems that will automatically cut to maximize value (grade) rather than volume. The value of a board can often be increased by properly

FIGURE 12.12.   Linebar resaw. Source: Photo courtesy of Corley Manufacturing, Chattanooga TN.

removing a defect from one edge or end. Excessive edging and trimming are costly, as they reduce lumber output.

Figure 12.13 illustrates the importance of properly edging and trimming hardwood lumber. This piece of basswood originally contained $0.031 \, m^3$ (13 board feet) of no. 1 Common grade valued at $144 \, m^{-3}$ ($340/1000 board feet). If edged and trimmed as shown, the volume of the piece would decrease to $0.028 \, m^3$ (12 board feet), but the grade would increase to FAS at $212 \, m^{-3}$ ($500/1000 board feet). The value of the piece in this

No. 1 Common basswood at $340/MBF
14 ft × 11½ in.

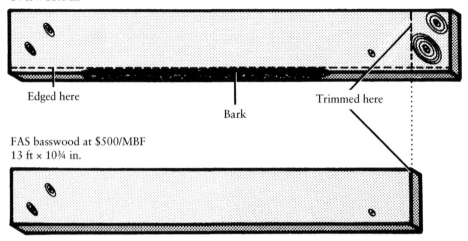

Edged here   Trimmed here

Bark

FAS basswood at $500/MBF
13 ft × 10¾ in.

FIGURE 12.13.   Value is increased by properly edging and trimming no. 1 Common basswood containing 13 board feet changed to FAS grade of 12 board feet, value increase of 36%. Note: Thousand board feet is abbreviated as MBF.

example is increased from $4.45 to $6.00, an increase of 36%. Most mills, both hardwood and softwood, attempt to increase lumber value through improved manufacturing.

Drying, Sorting, and Finishing

The drying process was discussed in Chapter 7. Before lumber is dried, it is usually sorted by grade, size, and species. To dry lumber uniformly and in a minimum time, it is desirable to dry only one thickness and species per kiln load. Mixed widths are acceptable because they have a minimal impact on the drying time as compared with thickness. With large-scale automated production and handling equipment, sorting by width is common in softwood production. In hardwood production it is common for species with similar drying characteristics to be mixed during kiln drying in order to increase kiln throughput. Automatic stacking equipment is used in many cases. Lumber packages ready for the kiln consist of alternate layers of lumber and *stickers* (also called spacer sticks or kiln sticks). After drying, packages are broken down and the stickers are removed.

Handling and sorting of lumber is often done twice. At the green chain (located after the trim saws in the mill) lumber is tallied and sorted by size and sometimes by species. Hardwood and appearance-softwood lumber are graded on the green chain. They are then often sold or brokered through a custom dry kiln. Structural softwood lumber is usually sorted by size and transferred to a dry kiln. After softwood lumber is dried and surfaced it is graded, finish trimmed, grade stamped, sorted, and packaged for shipment. A variety of lumber-sorting equipment is available for use at the green or dry end of the mill. The simplest system has lumber manually pulled and sorted as it proceeds down a green or dry chain. Most modern mills use mechanical sorters that can be controlled by a single grader/operator. Sorters are available to automatically sort lumber by length, width, or thickness.

The portion of a lumber manufacturing plant where cutting to final size and surfacing is done is called the *planer mill*. In hardwood mills shipping only rough lumber, no planer mill is required. Planing equipment commonly includes surfacers, matchers, and molders. A *surfacer* is a machine that planes lumber on two faces only (S2S). High-speed *matchers* used for softwood dimension have four or more heads to surface lumber on four sides (S4S). Some of these machines can operate at lineal speeds in excess of $7.6\,\mathrm{m\,s^{-1}}$ ($1500\,\mathrm{ft\,min^{-1}}$). With these machines, pieces are fed end-to-end and when they are running smoothly, there is a seemingly endless stream of wood passing through the machine. At production speeds, if for example, a planer is running 2×6 lumber, it produces $90\,000$ board feet per hour or one flat-bed truck load every 12–15 min.

Industrial Products

Products that are consumed for general shipping and handling as well as the construction or operation of railroads, powerlines, marine structures, roads and bridges, etc. are termed industrial. These products are typically not available to homeowners and do-it-yourself contractors. Wood utility poles and cross arms are generally the most sustainable and lowest cost solution for electrical distribution systems. Roundwood piling is a highly efficient means of building and other structural foundation development, particularly for heavy structures like bridges and in areas that require strong anchorage such as in coastal areas. These products require high-value forest resources which can be an lucrative opportunity for timberland managers. Railroad ties, most commonly red oak, support the rail lines that move freight, in a highly energy-/fuel-efficient manner, throughout North America. Dragline, crane, oilfield, and other mats are used in the development and maintenance of waterways, energy transmission, etc. These mats may be laminated from low-grade lumber or sawn directly from large but relatively low-grade logs. Concrete form boards, ship/sea walls, and crating similarly have minimal appearance requirements. All of these products help diversify the forest products market and even out wide swings owing to economic fluctuations.

Improving Sawmill Efficiency

A mill that buys wood by log rule can compare total log volume input with lumber volume output to determine overrun. Over time, this process gives an indication of any change in the performance of the key operators in the mill or machine centers, or in the quality of the logs, which is critical piece information for the mill's business decision makers. As noted earlier, the most common measure of mill efficiency is the LRF.

There are a number of possible means of improving the LRF of a sawmill. Among the most important are reducing kerf; reducing variability in thickness, which requires that lumber be sawn oversize; and making optimum decisions about how to cut each log and to accurately position it according to the decision made. Edging and wane allowance are also important, especially when sawing small-diameter logs.

Reduction of kerf losses in a sawmill is accomplished primarily by minimizing the saw cuts made on the headsaw and breaking down cants on smaller secondary saws that have thinner kerf. Modern rotary gang saws typically have kerfs of about 3.2 mm (1/8 in.) and produce smooth, accurately sawn lumber.

Another important means of increasing the LRF is by reducing the variability of lumber thickness. If a headrig has a variability of sawing thickness of 6.4 mm (±3.2 mm) and this can be reduced in some way to 3.2 mm (±1.6 mm), there can be a minimum saving of 1.6 mm of wood each time a saw cut is made. This saving results from setting the saw 1.6 mm over the desired thickness rather than 3.2 mm, which would be required with the greater variability. Proper selection and maintenance of equipment is very important in attaining this goal.

The greatest advancement in increasing the efficiency of sawmills in recent years has been the application of electronic log scanners and computers for the measurement of logs and placement of the first saw cut. The results for the headrigs shown in Figure 12.14 came from a Canadian study of improved sawing decisions from computer control. Two different models of a Chip-N-Saw, a chipper canter of the type shown in Figure 12.8, were used. Note in Figure 12.14 that the LRF was increased by 18–35 board feet m^{-3} (0.5–1.0 board foot ft^{-3}) by using a computer-controlled system.

The importance of making the first cut in a log at the proper position may not be immediately apparent. Once the first saw cut is made, however, the location of subsequent cuts is cast. An illustration of the difference in lumber yield that a shift in the first cut can make is shown in Figure 12.15. On this small log the lumber yield could increase from 0.075 to 0.094 m^{3} (32–40 board feet) by moving the cut 5.1 mm to the right. The data in this figure is from Hallock (1973), who did much work to demonstrate the importance of properly locating the first saw cut on a log. He referred to this as the *best opening face*, or the BOF.

A scanner measures the length, diameter, and shape of a log or cant by passing it through a light or laser beam. Some scanners operate with a light receiver that senses the width of the object being placed between the light source and the receiver. Other scanners take measurements from the light reflected from the object being measured, whether it is a log or a cant. Figure 12.16 illustrates an oak cant being scanned and positioned on the saw carriage. Such scanners are now commonplace in softwood mills and are used to a

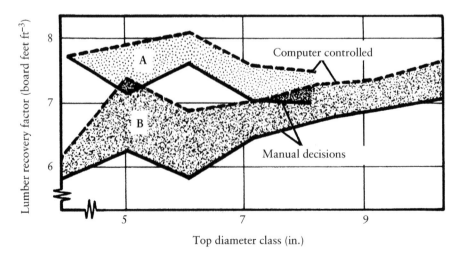

FIGURE 12.14.   LRF for two chipper-canters, one operated by manual decisions, the other by use of electronic scanning and computer control.

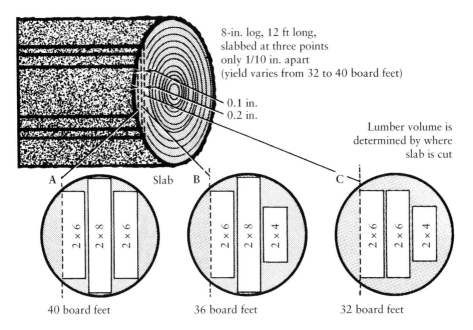

8-in. log, 12 ft long,
slabbed at three points
only 1/10 in. apart
(yield varies from 32 to 40 board feet)

0.1 in.
0.2 in.

Lumber volume is
determined by where
slab is cut

A

Slab

B

C

2 × 6
2 × 8
2 × 6

2 × 6
2 × 8
2 × 4

2 × 6
2 × 6
2 × 4

40 board feet

36 board feet

32 board feet

FIGURE 12.15. Importance of first saw cut in a log.

FIGURE 12.16. An oak cant being laser scanned and positioned on the saw carriage to maximize lumber recovery. Source: Photo courtesy of Corley Manufacturing, Chattanooga TN.

limited extent in hardwood mills. Other types of scanners such as ultrasonic, X-ray, laser, dielectric, and combinations thereof are used to a lesser extent.

Currently, the application of scanners and computers to sawmills is aimed at maximizing the volume of lumber. Sawmill optimization software computes and drives the highest volume and value solution from each log or cant. Value is determined using a combination of volume, grade, size, and market price. Figure 12.13 demonstrates that the piece with the highest volume does not necessarily have the highest value.

Importance of Standards and Grades for Proper Lumber Selection and Use

Those involved in the manufacture, sale, or use of lumber for secondary products must understand lumber grades if they are to utilize the material in the most-efficient manner. Using the wrong species, grade, or size for a job can be costly to the homebuilder or furniture manufacturer. It can also result in using more lumber than necessary or in unsatisfactory performance of the final product.

The grades used in the United States and Canada for hardwood lumber are written by the National Hardwood Lumber Association (2015). These rules are also widely used in other countries that export to the United States. A number of softwood lumber trade associations are responsible for preparing and administering the grade rules applicable to their particular species. Softwood grade rules conform to PS 20-15, a product standard developed under the jurisdiction of the US Department of Commerce and administered by the American Lumber Standards Committee (ALSC). The grades of dimension lumber for softwood associations conform to the National Grading Rule for Softwood Dimension Lumber prepared under the supervision of the ALSC.

Hardwoods

The standard grades of hardwood lumber are shown in Table 12.5. These grades are based upon the percentage of the total area of the face of a board that is usable as clear parts. For grading, the inspector first determines and evaluates the poorest side of the board. Then by visual judgment, the proportion and size of clear, rectangular cuttings from the piece is estimated. These cuttings must meet the minimum size requirements shown in Table 12.5. If over 83% of the board area is in these usable cuttings and the board is of sufficient size, the board is an FAS; if it yields between 66 and 83%, it is a no. 1 Common, etc. Experienced inspectors make these estimations at a rapid rate, taking only a few seconds per board. Grading is considered accurate if upon reinspection the value of the lumber is not found to differ from the original by more than 4%.

When purchasing hardwood lumber, a buyer selects the grade based upon the size of pieces (dimension) that they must cut from the lumber. The smaller the pieces required is, the lower the grade of lumber that can be used. Generally, plants obtain much higher yields of *usable cuttings* from the lumber than is indicated by the required yield percentages for that grade, as shown in Table 12.5. This results from the fact that mills are not restricted to the size of cuttings used in the grading rules and the fact that the required yields for the grades are minima and not averages.

TABLE 12.5. Some characteristics of the NHLA hardwood lumber grades.

Grade name	Required yield	Minimum allowed board size (in. × ft)	Minimum allowed cutting size (in. × ft)
FAS (firsts and seconds)	83.3%	6 × 8	4 × 5 or 3 × 7
F1F (first one face)	83.3% on face and 66.7% (no. 1C on back)	6 × 8	4 × 5 or 3 × 7 (FAS side) and 4 × 2 or 3 × 3 (no. 1C side)
Selects	83.3% on face and 66.7% (no. 1C on back)	4 × 6	4 × 5 or 3 × 7 (FAS side) and 4 × 2 or 3 × 3 (no. 1C side)
No. 1 Common	66.7%	3 × 4	4 × 2 or 3 × 3
No. 2A	50%	3 × 4	3 × 2
No. 2B	50%	3 × 4	3 × 2
No. 3A	33.3%	3 × 4	3 × 2
No. 3B	25%	3 × 4	1.5 or wider and contain 36 in.^{2}

Inexperienced hardwood lumber buyers tend to buy higher grades than necessary and as a result needlessly increase raw material costs. To approach the use of lumber grades scientifically, many private firms and public research agencies have developed tables of expected yield for various size cuttings obtained from different grades and species.

There is no moisture content standard for hardwood lumber; i.e. there is no specific moisture content maximum for lumber sold as dry or kiln dried. A purchaser should specify the moisture content requirement when buying such lumber.

Many uses of hardwoods do not include cut-up operations like those in furniture or flooring plants. Pallet plants usually purchase lumber pre-ripped to the width needed. Much of the material used in pallets is of grade no. 3 Common or is purchased ungraded. This material is often sold by groups of species of similar density. The higher-density woods are preferred for most pallets where strength and nail-holding power is needed. Low-density woods are selected where ease of nailing and shipping weight are important.

Hardwoods are not widely used for *construction lumber*, e.g. as joists, rafters, studs, and truss members. Some lower-density hardwoods, however, are suitable for this purpose and may find greater use in the future. For example, aspen has been used to produce 2 × 4s graded similarly to softwoods.

Softwoods

Most softwood lumber is graded by visual inspection. Each piece is inspected on both wide faces. In a few seconds experienced graders assess the knot size and location, the slope of grain, freedom from decay, and other characteristics that determine grade. Dimension lumber to be used in engineered structures can also graded by machine. This is termed *machine stress-rated* (MSR) *lumber*. The main use of MSR lumber is in floor trusses, trussed rafters, and glulam beams. The major components of a machine that stress grades lumber are shown in Figure 12.17. The machine measures the stiffness of lumber by flexing it in the flatwise direction and measuring the force required to do so. The variability in strength of a grade of MSR lumber is less than for visually graded lumber, which is an advantage when high strength is needed.

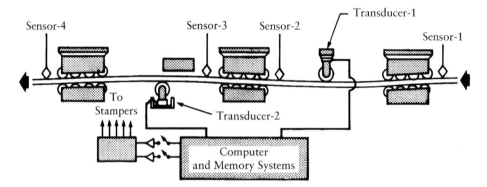

FIGURE 12.17.   Main elements of a machine that mechanically stress grades lumber.

TABLE 12.6.   Some of the grades of softwood dimension lumber that
are assigned allowable working stresses.

| | Grades | |
Nominal size	General use	Structural use
2–4 in. thick and 2–4 in. wide	Construction	Select structural
	Standard	No. 1
	Utility	No. 2
	Appearance	No. 3
		Stud
2–4 in. thick and 5 in. or wider	Select structural	Select structural
	No. 1	No. 1
	No. 2	No. 2
	No. 3	No. 3
	Appearance	

Source: US Department of Commerce (2015). American Softwood Lumber
Standard. PS. 20-15.

Technology is also available to machine grade softwood lumber via X-rays, lasers,
digital photography, ultrasound, et al. For commercial applications, usually a combination
of at least two of these technologies is used, such that both internal and external defects
are identified. Although still a relatively new technology, current automated grading systems are operable at the commercial speeds required of high-production mills and are
ruggedly constructed to survive the routine abuse associated with lumber production.

The major use of softwood lumber is for construction. A few grades that are intended
for cut-up operations to produce window and door parts are called *shop grades*. The
largest portion of construction lumber goes into dimension, i.e. lumber 50–125 mm
(2–5 in.) thick. Some of the softwood lumber grades that are assigned allowable stress
values are shown in Table 12.6, with the highest grade in each category listed at the top.
(Recall that the use of the allowable stress values was discussed in Chapter 9.)

The many grades and species of softwood dimension lumber can cause confusion in
selection of building materials. For instance, there are 11 grades of softwood 2×4s, so if
a mill produces lumber from three species groups it will have 33 different categories of

material for just this one size. Most manufacturers and building materials dealers restrict themselves to a limited number of grades and species. This reduces inventory problems and confusion of the customer/builder. It is common practice to sell mixed grades of dimension, for example, no. 2 and Better or Standard & Better.

Three moisture content levels can be specified for softwood structural lumber: (i) *S-Grn* or *Grn* – the lumber was green or above 19% MC when surfaced; (ii) *S-Dry* or *KD 19* – the lumber was surfaced when at 19% MC or less; and (iii) *MC-15* or *KD 15* – the lumber was surfaced when at 15% MC or less. These categories are used for both boards and dimension. Most structural softwood lumber produced in the United States is KD 19. Timbers are ordinarily sold only in the green condition.

Most softwood lumber, particularly dimension, is grade-stamped at the mill after it has been dried and surfaced. Building codes and most mortgage lenders require the use of grade-stamped lumber for framing. Grade stamps for lumber indicate the grade, the species or species group, the moisture content, the supervisory association, and the number that designates the mill at which the lumber was produced. An example of such a grade stamp is shown in Figure 12.18. A grade stamp for MSR lumber is illustrated in Figure 12.19. The elements MSR stamps are similar to those of visually graded pieces but MSR lumber is designated by allowable bending strength and stiffness values.

The sale of softwood lumber species in various regions follows traditional marketing patterns developed over time. Builders and building contractors become accustomed to certain species and grades and sometimes it is difficult to change their preferences unless significant cost savings are possible. Although mechanical property differences exist among different species, there is no reason why any species of softwood dimension will

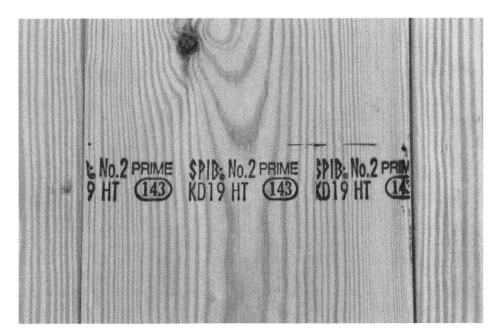

FIGURE 12.18.   A grade stamp on a 2 × 8 indicating that the piece of lumber is no. 2 grade, Prime, southern pine, kiln-dried to 19% MC before surfacing, and qualified as heat treated. The stamp also indicates the grading agency and mill identifier (143). Source: Photo courtesy of Frank C. Owens.

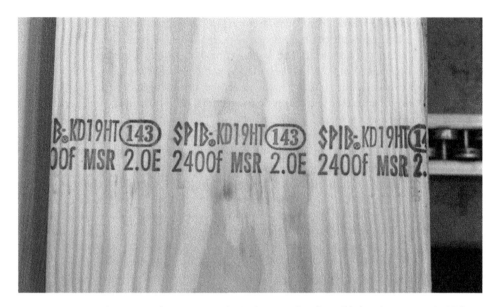

FIGURE 12.19.   A MSR grade stamp on a 2×8 showing the allowable bending stress (2400 f) in psi and modulus of elasticity code (2.0 E) in psi $\times 10^6$. The stamp also indicates the grading agency, mill identifier (143), moisture content at time of surfacing, and that it its qualified as heat treated. Source: Photo courtesy of Frank C. Owens.

TABLE 12.7.   Factors to compare the bending strength and stiffness of various structural dimension lumber sizes.

Nominal size	Depth (in.)	Depth squared (in.^{2})	Depth cubed (in.^{3})
2×4	3.5	12.3	42.9
2×6	5.5	30.3	166
2×8	7.25	52.6	381
2×10	9.25	85.6	792
2×12	11.25	127	1424

> Note: Bending strength (MOR) is a square function of beam depth. Bending stiffness (MOE) is as cubic function of beam depth. These values assume the lumber is used in "joist" or "edgewise orientation not "plank" or "flatwise" orientation.

not do the job for which it is intended if the lumber is properly manufactured and the structure is designed with the properties of that species and grade in mind. In some building construction situations, it is possible to use the next larger size of dimension rather than changing to a higher grade or a stronger species. Such a change is often less costly and results in a stronger and stiffer structure as compared with that made by using the smaller size of the higher grade.

When using dimension lumber for construction, it is often helpful to compare the strength and stiffness of different lumber sizes. Two rules of thumb can be very useful in this regard: (i) *the strength of a beam is proportional to the square of its depth*; and (ii) *the stiffness of a beam is proportional to the cube of the depth*. For example, from Table 12.7, the strength of a 2 × 8 can be compared to a 2 × 10 by comparing the squares of their actual depths: 52.6–85.6. The 2–10 can safely support 85.6/52.6 or 1.6 times as great a load as the 2 × 8. In terms of stiffness, the 2 × 10 is 791.5/381.1 or 2.1 times as stiff as the 2 × 8. This means it will deflect 1.0/2.1 or only 48% as much under a given load. Such comparisons assume that the span and allowable stress values for the sizes being compared are the same. The large increases in strength and stiffness obtained by increasing the depth of structural members by one increment should be understood.

References and Supplemental Reading

Aune, J.E. and Lefevre, E. (1974). Chipping headrigs: do they achieve maximum recovery? *Can. For. Ind.* 94 (8): 70.

Barrett, G. 2004. The U.S. Hardwood Industry Today and Tomorrow. Hardwood Manufacturers Association National Convention.

Cohen, D.H. and Sinclair, S.A. (1990). The adoption of new technologies: impact on performance of producers of softwood lumber and structural panels. *For. Prod. J.* 40 (11/12): 67–73.

Denig, J. (1993). *Small Sawmill Handbook*. San Francisco: Miller Freeman.

Donnell, R. (2004). *Top 200 U.S. Softwood Sawmills. Timber Processing*. Montgomery, AL: Hatton-Brown.

Evans, D.S. (ed.) (1993). *Terms of the Trade*, 3e. Eugene, OR: Random Lengths.

Foreign Agricultural Service. 2006. USDA–FAS Commodity Statistics, (www.fas.usda.gov).

Hallock, H. (1973). *Best Opening Face for Second Growth Timber. Modern Sawmill Techniques*, vol. 1. San Francisco: Miller Freeman.

Hallock, H. and Lewis, D.W. (1974). *Best Opening Face for Southern Pine. Modern Sawmill Techniques*, vol. 2. San Francisco: Miller Freeman.

Howard, J.L. and Jones, Kwameka C. 2016. U.S. timber production, trade, consumption, and price statistics, 1965–2013. Research Paper FPL-RP-679. Madison, WI. USDA, For. Serv., For. Prod. Lab.

HRM. 2017. Hardwood Consumption Estimates 2017. Hardwood Market Report. Memphis, TN.

National Hardwood Lumber Association. 2015. NHLA Rules for the Measurement & Inspection of Hardwood and Cypress. Memphis, TN.

Page, R., and Bois, P. 1961. Buying and selling southern yellow pine sawlogs by weight. GA For. Res. Counc. Rep. 7.

Page, R. and Bois, P. (1993). Sawmill production in the United States – 1991. *For. Prod. J.* 43 (3): 19–21.

Schuler, A. 2005. U.S. softwood lumber supply – imports absorbed almost all the growth in demand. In: Davies, D. Globalization, consolidation, and innovation: evolving markets and changing competitive landscapes. (http://www.interfor.com/pdf/Intl_Wd_Blding_Forum_Sept14-05.pdf).

Spelter, H., McKeever, D., and Alderman, M. 2007. Profile 2007: Softwood Sawmills in the United States and Canada. USDA For. Serv., For. Prod. Lab. FPL-RP-644.

US Department of Commerce. 2015. American Softwood Lumber Standard. PS. 20–15.

Wade, M.W., Bullard, S.H., Steele, P.H., and Araman, P.A. (1992). Estimating hardwood sawmill conversion efficiency based on sawing machine and log characteristics. *For. Prod. J.* 42 (11/12): 21–26.

Wagner, F.G., Gorman, T.M., Pratt, K.L., and Keegan, C.E. (2002). Warp, MOE, and grade of structural lumber curve sawn from small-diameter Douglas-fir logs. *For. Prod. J.* 52 (1): 27–31.

Wagner, F.G., Steele, P.H., Kumar, L., and Butkovic, D. (1991). Computer grading of southern pine lumber. *For. Prod. J.* 41 (2): 27–29.

Williams, D.L., and Hopkins, W.C. 1968. Conversion factors for southern pine products. LA State Univ. Agric. Exp. Stn Bull. 626.

Williston, E.M. (1976). *Lumber Manufacturing: The Design and Operation of Sawmills and Planer Mills*. San Francisco: Miller Freeman.

13

Structural Composites

Structural composites have revolutionized the nature of wood construction. During the last 70 years, the use of composite lumber and panels has grown from limited applications in specialty markets to full integration into commodity and specialty markets such as framing, subflooring, roof and wall sheathing, corner bracing, concrete forming, and others. This chapter discusses both structural panels and composite lumber. The uses of these are significantly different; however, they share a significant amount of common manufacturing technology. Crosslaminated timber (CLT) has characteristics of both panels and lumber.

Figure 13.1 illustrates an updated rendition of Marra's "Non periodic Table of Wood Elements" (Marra 1979). The table illustrates products associated with increased degrees of refinement or *comminution*. Note that as the level of refinement increases, wood or bio-based raw material quality can decrease. Through application of composite products technology, low-value and small raw materials can be utilized to make high-value large-scale products. Nanotechnology may further elevate the performance and economic benefits from wood and biobased composites in products of homogeneous furnish and in those that marry biobased and nonbiobased materials, and in the performance of adhesives and coatings.

Composite wood products, in general, extend and increase the value of the forest resource. Through time, wood-based composites have been developed in response to a wood resource that is tending toward faster growth rates, shorter growth/harvest rotations, smaller-diameter trees and increased proportions of juvenile wood, and lesser strength and stiffness properties. The development of such products has also sought to address increasing market demand for stronger, straighter, and more durable products that are cabable of spanning greater distances and are able to compete structurally and economically with nonwood alternate materials.

Composite lumber combines the the natural strength properties of wood with modern engineering and production technologies to create resource efficient structural products. A group of these are shown in Figure 13.2. These engineered wood products

Forest Products and Wood Science: An Introduction, Seventh Edition. Rubin Shmulsky and P. David Jones.
© 2019 John Wiley & Sons Ltd. Published 2019 by John Wiley & Sons Ltd.
Companion website: www.wiley.com/go/shmulsky

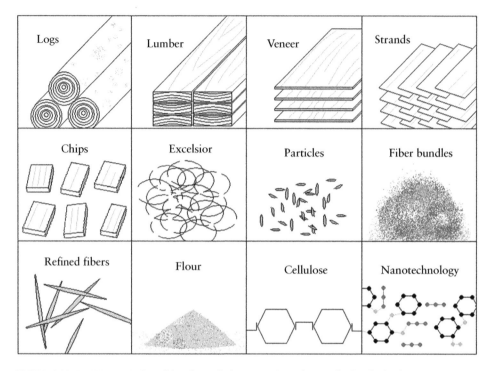

FIGURE 13.1. Nonperiodic table of wood elements. Note that as the level of refining increases, the acceptable quality and cost of raw material decrease.

can outperform solid-sawn lumber in a variety of ways. Structural composites are produced from both the same forest resource as solid-sawn lumber as well as smaller-diameter trees. Fingerjointed lumber uses short lengths of board, called *shook*, as a raw material. Glue-laminated timber utilizes finished lumber as its elemental constituent. Laminated veneer lumber (LVL) uses veneer from relatively high-grade trees. Other composite lumber products rely on low-grade forest resources or by-products from various other wood production operations. In each of these products, natural wood defects are more uniformly distributed than in solid-sawn lumber, which increases the uniformity of their strength properties. In lumber manufacturing, producers chase yield. That is, they benefit when they can squeeze the most board feet (rectangular pieces) out of each log (truncated cylinder). With composites, yield is typically higher and more consistent and manufacturers chase production. Product size is restricted only by manufacturing and handling equipment, not tree size. Many of the structural composite lumber products complement each other in construction (Figure 13.3). This factor greatly enhances potential architectural design and engineering freedom. Research and development of these types of building products is ongoing. Light weight, reduced warp, ready availability, predictable performance, size uniformity, aesthetic qualities, and economics work together to pull increasing amounts of composite lumber products into the market.

In the future, increased development and adoption of wood + nonwood hybrid composites is assured as the market strives for lower-cost, high-performance structural materials. As mentioned in Chapter 9, structural materials are ultimately evaluated based

FIGURE 13.2.   A sample of structural wood composite lumber products; from the left: glulam, PSL, I-joist, LVL, LSL.

on unit performance per unit cost, that is, MPa $\$^{-1}$ (psi $\$^{-1}$) for strength and stiffness. As with solid wood products, structural wood and bio-based composites compare favorably against nonwood and non-bio-based alternatives.

There are two major types of wood-based structural panels, plywood and *oriented strandboard* (OSB). Structural panels are produced from a variety of raw materials varying in size from the raw material used for sawn lumber products or plywood to small bolts suitable for pulp. Like structural composite lumber, structural panels feature more uniformly distributed defects, uniformity, and a high degree of size flexibility. These factors enhance architectural design and engineering freedom. Large panel size, $3\,m^2$ ($32\,ft^2$), facilitates faster construction rates than were ever achievable when lumber was used as roof, floor, and wall sheathing. High racking strength is another advantage common to plywood and OSB. Figure 13.4 shows a home of light-frame construction and plywood sheathing hit heavily by an earthquake in Anchorage, Alaska, in 1964. The home did not collapse but retained its shape as a rectangular solid because of the racking strength of the plywood nailed to the lumber framing. Additional features include light weight, reduced warp, ready availability, predictable performance, size uniformity, pleasing aesthetic qualities, and favorable economics.

What follows is a discussion of the manufacturing and use of some of the most commonly available commercial products. Examples of some new products that are moving toward commercialization are included.

FIGURE 13.3.   Wood frame construction with a combination of engineered wood I-joists and parallel strand lumber. Source: Photo courtesy of Trus Joist, A Weyerhaeuser business.

FIGURE 13.4.   A plywood-sheathed home that remained structurally intact during the Alaskan earthquake of 1964.

End-jointed Lumber

End jointing is the process of connecting two pieces of lumber end to end. Because the end grain of wood is composed of mostly open pores and little surface area, very little tensile strenth can be developed by simply butt jointing the ends of wood. As a means of overcoming this hurdle, scarf jointing was developed (Figure 13.5). To scarf joint wood, a low-angle cut is machined into the ends of lumber, with mated pieces then glued together under pressure. The result is longer lumber than could be otherwise achieved, with tensile strengths on the order of 90% of that of the solid lumber. To minimize the waste associated with endjointing, *fingerjointing* has been developed. In this case, the low-angle cut is folded back upon itself numerous times, giving the appearance of multiple fingers (Figure 13.5). Currently fingerjointed lumber is used for structural members such as studs, composite laminations in glue laminated timber, and nonstructural molding and trim. In the United States, short trim blocks or other *offal* is defected (defects chopped out) and recovered as fingerjointed lumber. In other parts of the world, lumber is often defected and reassembled as clear fingerjointed material. This process works especially well with radiata pine lumber in which short clear lengths of wood are separated by knot *whorls*.

Glue-laminated Timber

Glue-laminated timber (glulam) is produced by face-laminating lumber to form beams. Individual lamina are generally between 19 and 38 mm (0.75 and 1.5 in.) thick. Compared with solid-sawn timbers, wider, deeper, longer, and stronger beams can be manufactured

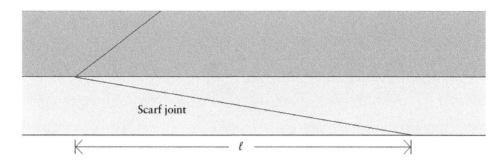

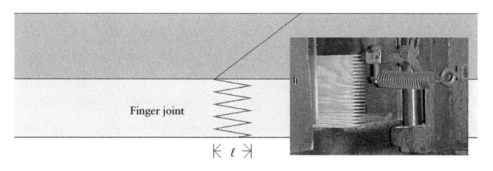

FIGURE 13.5. Schematic of endjointing techniques. Each involves the machining of mating low-angle surfaces to facilitate wood bonding. Scarf jointing (upper frame) evolved into fingerjointing (lower frame) as a means of reducing waste and simplifying the clamping procedure. Inset illustrates a piece of shook emerging from the finger profile cutter.

with glulam technology. Applications for engineered beams include structures such as field houses, sports arenas, bridges, warehouses, residential and commercial buildings, churches, transmission poles, and farm buildings (Figure 13.6). By region, Europe accounts for approximately half of production volume, North America nearly 30%, and Japan close to 20%. In the United States the majority of commercial glulam is manufactured with structural softwood species, mainly southern pine and Douglas fir; however a variety of other softwood and hardwood species are incorporated into glulam production.

Rectangular, commodity-type glulam beam production is highly automated. Curved glulam production is labor intensive. First, high-grade lumber is machine stress rated and sorted according to bending stiffness and stress. Such sorting allows the strongest material to be located on the outer faces of the laminated beams (tension and compression). Weaker material is efficiently placed closer to the neutral axis where bending stresses are lower. Next, individual boards are end-jointed (fingerjointed) to form continuous, full-length lamina. Following fingerjointing, the individual lamina are proof tested and then face planed to remove any adhesive squeeze-out from the fingerjointing process, to produce uniformly thick laminations for the beams, and to provide fresh surfaces for gluing. Once planed, a layer of cold curing waterproof adhesive (generally a phenol–resorcinol–formaldehyde-type resin) is applied to each lamina. Laminations are then pressed together. Standardized rectangular section beams can be produced in

FIGURE 13.6. A commercial building designed to highlight its glue-laminated timber frame.

batch- or caterpillar-type presses that are heated with radiofrequency energy. Such technology can heat and cure the adhesive for a 100 mm (4 in.) thick beam in about 7 min. This rapid processing allows for higher production and reduced manufacturing costs as compared with systems using cold-setting resins. Also, because heat is used to cure the adhesive, phenol formaldehyde or melamine formaldehyde can be used instead of the more expensive resorcinol. Cambered, arched, and tapered beam and column designs are pressed together in clamps on a working floor system. Once glued, the timbers are surface planed, sanded, drilled, or otherwise worked to meet design specifications for the particular application. In addition to a wide variety of custom sizes and shapes, many standard-sized beams are available.

Structural glulam beams can be pressure treated with preservatives for use in decay-susceptable situations. This feature is especially important for bridge timbers and transmission poles. By completing all of the required machining and treating in the factory, these products outperform field-cut and field-treated timbers. In the case of transmission poles up to about 34 m (110 ft) long, glulam products can be delivered within 1 or 2 months of an order as compared with similarly sized solid wood poles which can take more than a year between an initial order and final delivery. In cases where decay or severe weather destroys high-tension line poles, replacement with glulam poles is often a cost-effective and timely alternative.

Glulam timber can also be combined with polymer composite reinforcement (Dagher and Bragdon 2001). Fiber-reinforced composite technology allows the mating of dissimilar materials such as fiberglass, carbon, and polyester fibers to glulam timbers. In this process, high-tensile-strength synthetic fibers are laminated to the tensile faces of glulam beams. These lamination processes yield stronger, stiffer beams that require less wood fiber and have smaller cross-sections. Also, the inclusion of high-strength reinforcement lamina can change the failure mode in the beams from axial tension to axial compression, which improves design safety. This type of composite engineering is shared by the military and by aerospace, auto racing, and performance marine industries.

The fire performance of glue laminated timber is generally very good. Because of the inherent large sizes, these heavy timbers char on the outside and do not support full exothermic combustion. Heavy timber frames of buildings and bridges often survive fires which could easily burn lighter wood frames or weaken steel members to the point of failure.

On a weight-to-weight basis, the cost of glulam is often somewhat higher than that of solid-sawn wood. This cost difference is attributable to many factors. Foremost of these is that the high-value raw material of glulam (high-strength structural lumber) is expensive. Lumber must also be transported from the sawmill to the glulam factory, thereby adding an intermediate shipping cost. Next, the process of producing custom, heavy-laminated timber requires significant capital and operating investment. Costs are better controlled when standard glulam sizes, which lend themselves to automated production, are specified. The structural adhesive incorporated in glue-laminated construction is one of the most costly resins used in the wood industry, and relatively high spread rates are required. Approximately 340 g of adhesive mix is required for every square meter (70 lb per 1000 ft^2) of single glue line. This is about twice the requirement of plywood or LVL. There are also substantial engineering costs associated with glulams intended for unique applications; much of the cost is based on the higher structural performance of glulam versus solid lumber and the large scale of members. Despite these costs, design freedom, warmth of appearance, long-term performance, environmental soundness, and availability allow glue-laminated timber to vigorously compete not only with solid-sawn material but also with other structural products such as concrete and steel in many applications.

Cross-laminated Timber

Similar to glulam, CLT utilizes structural lumber as a raw material. Similar to plywood, the lumber that is used to build up each layer of the panel is oriented at 90° to the adjacent layer(s) (Figure 13.7). In this manner, large panels can be manufactured quickly.

FIGURE 13.7.   A stack of three crosslaminated timber panels. Each panel comprises three layers or plies of structural lumber.

CLT construction is more costly than stick- or factory-built wood construction. However, its potential for rapid and tall (in the 6–10 storey range) construction make it an appealing alternative to concrete. Because it is lighter than concrete, less equipment is needed during building construction, which reduces cost, risk, and time to completion. With an increasing number of production facilities, CLT is becoming an attractive and sustainable alternative for mid-rise home and light-commerical construction.

For manufacturing, kiln-dried, graded structural graded lumber is purchased. Adhesive is applied and the lumber is pressed together to make large panels. Once cured, the panels are removed from the press. The large panels can then be routed for windows, doors, and other fenestration as needed. Also holes and chases for wiring, plumbing, or other utilities can be machined in the factory on a computer-controlled router or saw. The issue of panel durability remains not yet fully investigated. While architectural designs for durability are an important part of the long-term performance equation, so too is wood protection and preservation. To succeed, CLT for use in exterior structural situations will need to be treated against decay and insects just like all wood products.

Structural Plywood

Structural plywood is a panel product of peeled veneer layers glued together so that the grain direction of some wood veneers runs perpendicular to and others parallel to the long axis of the panel. In most types of plywood, the grain of every other layer is applied parallel to the first; the grain orientation of adjacent veneers lies at right angles. Therefore, to maintain a balance from one face of the panel to the other, an uneven number of veneers is often used. Some plywood, however, is produced with an even number of veneers, two examples being softwood plywood made up of four or six *plies* (veneers) with two of the veneers applied parallel to form a thick center core. This type of construction yields a panel with excellent dimensional stability both along and across the panel's long axis. Panels have significant bending strength properties along two major axes, which is favorable for sheathing applications. Waterproof structural plywood has been available since ~1940. A variety of plywood panel types are illustrated in Figure 13.8.

Southern yellow pine species (termed southern pine in the trade) and Douglas fir are the two main species groups used for the manufacture of structural plywood in the United States. However, other major softwoods including the true firs, western hemlock, and western pines are also used. Some hardwoods such as yellow poplar and sweetgum are also utilized.

Structural panels are not sold by species group. Instead, they are classified by their intended structural application such as sheathing, siding, concrete form, or marine-grade plywood. In the case of plywood intended for construction, panels are graded by a *span rating*, which indicates the maximum spacing between framing supports. Other panels are classified by a species group that corresponds to the species of the face and back veneers. The grouping of species for plywood according to United States product standard PS1-09 is shown in Table 13.1.

Many hardwood species, both domestic and foreign, are listed in Table 13.1. Apitong and keruing are a group of species from the genus *Dipterocarpus* originating in the Philippines, Malaysia, and Indonesia. *Lauan* is a group of species once marketed as Philippine mahogany, a misnomer because they are not true mahoganies. *Meranti* refers to many of the same species as lauan but originating in Malaysia or Indonesia rather than in the Philippines.

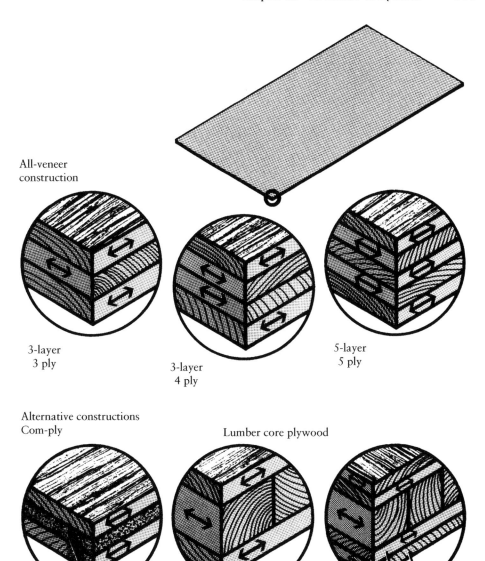

All-veneer
construction

3-layer
3 ply

3-layer
4 ply

5-layer
5 ply

Alternative constructions
Com-ply

Lumber core plywood

Particleboard

3-layer

5-layer

Face

Crossband

FIGURE 13.8. Various types of plywood construction.

Plywood Production

Peeler logs from which veneer and subsequently plywood and LVL are produced have strict size, form, and quality requirements. As such, their prices are relatively high. Production facilities consume roughly 1500–2000 tons (1650–2200 short tons) of logs per day. Logs range from full tree-length to mill-specified sizes. To begin production, all

TABLE 13.1. Grouping of species for structural plywood.

Group 1	Group 2	Group 3	Group 4	Group 5
Apitong	Cedar, port	Mengkulang	Aspen	Basswood
Beech, American	Orford	Meranti, red	Bigtooth	Poplar, balsam
Birch	Cypress	Mersawa	Quaking	
Sweet	Douglas fir*ᵃ*	Pine	Cativo	
Yellow	Fir	Pond	Cedar	
Douglas fir*ᵃ*	Balsam	Red	Incense	
Kapur	California red	Virginia	Western red	
Keruing	Grand	Western white	Cottonwood	
Larch, western	Noble	Spruce	Eastern	
Maple, sugar	Pacific silver	Black	Black (western poplar)	
Pine, southern	White	Red	Pine	
Caribbean	Lauan	Sitka	Eastern white	
Loblolly	Almon	Sweetgum	Sugar	
Longleaf	Bagtikan	Tamarack		
Ocote	Mayapis	Yellow poplar		
Shortleaf	Red			
Slash	Tangile			
Tanoak	White			
	Maple, black			

Source: APA (2010).
*ᵃ*Varies by growth region.

logs are debarked. For plywood and LVL, logs are cut to length on a slasher saw into bolts, ~2.64 m (104 in.) long. This length leaves ~100 mm (4 in.) at each end for final trim allowance of the finished 2.44 m (96 in.) long panels. The bolts or peeler blocks are subsequently heated and peeled on lathes. Short or crooked sections not fit for peeling are chipped for pulp or sawn to lumber. The small-diameter stems and cores left after peeling can also be sawn into lumber, or converted into posts or landscape timbers. Peeled veneer yield is generally between 60 and 70%. Bark, wood veneer round-up, and core material do not go into the finished product.

Heating the Blocks

Most softwood veneer blocks are heated prior to peeling. Heating softens the wood and knots, thereby reducing cutting power consumption and producing a higher volume of smoother, higher-quality veneer. Steaming, soaking in hot, slightly alkaline water, spraying with hot water, or combinations of these methods are all used in various situations to obtain the increased wood temperature required. The objective is to heat logs to a suitable temperature as deeply as veneer will be cut. Most softwoods are heated to a core temperature of 50–60 °C (120–140 °F). The exact heating time required depends upon the density of the wood, the diameter of the block, the bath temperature, the initial wood temperature, and the temperature required for a satisfactory cut.

Peeling Veneer

Virtually all veneer in structural veneer is peeled on lathes. Rotary lathes are designed to produce continuous ribbons of green (undried) veneer that is subsequently clipped to recover usable veneer widths. From each block, a small center core is ejected and further processed into such things as lumber, two-sided landscape timbers, or pulp chips. As shown in Figure 13.9, the pressure bar and knife assembly move forward simultaneously as each block is turned on the lathe. Also shown is a veneer slicer used for manufacturing lower-production, higher-quality veneer generally utilized in nonstructural panels. Careful adjustment of the cutting angles and the horizontal and vertical gap between the pressure bar and the knife edge is necessary to obtain the proper peel thickness. The proper pressure must be developed by the pressure bar to reduce checking of the veneer as it is severed from the log and to minimize roughness. The use of a powered roller on the tip of the *pressure bar* improves veneer quality. The veneer then springs back in thickness once the cut is made. Most structural veneer is peeled to a target thickness of 2.5–4.2 mm (1/10–1/6 in.).

Close examination of veneer shows hairline fractures running parallel to the grain, called *lathe checks*. These are a result of peeling a flat-surfaced veneer from a round tree stem. The face of veneer from the outside of the round bolt is called the tight side because it has no such checks. The opposite face, on which lathe checks appear, is called the loose side. For the face plies on panels, the loose side of the veneer is oriented inward toward the glueline, which works to repair the checks and improve appearance. If a sanded grade of plywood is painted and subjected to weathering, lathe checks may open and appear as checks in the finish. For this reason, plywood for use as siding or for other exterior finish applications is generally manufactured to have a rough-textured surface where such checks will add to the textured appearance.

One of the keys to the high-speed production of veneer in a modern plywood plant is a computer-controlled scanner-operated *charger* that automatically locates the geometric center of each block and then quickly loads (charges) the lathe. The positioning devices are called *optimizers*. A veneer lathe and charger are shown in Figure 13.10. Equipment of this type can charge the lathe with a small-diameter block, round the block, peel the block to a 90–120 mm (3.5–4.75 in.) diameter core, and eject the core in about 10 s. As the core is ejected, the charger has another bolt ready for the next cycle. This equipment is abso-lutely necessary in mills that peel relatively small average diameter blocks (~215 mm) because high piece counts per hour are needed to maintain profitable production.

An innovative peeling method was introduced in the late 1980s. In this case a "spin-dleless" or "centerless" lathe uses a series of rollers rather than end chucks to support the block during peeling. This allows the peeling of cores down to diameters of 50 mm (2.0 in.) or less. The increased veneer yield is especially important to mills that rely on small-diameter logs. Wood quality from these machines, however, is lower than from larger chuck-driven lathes. Wood in the inner core is primarily juvenile wood, has the highest occurrence of knots, and tends to be excessively wavy – an artifact of producing flat veneer from small-diameter stems.

Veneer Clipping

In high-speed plywood mills a series of trays is generally used to handle the ribbon of veneer, which is peeled at a rate of 1.5–4.0 m s^{-1} (300–780 ft min^{-1}). The trays are long enough, about 36 m (120 ft), to handle the veneer that comes from an entire 380 mm (15 in.) diameter block. Figure 13.11 shows a *tray storage* system located between the lathe and the clipper.

Clippers are high-speed knives that chop the veneer ribbons to usable widths. In struc-tural veneer mills, veneer is clipped automatically at speeds up to 7.5 m s^{-1} (1450 ft min^{-1}).

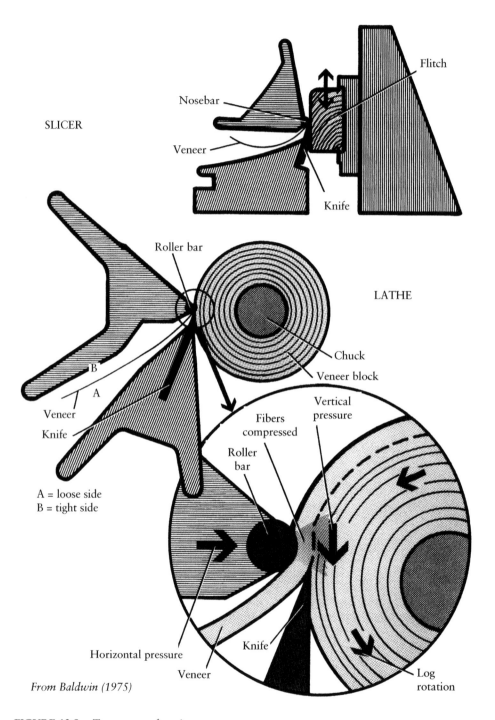

SLICER

LATHE

A = loose side
B = tight side

From Baldwin (1975)

FIGURE 13.9.   Two means of cutting veneer.

FIGURE 13.10. Step feeder, lathe charger, and a small-log veneer lathe.

FIGURE 13.11. A typical veneer tray storage system as seen looking back toward the lathe. The veneer being discharged is moving to the clipper.

Clippers cut veneer to widths of about 1.37 m (54 in.) to provide an allowance for drying and trimming to the 1.22 m (48 in.) finished panel width. Open flaws in the ribbon are detected and removed by the clipper. Broken or short veneer produced during round-up can be trimmed and clipped for core stock, burned for energy at the mill, or used in other wood composites.

Drying

The fundamentals of drying were outlined in Chapter 7. Many types of dryers have been used in structural veneer mills: forced air roller-restraint, platen type heated by steam, and radio frequency (Sellers 1985). The majority of structural veneers are dried in the forced air roller-restraint type of dryer. The operation of one such dryer is illustrated in Figure 13.12 showing how hot air at speeds up to 20 m s^{-1} (4000 ft min^{-1}) is impinged on the veneer. This removes the boundary layer of moist air that can act as an insulator in dryers with low-velocity air circulation. Dryer temperatures are generally not more than 205 °C (400 °F) and residence time in the dryer is typically 8–10 min. Ideally, final veneer moisture contents range from 3 to 6% (Sellers 1985).

A significant problem encountered in veneer and strand drying is the generation of air pollution emissions that contain hydrocarbon-based volatile organic compounds and hazardous air pollutants such as methanol and formaldehyde. Because of strict and costly pollution regulations, mill production is often limited by generation of air pollutants. Current control technology uses wet electrostatic precipitation to remove particulate matter from the dryer stack before the volatile organic compounds and hazardous air pollutants are destroyed by post-dryer combustion or biofiltration. Such end-of-pipe emission control technologies are costly.

Veneer Grading

Most structural veneer is first graded ultrasonically. The highest grades (stiffest, strongest, and densest) are routed to LVL production. The lesser grades are then visually graded for use in plywood. Grades are based generally on the occurance of knots and their sizes

Jet tube

Veneer surface

FIGURE 13.12.   Principle of the jet (impingement) dryer.

across the sheets of veneer. After grading, high-value veneer can be upgraded by sewing or taping splits, plugging and patching holes, and filling cracks.

Lay-up

Assembly-line-type processing technology has mechanized most production operations. Panels are built up, consolidated, pressed, and trimmed with minimal manual labor. The application of adhesive to veneer in automated systems is commonly done with spray or curtain coaters. Both of these methods are well suited to automated lay-up systems. In these systems veneer travels on a belt under a liquid spray or curtain. *Spray coaters* use low-pressure nozzles to apply liquid droplets. In excessively dry processing environments, spray coaters are less favorable because droplets can dry out before they get to the hot press. *Curtain coaters* consist essentially of a box with a slot in the bottom through which adhesive flows in a continuous sheet or curtain. With curtain coaters special care must be taken to adjust the spread rate; misapplication is costly. Other application methods are used to a lesser extent. To build thicker panels, additional layup stations can be added to the assembly line.

The assembly of veneers into plywood panels is highly mechanized in large mills. Modern modular assembly systems contain a station for each successive layer of veneer in the product. Systems that are nearly automatic are widely used, but the variability in size of veneer makes some manual input necessary. For example, the full-size face sheets of veneer may be handled by machine, but the narrower strips used to make up the core may be handled manually. Systems such as stapling, sewing, taping, hot gluing, or otherwise connecting the strips of veneer for the core are widely used. Veneer sheets that are so created can then be handled by machine rather than manually.

Pressing

Most plywood mills prepress batches of layed-up panels in a cold press prior to final pressing in the hot press. This process permits easier loading of the hot press and reduces shifting of the veneer, which can result in down-graded panels. Hot pressing to cure the plywood gluelines is done in hydraulically powered heated presses like the multiopening presses shown in Figure 13.13. Such presses can produce 20–40 1.22 by 2.44 m (4 by 8 ft) panels with each press cycle, which takes about 2–7 min. The purpose of the press is twofold: to bring the layers of veneer tightly together and to heat the adhesive to the temperature required for resin polymerization. The *phenol–formaldehyde* (PF) based resins used in most structural veneer-based products typically require temperatures of about 115 °C (240 °F) at the innermost glueline for ~90 s to cure properly. For plywood, heated platens are used to transfer the heat energy to the wood composite system. Steam, electricity, and heated oil are all used as heating media. During the hot pressing and hot stacking immediately thereafter, the phenol and the formaldehyde bind with each other to form a thermoset plastic. Their chemical reaction is controlled such that upon completion there is no free formaldehyde in the panels which could otherwise later off-gas.

High-quality smooth veneer with uniform thickness requires less pressure than lower-quality veneer. Pressures used in the press cycles vary from about 750 kPa (110 psi) for low-density woods to 1500 kPa (220 psi) for denser species.

FIGURE 13.13. Two 24-opening 1.22×2.44 m (4×8 ft) plywood hot presses. The press openings are loaded by an elevator seen in front of one of the presses.

Panel Grades and Uses

There are five primary considerations when selecting plywood panels for a specific use: (i) durability of the glueline needed to avoid delamination; (ii) strength requirements for panels to be used structurally; (iii) quality needed on the faces to accomplish the appearance desired; (iv) special requirements such as fire or decay resistance; and (v) market cost differences.

For structural plywood manufactured under APA–The Engineered Wood Association standards, the durability of the glueline is specified as being either *Exterior* or *Exposure 1*. These two designations indicate not only the durability of the glueline but also the grade of veneers that must be used in laying up the panel. All structural plywood produced today in the United States has an Exterior-rated adhesive (phenol formaldehyde) regardless of whether it is classified as Exterior or Exposure 1.

The Exterior durability classification is intended for applications that may involve permanent exposure to weathering such as siding and concrete forming. The Exposure 1 durability classification is intended for uses that may involve temporary exposure to weathering such as during a construction delay. Exposure 1 panels should ultimately be protected from permanent exposure to weather.

Panel grades are generally identified in terms of the veneer grades of the two surfaces, e.g. A–B or C–D, or by the intended use, e.g. underlayment or rated-sheathing. The veneer grades define both the appearance of the growth characteristics (knots and holes) and the amount of veneer repair that may be made. The main grades of veneer and the characteristics are outlined in Table 13.2.

TABLE 13.2.   Relationship among block diameter, core diameter, and theoretical percentage veneer yield.

Block diameter (cm)	Veneer yield (%)		
	Core diameter		
	8 cm	10 cm	12 cm
20	84	75	64
25	90	84	77
30	93	89	84
35	95	92	88
40	96	94	91
45	97	95	93

Source: Adapted from Williams and Hopkins (1968).

There are also differences between the grades of sanded structural plywood and the unsanded sheathing/structural panels. Although oriented strand board has largely displaced plywood as a structural sheathing in housing construction, the residential construction market still accounts for about one-third of plywood market demand in the United States. Remodeling is another major market. By far the greatest volumes of unsanded structural plywood used in construction and remodeling are used for subfloors and roof and wall sheathing. Sanded grades are used for paneling and siding among other things. The grades C–C and C–D are commonly used in these applications. Marine, industrial, and export are additional markets where the unique and positive attributes of plywood fare well.

Panels are designated by the spans (in inches) over which they can be used as roof sheathing and subflooring in residential construction, e.g. 32/16. A 32/16 panel can be used as roof sheathing to span rafters up to 810 mm (32 in.) apart or as subfloor over joists spaced 405 mm (16 in.) apart (distance from the center). Current practice in the United States is to only include imperial units (and not metric measures) on the grade stamps. A typical registered grade stamp of Timber Products Inspection, a third-party inspection agency, is shown in Figure 13.14. In addition to the span rating, grade stamps indicate the durability classification, the thickness of the panel, the mill number (318 is shown), and the product standard. Only association member mills that participate in their quality inspection and testing programs can use these stamps. Other agencies have their own grade stamps. Each agency has its own quality verification programs and each stamp contains similar information.

Another major market for softwood plywood today is the industrial sector, including such applications as furniture frames, truck trailer linings, RV floors, agricultural bins, shipping containers, and pallets. Nonresidential construction, including panels for concrete forming, is a lesser market. Two approaches are used to overcome surface checking of plywood in exterior situations. One method is to *overlay* the plywood with a layer of medium- or high-density resin-impregnated wood fiber sheets. These wood fiber overlays hold paint well and provide a check-free base. The second approach is to provide a rough or textured surface that is not downgraded by checks. These rough, natural surfaces are receptive to stains and are popular as exterior siding. Figure 13.15 shows some of the softwood plywood surfaces that are manufactured.

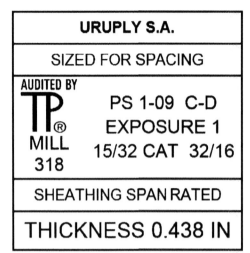

URUPLY S.A.
SIZED FOR SPACING

AUDITED BY
TP ®
MILL
318

PS 1-09 C-D
EXPOSURE 1
15/32 CAT 32/16

SHEATHING SPAN RATED
THICKNESS 0.438 IN

FIGURE 13.14.   Sample grade stamp for softwood plywood. Among other things, this stamp indicates the auditing agency (TP, Timber Products Inspection, Inc.); mill number 318; product standard to which the panel conforms (PS 1-09); grade of face and back veneers (C and D); adhesive exposure rating (Exposure 1, waterproof); thickness class [15/32 in. (0.438 actual)]; and span rating 32/16. Source: Courtesy of Timber Products Inspection, Inc.

Oriented Strandboard

OSB was first developed and marketed as an improved form of Canadian waferboard in the early 1980s. OSB shares many characteristics with plywood but it is manufactured from a lower-value forest resource and it is more resource utilization efficient. Raw material costs for OSB, and strand-based composites are generally lower and typically follow pulp wood prices. Logs or short bolts are relatively small in diameter and are processed to create millions of thin strands which are subsequently layered together. As such, OSB's production costs are lower and, in general, it can compete favorably with plywood on a direct cost basis. That said, to successfully meet the various mechanical properties and quality tests, more wood and more adhesive may be used, which are costly. Thus making OSB is a constant balancing act between reducing costs and making quality products.

Most panels are usually manufactured in standard 1.22 × 2.44 m (4 × 8 ft) sizes; however, other custom sizes are available. In the United States, most plywood and OSB is manufactured within a thickness range of 6.3–19 mm (0.25–0.75 in.).

Early OSB mills primarily utilized aspen because of its low density, low cost, and wide availability. Following development with aspen, a variety of other species have since been incorporated. Currently, mills can use almost any low- to medium-density species that is widely available. Southern pine, spruce, birch, yellow poplar, sweetgum, sassafrass, beech, and others are incorporated into OSB. Species of relatively high density, such as beech, are often mixed with species of relatively low density, such as aspen, to maintain acceptable board properties.

Strand yields, however, run about 85–90% with losses coming mainly from bark and strands that break (fines) during processing. To improve yield, OSB mills in colder climates also heat blocks prior to stranding. Either long or short logs can be utilized, depending on a specific mill's operation.

Mills often procure market logs from a radius approaching 200 km (125 miles). Where mixed species are used it is advantageous to maintain a uniform mix of species and wood moisture content in order to minimize processing adjustments in the mill. Truck or railcar loads of roundwood are most commonly purchased by weight but

(a)   (b)

(c)   (d)

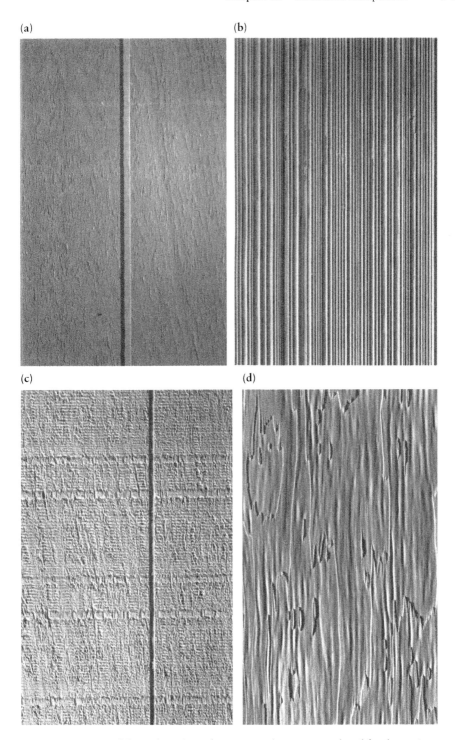

FIGURE 13.15.   Some of the wide variety of textures and patterns produced for decorative softwood plywood.

FIGURE 13.16.   OSB log yard crane. Source: Photo courtesy of Mississippi State University.

sometimes by bulk volume. Truckloads of logs, about 20 tons (22 short tons) each, are handled by fork lift or cranes capable of lifting entire loads with a single pick (Figure 13.16). In locations where risk of decay is high and logs must be stored for extended periods, water spray storage is employed; keeping logs continuously wet has been found to provide short-term protection against stain and decay. At certain times of the year, log yard inventories may reach $150\,000\,m^3$ (5.3 million ft^3). Daily roundwood volume consumed by commercial mills is $\sim\!2400\,m^3$ ($85\,000\,ft^3$) with daily product output for a single mill on the order of $1000\,m^3$ ($35\,000\,ft^3$).

Producing OSB Strands

The mechanical bending strength in composite wood panels comes largely from the unbroken, elongated, wood pieces. Therefore, to engineer mechanically strong strand-based panels, relatively long elements of wood should be incorporated. The ideal particle for developing strength, dimensional stability, and product uniformity is a thin flake of uniform thickness with a high length-to-thickness ratio. Roundwood is required to produce such strands. To produce strands, disk stranders are commonly utilized. First, a group of logs is advanced into a strander and clamped (Figure 13.17). Disk stranders have a series of $\sim\!50$ radially oriented knives inserted into large diameter ($\sim\!2.5\,m$, $\sim\!8\,ft$) steel disks. As the cutting disk rotates, it advances into the side of the clamped bolts or tree length stems, thereby cutting long strands, parallel to the long axis of the roundwood. Scoring knives inserted onto the long cutting knives control the strand lengths. A strand length of $15\,cm$ ($6\,in.$) is commonly used so a $60\,cm$ ($24\,in.$) long knife will

FIGURE 13.17.   A 750 kW commercial disk flaker advancing into a bundle of clamped tree-length stems. Source: Photo courtesy of Mississippi State University.

produce four 15 cm (6 in.) long strands simultaneously. Strand thickness can be manipulated by adjusting the speed at which the disk advances into the roundwood. Target thickness is generally on the order of 0.75 mm (0.030 in.). Strand widths are generally between 2 and 5 cm (0.8 and 2.0 in.). These geometries have been developed through commercial-level trial and error. They balance board performance with productivity and yield. To maintain adequate production, many mills have two or three stranders. This situation allows continuous green strand production on one (or two) machine(s) while another is being maintained. Occasionally, mills have separate stranders for producing face and core strands. Each strander can reduce a ton of roundwood to strands in about 90 s. Once the strands have been produced, they are conveyed to green storage bins to await drying.

Strand Drying

Good adhesion between strands in OSB production generally requires a wood moisture content from 2 to 6%, depending on the type of resin. Strand residence time in the dryer is generally between one and five minutes. Both rotary-type (somewhat like a tumble dryer for clothes) and flat-screen-type dryers are used. Hot air entering the dryers can be as hot as 540 °C (1000 °F). As the air and strands move through the dryer, the moisture evaporating from the strands cools the air. Exiting air temperatures are generally about 200 °C (392 °F). These temperatures maximize production volume without appreciably degrading the strands. Flat screen dryers are the more modern of the dryer types. As compared with dryers using older technology, these typically retain better strand surface quality, provide less strand "curling," break fewer strands, and generate fewer fines which absorb much resin without improving board properties.

Strand Blending and Mat Forming

Analogous to the layup process for plywood are the blending and forming processes for OSB and strand-based composites. The addition of resin adhesive and *wax* or other sizing agents to the strands is called *blending*. The wax (ranging from about 0.25 to 2% by weight) is added to provide some water repellency to panels. The quantity of resin solids (ranging from 2 to 5% by weight) for OSB is low considering the large surface area of the strands. When plywood veneers are glued, a continuous layer or film of adhesive is applied between the layers. In OSB and strand-based composites, however, the adhesive coverage is dispersed. Coverage consists of discrete resin droplets. This coating is somewhat analogous to spot welding metal.

Blending of the furnish for the surface and core layers of OSB is often done separately such that more resin, wax, and moisture can be added to the surface layers if desired. The resin and wax are usually added to the blender through spray nozzles or simple tubes, or on rotating centrifugal applicators that atomize the materials. Finely atomized particles form a resin-fog inside the blender. Historically, resin was applied as a liquid. Currently, processing technology allows for adhesive blending as a powder. This process eliminates the addition of excess water to the wood from the resin, lowers the shipping cost of resin, and increases resin shelf-life. Both types of blending are used commercially. In some cases, such as in producing exposure- or exterior-type composites, powdered or granular borates or other fungicides can be added to the furnish during blending.

The process of depositing blended strands into the form of a mat is termed *forming*. The objective of formation is to provide as uniform a mat as is possible, i.e. to produce a uniform mat weight across the area of the board. Multiple forming heads are used to deposit oriented strands onto a moving belt. Electrostatic flake orientation, in which strands must pass through an electrical field, has been used with limited success. Theoretically, because strands are somewhat polar, they align themselves accordingly and form successive strand layers in the panel. Commercially, this technology has not proven reliable. Mechanical orientation has proven to be the manufacturing technology of choice. Using this technology strands drop through an orienter which, for the most part, aligns them within about 15° of their intended axis. To align the strands parallel to the belt direction the strands drop through a series of spinning disks, spaced ~10 cm (4 in.) apart. The disks are about 40 cm (16 in.) in diameter. This apparatus looks similar to an agricultural harrow which has several smooth and evenly spaced disks mounted on a common arbor (Figure 13.18). To align the strands perpendicular to the forming belt direction (the core layers of OSB) a different machine is used. There strands fall onto a large rotating fluted cylinder that looks similar to the paddlewheel on a stern-driven riverboat. Strands self-align in the flutes of the wheel (perpendicular to the forming belt direction) and land on the belt, duly oriented, as the cylinder rotates. Most OSB is produced with four forming heads; the first forms the bottom face (~25%, parallel), the next two form the core (~50%, perpendicular), and the last forms the top face (~25%, parallel). This process can be considered a four-layer construction. Forming the face and core at these ratios provides a dimensionally balanced panel. Final board thickness, dimension, and density are determined during this stage. Process control continually monitors and adjusts the former and belt speeds to create the desired mat thickness and density. It requires a loose mat, ~100 mm (4 in.) thick, to produce a finished 12.7 mm (0.5 in.) thick panel (Figure 13.19).

FIGURE 13.18.   One type of former and strand orienting machine.

FIGURE 13.19.   Loosely formed OSB mat ~14cm (5.5 in.) thick that will be condensed to form a 11.1 mm (7/16 in.) thick panel. Source: Photo courtesy of Mississippi State University.

Natural wood characteristics such as knots and holes, which affect the strength of veneer and therefore structural plywood, are essentially absent in OSB. While a sheet of structural plywood typically has between four and seven discrete layers, a comparable sheet of OSB has more than 50 discrete plies of strands. Strand-based composites are more homogeneous, which increases their inherent uniformity. Mechanical strength properties of OSB are typically the same or higher than those associated with structural plywood.

The loose-strand mat is either cut with a flying cutoff saw (in the case of batch presses) or continues directly to the press (in the case of continuous presses). With continuous presses the loose-strand mat is cold pressed via converging belts as it enters the hot press. This step is not necessary with batch presses because there the discrete sections of the loose mat are loaded into a press charger.

OSB Hot Pressing

For batch-pressed OSB, the discrete sections of the loose strand mat are transferred from the press-charger to the press simultaneously. Multiopening presses like those shown in Figure 13.13 are used throughout North America. Large presses have platens measuring up to about 3.66 × 12.2 m (12 × 40 ft). Large presses are favorable because they greatly increase production capacity. Rapid loading of 2.74 × 7.62 m (9 × 25 ft) panels into the indexed charger of a multiopening press is shown (Figure 13.20). Modern, large single-opening presses are also used in some mills, primarily in Europe. Continuous presses are fed directly from the forming line. There, the loose-strand mat passes between converging belts until it reaches its target thickness, thus causing substantial mechanical pressure to build within the mat. At that point heat is applied and the thermoset adhesive cures.

For most OSB sheathing and subflooring-type products which range in thickness from 11.1 to 19.1 mm (0.44–0.75 in.), heated press platens are used to cure the adhesive.

FIGURE 13.20.   A 16-opening 2.74 × 7.62 m (9 × 25 ft) OSB hot press charger. Each of the 16 billets will be cut into six, 1.22 by 2.44 m (4 × 8 ft) boards. Source: Photo courtesy of Mississippi State University.

These platens conduct heat from the press to the mat to cure the adhesive. Platen temperatures range from ~175 to 205 °C (350–400 °F). Press times are generally between three and six minutes depending on board density and thickness. Because conductive heat transfer is approximately a function of the square of panel thickness, the press time for a 38 mm (1.5 in.) thick board is about four times that of a similar density 19 mm (0.75 in.) thick board. This relationship makes conventional hot pressing of thick composite panels slow and costly. Long hot press cycles can damage board surfaces and substantially impede production volumes. To accelerate the cure of thicker mats, both high-frequency curing and steam injection are used. In either case, batch or continuous, the key to high production is short press cycles. During pressing, maximum platen pressures are generated while the mat is consolidated to the desired thickness. These initial pressures are on the order of 4800–5500 kPa (700–800 psi). Throughout the cycle, only enough pressure is maintained to sustain the desired thickness, which means that hydraulic gauge pressure drops significantly. The parameters of press temperature, closing time, and adhesive cure rate are constantly being refined and manipulated to decrease press cycle time and improve production rates.

Because the press is one of the most expensive pieces of equipment in the plant, manufacturing processes are designed to keep the press operating continually. Production bottlenecks ahead of the press must be eliminated by both competent design and diligent operations engineering. As the multiopening press opens, the mats on the press charger are loaded into the press. At the same time the pressed panels are unloaded on the other side. Also, the forming process is difficult to start and stop, so once the forming line is running well it is best to keep it going. To maximize return on investment most mills operate 24 h per day, 365 days a year, with regularly scheduled down-time periods for maintenance.

OSB Applications

The major use of OSB is for structural panels, primarily roof and wall sheathing and subflooring, in light-frame construction (Figure 13.21). Required board properties are determined only by the specific end use of the composite board. I-joist web stock is another major end uses for OSB. Each application requires different board performance attributes. Thus, processing conditions for each product are manipulated and controlled to produce specific board products tailored to specific applications. Similar to structural plywood, OSB is grade stamped. The grade stamps contain information about its manufacturing, physical and mechanical properties, as well as its acceptable use categories (Figure 13.22).

Because OSB is formed in layers it is possible to manipulate the proportion and type of resin, moisture, processing parameters, and flake geometry throughout the board to provide superior strength. The major strength advantage of OSB is derived from the parallel orientation of the strands. Note the 3:1 or 4:1 strength ratio between parallel and perpendicular orientation. Randomly oriented waferboards typically have bending MORs of about 17 MPa (2500 psi), while parallel-to-grain MOR for sheathing-grade OSB is generally at or above 29 MPa (4200 psi). The density of sheathing-grade OSB is usually kept between 640 and 670 kg m^{-3} (40–42 lb ft^{-3}), depending on individual mill processing parameters. Wood species with densities from about 370 to 480 kg m^{-3} are usually used to meet board density requirements.

FIGURE 13.21.   OSB being used as wall sheathing on a residence.

Laminated Veneer Lumber

Similar to glulam, this structural composite utilizes more or less continuous lamina along its length. Similar to plywood, LVL is made from sheets of veneer that are face laminated to build up member depth (Figure 13.23); however, with LVL all of the veneer sheets are oriented parallel to the long axis of the product. The density of LVL nearly matches that of its wood constituents, that is, there is little "densification" during pressing. Veneer thicknesses range from ~1.5 to 6 mm (0.06–0.25 in.).

FIGURE 13.22. Sample grade stamp for OSB. Among other things, this stamp indicates the auditing agency (TP, Timber Products Inspection, Inc.); mill number 361; product standard to which the panel conforms (PS 2-10); that it is tongue and grooved, cut 1/8 in. scant to leave room for expansion at edges; adhesive exposure rating (Exposure 1, waterproof); thickness class (23/32 in. [0.703 actual]); and floor span rating 24 in. on center (24 oc). Source: Courtesy of Timber Products Inspection, Inc.

The strength of LVL is similar to that of the highest grades of solid lumber because the grain direction of each lamina is parallel to the board axis and the wood defects in the veneer sheets are randomized, so the strength properties are more uniform. LVL is usually manufactured at a 38 mm (1.5 in.) thickness, to match the standard thickness of solid-sawn dimension lumber, or 44 mm (1.75 in.). For the most part, LVL production parallels that of structural plywood. Round logs are slashed and heated, then veneer is peeled, clipped, dried, and ultrasonically graded. This process allows the most efficient use of the veneer. From the sorted veneer, products with predictable and relatively uniform strength properties are readily achievable. Following veneer grading, large billets (typically 1.22 m wide by 24.4 m long) of LVL are layed up and pressed. This process is highly automated. LVL billets are produced by applying layers of veneer and adhesive sequentially to a moving belt. Modern modular assembly systems contain a station for each successive layer of veneer in the product. Most mills prepress the LVL billets in a cold press prior to final pressing in the hot press. Pre-pressing is done at room temperature and relatively low pressure. As with plywood, the purpose is to allow the wet adhesive to tack the veneer layers together.

FIGURE 13.23.   Structural laminated veneer lumber (LVL), 38 mm (1.5 in.) thick. Note the crush-lap veneer joint near the center.

LVL hot presses fall into two major categories: single opening and continuous. Single-opening presses produce large billets, ~1.22 m (4 ft) wide and up to 24.4 m (80 ft) long, one at a time. Continuous presses have caterpillar-type track systems that continually press the material as it moves through the press. Owing to the greater thickness of LVL, compared with plywood, and the nature of heat transfer, it takes 15–20 minutes to conventionally hot press LVL. High-frequency energy pressing is commonly used to decrease the press time; using this technology, tuned electric waves are passed through the wood to selectively excite water molecules in the resin, similar to the action of a microwave oven. This type of energy transfer can reduce press cycle times to 4–6 min. In each product and processing case, resin systems must be carefully tailored to the specific conditions of the plant.

The exact pressing conditions are designed to bring the veneer surfaces tightly together without overcompressing the wood. Some compression and crushing of overlapping lamina at the veneer joints is a necessary part of manufacturing. Pressures used in the press cycles vary from about 750 kPa (110 psi) for low-density woods to 1400 kPa (200 psi) for dense species. Cured billets are ripped into a variety of widths, often to compete directly with high-grade solid-sawn dimension lumber.

LVL Species, Grading, and Use

Because southern yellow pine and Douglas fir are the two main species used for the manufacture of structural veneer in the United States, these are also the two primary species consumed in LVL production. Other softwood species such as fir, spruce, and hemlock and hardwoods such as gum and yellow poplar are also used.

LVL is rated and sold based on mechanical strength and stiffness properties. The mechanical properties are influenced mainly by the raw material properties and manufacturing parameters. Because established mechanical strength properties are specified for each billet, LVL applications can be designed following engineering *allowable stress design* practices.

LVL is more costly than solid-sawn lumber but it has several technical advantages that make it a popular product for applications where straightness, high strength, and/or long lengths are needed. One advantage of LVL is the fact that it has a uniform and well-controlled moisture content. The orientation of wood elements in LVL is quite similar to that in solid wood and therefore it shrinks and swells with moisture content change similar to wood in solid form. However, because of its multiveneer construction it is less likely to warp with moisture content changes. Dispersion of strength-reducing growth characteristics such as knots and slope of grain throughout the piece (randomized in each veneer lamina) results in design strength properties that are higher and more uniform than most solid-sawn lumber. Typical axial design bending stress values for strength and stiffness are ~22 and 13 800 MPa (3200 and 2 000 000 psi), respectively. The major applications of LVL are flanges in wood I-beams (about 50% of LVL production), headers over garage doors and large windows, structural members where safety is critical such as in scaffold planks, built up girders, and other high-stress applications. Nonstructural LVL is produced with a colorless adhesives for door and window parts, other millwork, and furniture. Like plywood, LVL can be preservative treated for use in decay-prone situations.

While LVL is one of the highest performing composite lumber substitutes, its future is closely tied to that of structural plywood. LVL is manufactured from high-grade structural veneer which often comes from plywood plants that grade and separate the highest strength veneer specifically for LVL.

Strand-based Composite Lumber Products

Strand-based composites include *parallel strand lumber* (PSL, made from long veneer strands), *laminated strand lumber* (LSL) and *oriented strand lumber* (OSL), both made from thin wood strands or flakes. The thin strands used in manufacturing LSL are made from carefully refined small-diameter, low-quality roundwood. The refined strands are then parallel laminated in order to produce elongated beams or billets with high uniaxial performance.

These products have achieved nontrivial market development and growth during the past two decades. Strand-based structural lumber substitutes and engineered wood I-joists have brought viable alternatives to solid-sawn lumber and LVL. Development of these and other new building products is ongoing and highly competitive.

Parallel strand laminating technology has been used successfully to manufacture composite "lumber-like" products. As with OSB, strands for this product are manufactured, dried, and blended with resin. The dry strands are then aligned, and formed on a continuous belt. In this case, however, all of the strands are oriented parallel to the long axis of the product. While this design loses the transverse dimensional stability associated with OSB, it maximizes the product bending strength. As such, some types of strand-based composite lumber can compete directly with solid-sawn structural framing lumber.

Parallel strand lumber is produced mainly from long oriented strands of Douglas fir or southern pine structural veneer. Veneer is sourced from structural plywood or LVL mills. The first or initial veneer removed from logs as they are rounded-up on a lathe (as the logs' taper is removed) does not come off in full-length pieces. Rather it comes off in short and irregular pieces, which are difficult to incorporate into plywood or LVL. This veneer, however, is generally of excellent wood quality (from the mature part of the tree), has no heartwood, which makes it readily gluable and treatable with preservatives, and has few defects such as knots. This veneer is processed through a clipper and a trimmer to make the long strands necessary for PSL. Strands are ~20 mm (0.8 in.) wide, 4 mm (0.16 in.) thick, and up to 1 m (39 in.) long. Strands are then dried to a moisture content of 3–5%. PF-based adhesive is then applied to the long strands. High levels of resin solids, up to about 15% by weight, are incorporated, which gives PSL excellent dimensional stability and durabilty. Billets are pressed with high-frequency (radio or microwave) energy to cure the phenolic resin adhesive in a continuous-type caterpillar press that slightly increases the density of the wood. The high-frequency energy allows thick billets, up to ~25 cm (10 in.), to be cured in minutes. The billets can then be sawn into lumber of virtually any size and length needed. This composite lumber product provides an alternative for structural softwood lumber of large size in applications where high uniform strength and reliability are essential. An architectural atrium designed using PSL is shown in Figure 13.24. The strength of PSL is superior to the best grades of Douglas fir and

FIGURE 13.24.   Vaulted atrium timber framed with parallel strand lumber. Source: Photo courtesy of Trus Joist, A Weyerhaeuser business.

southern pine lumber. Because there are numerous small voids in the finished product, full treatment with preservatives is possible. This characteristic allows PSL to be used in severe exterior conditions such as bridge stringers, cross and switch ties, and wood trestle members.

LSL is similar to PSL except that it is made with thinner and wider strands and it uses different resin binder. Aspen, yellow poplar, basswood, or other low-density hardwood flakes are produced on a modified disk flaker. Strand-production equipment is similar to that used in the manufacture of OSB; however the strands are about twice as long, about 30 cm (12 in.). Green strands are dried and then screened. Broken or short strands are continually removed from the process. The strands are then blended with resin and formed, like OSB, into a loose mat. This product is manufactured with pMDI (polymeric diphenylmethane diisocyanate) resin adhesive to provide a waterproof, fast-curing, light-colored product. The use of pMDI reduces press cycle time as compared with phenolic adhesives. The billets, ~25–150 mm (1–6 in.) thick, are cured in steam-injection presses. Steam injection further reduces press cycle time and minimizes density gradients through the billets' thickness. Because diphenylmethane diisocyanate (MDI) utilizes moisture in its curing reaction, strands can contain relatively high levels of moisture, on the order of 15%, which enhances dryer throughput 32 mm (1.25 in.) thick panels can be cured in ~60 s.

LSL is sold for a variety of applications requiring high-grade lumber. Different levels of structurally rated material are available to compete with solid-sawn dimension lumber (Figure 13.25). Also, nonstructural-rated material is available to be used in molding, millwork, furniture, and other secondary products.

OSL is produced from oriented flakes, in much the same way as the faces and core of OSB. In this case, though, all strands are oriented in one direction. OSL differs from LSL in that its flakes are shorter, about 15 cm (6 in.) long, the same length as those used in OSB. Specialized curved press platens allow for numerous architectural shapes to be produced (Figure 13.26). These products have a ready market in architectural windows and doors, furniture parts, and other specialty applications.

Crushed long-strand products are close to being commercialized. They have been researched in Australia, Japan, and the United States. Production is developed and commercialization depends largely on market demand. In this process, low-value logs are debarked and then processed through a tree crusher. Within the crusher, rollers either oscillate or contain grooves that separate wood along its weakest planes, similar to what can occur when a wet tree is loaded to the point of failure. This process mainly generates fractures within the weaker earlywood and leaves long latewood splinters or strands intact (Figure 13.27).

In one such process, the wood goes through a series of rollers that contain sequentially smaller grooves. The result is a loose fibrous mat of long, parallel-to-grain strands with most of the tensile strength undisturbed. Strand mats are then dried and coated with adhesive. Adhesive application is challenging because the strands are, to a considerable extent, held together in the loose mat, which is not conducive to the use of spray, roll, curtain, tumble, or other traditional resin application processes. The blended loose strand mats are then assembled into thick mats and pressed. Both high-frequency energy and steam injection have been used to expeditiously cure the thick billet. Because low-value raw material is used (thinnings or underutilized species) and there is no intermediate commodity product produced (veneer or strands), the production cost of this product is

FIGURE 13.25.   House frame constructed with laminated strand lumber. Source: Photo courtesy of Trus Joist, A Weyerhaeuser business.

somewhat lower than that of competing products. Crushed-strand lumber is adaptable to both softwoods and ring porous hardwoods. The strength properties of such composites rival those of high-grade structural lumber and LVL.

A major advantage of composite lumber products is that they allow the production of relatively large sizes of lumber from small logs. Therefore the environmental impact on mature forest resources is reduced. Additionally, there is a significant raw material cost advantage associated with using underutilized "low-value" small-diameter timber. Also, raw material utilization is high – typically 70% or more ends up in the primary product. Although it is more costly, composite lumber compares favorably with solid-sawn lumber in terms of strength, uniformity, and straightness. While the quality of solid-sawn lumber is determined to a great extent by the quality of the raw material, the quality of composite lumber is dependent upon the manufacturing process. Structural composite lumber products continue to increase in importance.

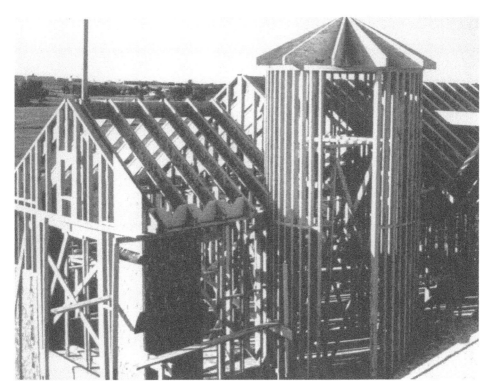

FIGURE 13.26.   Residential construction using engineered wood composites. Note that the roof structure on the tower utilizes LSL members more than 1.5 m deep. Source: Photo courtesy of Trus Joist, A Weyerhaeuser business.

FIGURE 13.27.   Tree that was catastrophically damaged in a storm; note the unbroken strands and splinters. (left) Small diameter tree stems being crushed along its weakest planes, note the long unbroken strands similar to those in the tree (right).

Basic Steps in Composite Lumber Maufacturing

At the mill, roundwood is either used immediately or inventoried depending on season and price considerations. Truck or railcar loads of roundwood are most commonly purchased by weight but sometimes by bulk volume. Like strand production in OSB, roundwood is required to produce structural strands. Tree length logs or roundwood bolts are debarked before being reduced to strands. Strands for strand-based composite lumber are longer than those used for OSB. These long and unbroken strands deliver the most favorable bending strength properties to the final products. A strand length of 30 cm (12 in.) is commonly used so a 60 cm (24 in.) long knife will produce two 30 cm (12 in.) long strands simultaneously. Strand thickness can be manipulated by adjusting the speed at which the disk strander advances into the roundwood. Strand widths are generally between 2 and 5 cm (0.8 and 2.0 in.).

Strand and particle drying was discussed briefly in the moisture chapter Adhesion of strand-based lumber products generally requires a wood moisture content from 2 to 6%, depending on the resin type used, so this range is the target of the dryer. Strand drying for these products is similar to that for OSB.

The fundamental aspects of long-strand blending are the same as those for OSB. Wax and adhesive are added to the strands to promote water repellancy and adequate adhesion, respectively. As with OSB, liquid adhesive is often centrifugally atomized into a fog and dispersed as an array of very fine discrete resin droplets.

Only structural waterproof resins are used for strand-based composite lumber. PF and MDI resins are the most common. The decision to use one versus the other depends on several factors including appearance, product application, production rates and wood moisture. The greatest factor that affects resin choice for most applications is economics.

In contrast to OSB, loose mats are formed with all of the strands aligned parallel to the long axis of the lumber product. Mat formation is similar to OSB in that the strands are typically dropped from above and deposited on a moving belt. Multiple forming heads build up the thickness of the mat with uniform density and closely aligned flakes; alignment is important, since misaligned flakes decrease the strength of the final product. The formers deposit the appropriate quantity of strands onto the forming belt. This assures that when the press closes, boards of the desired thickness and density are made. For batch pressing, as the loose mat is continuously formed, a flying cutoff saw cuts sections to be loaded into the press. Because the mats are not consolidated, care must be taken that material is not lost from the edges. After cutting, the large sections of loose mat are accelerated into the hot press. The action of the formers, cutoff saw, press, and unloader are synchronized by computers. This synchronization promotes maximum productivity through the pressing stage. In continuous pressing, the formed mat is consolidated as it is forced through converging rolls or belts. Then it passes through the press on heated belts or moving caterpillar-type tracks. Pressure is provided internally by the compaction of the wood in the press.

To accelerate the cure of thicker mats, either high-frequency curing or steam injection can be used. Similar to the high-frequency pressing of LVL, electrical energy can be used to rapidly cure thick, greater than 25 mm (1.0 in.), strand-based products. Electromagnetic waves use the dielectric properties of wood to heat the billets from the inside. Steam injection pressing utilizes small perforations in the press platens to inject dry, high-pressure steam into the mat to heat and cure the core quickly. Steam injection pressing creates a composite product with relatively uniform density through the thickness.

Engineered I-joists

From the discussion of shear and flexural stresses in the mechanics chapter, it is intuitive that strong and efficient beams can be engineered to provide maximum strength where stresses are the highest. Such is the case of steel and aluminum I-beams. The production of engineered glued wood beams with efficient I- or box-sections dates back to the 1940s (Hansen 1948), the era in which synthetic water proof adhesives were first developed. Early beams were factory made by special order or fabricated on-site. To efficiently utilize material, scarf-jointed solid members were used for the tension and compression elements. Early engineered I-joists had solid wood flanges (compression and tension), similar to large I- and box-section beams which were designed and fabricated at that time for commercial and industrial buildings. Structural plywood was used for the web elements of those members. Around 1970, the US manufacturer Trus-Joist Corporation began commercially manufacturing a universal type of composite wood I-joist (Smulski 1997). Although I-joists were first developed utilizing structural plywood for joist web material, OSB is now used almost exclusively. Uniformity, availabilty, cost, and performance drove this change. Currently, engineered wood I-joist production provides a significant market for its constituent materials (OSB, LVL, high-grade structural lumber).

Manufacturing involves preparation of both the flange and web stock followed by assembling them into the composite system. Flange stock can consist of solid-sawn lumber, LVL, or other structural composite lumber substitutes. This material is ripped to length, fingerjointed (if necessary), and grooved to receive the web stock. Specially produced OSB is ripped and fingerjointed before it is merged with the flange stock. Structural adhesive is applied to the grooved flanges as they are continuously pressed together with the webstock. Both phenol and phenol–resorcinol-based adhesives have been used in this manufacturing process. Polyurethane adhesive has the potential for use in this application and may be commercialized in the near future. Because manufacturing is done in a continuous press system, virtually endless lengths are possible. Shipping and handling restrictions, however, generally limit the usable length to ~24 m (80 ft). Because LVL and OSB are major components, I-joist mills are often located on shared properties with LVL or OSB mills. This location strategy eliminates the shipping cost associated with transporting at least one of the primary constituents. Also, it usually facilitates better communication between the two mills, which can lead to better coordinated and more efficient production schedules for the two plants.

The sectional configuration of I-joists makes it possible to increase the strength-to-weight ratio as compared with that of solid wood. In addition to light handling weight and reduced shipping charges, these products are highly resource efficient. The ability to manufacture beams with deep sections greatly increases their stiffness to weight ratios. Thus, longer spans and more open floor plans are available than when only solid wood framing members are available (Figure 13.28). For all of these reasons, market penetration of I-joists has been excellent over the past two decades.

FIGURE 13.28.   Engineered I-joists over large open clearspan. Source: Photo courtesy of Trus Joist, A Weyerhaeuser business.

There are currently numerous manufacturers of wood I-joists. Each manufacturer conducts strength testing and provides building code-approved products. Because there are many manufacturers of wood I-joists, there has been a desire among builders and other consumers to standardize these products and to some extent this has been achieved.

Two variants of the most common form of wood I-joists are the glue-laminated I-joist and the LVL I-joist. In each case, a structural wood composite is used as a raw material, either glulam or LVL. The raw material is then sawn appropriately and glued together into an I-type section, to increase structural efficiency. While manufacturing of these products is relatively costly, structural performance is very high.

Another commercially available laminated structural composite lumber product is construction mats from low-value sweetgum and other hardwood lumber (Figure 13.29). Historically construction, crane, and dragline mats were manufactured from solid oak timbers, ~30 cm (12 in.) deep. The composite action associated with glue lamination, however, allows for a 15 cm (6 in.) deep sweetgum mat to achieve the same bending strength characteristics as a 30 cm (12 in.) deep oak mat. The end result is an industrial material that uses about 50% less wood, from a lower grade, and from less-valued species. Innovation such as this will largely ensure structural wood production well into the twenty-first century.

(a)

(b)

FIGURE 13.29. Industrial laminated mats produced from low-grade sweetgum lumber. Upper frame illustrates wood mats applied as a temporary access road through an environmentally sensitive area. Lower frame illustrates a crawler construction crane atop 38 cm (15 in.) of laminated wood mats.

Adhesives

Thermosetting synthetic resin is used exclusively for the manufacture of structural panels and structural lumber composites in North America. Resorcinol formaldehyde, phenol resorcinol formaldehyde and melamine formaldehyde are used for glue-laminated timber. PF is the primary adhesive type used in manufacturing structural plywood and LVL. Both phenolic and isocyanate (MDI) based resins are used for OSB and LSL. Phenol formaldehyde is used for parallel strand lumber.

Both PF and MDI resins form waterproof bonds necessary for the use of structural wood composites. While both PF and MDI are derived from petroleum and thus have prices that are tied to the price of oil, PF is widely available at about one-fourth the cost of MDI. PF is also available in powdered form, which is a significant benefit in strand blending. Advantages of MDI are that it is clear-colored, which gives finished products a brighter surface than PF-bonded products (PF is dark red, brown, or black), it does not contain water and can thus tolerate higher moisture content strands, which increases dryer production and cure time for MDI is faster than for PF. There are exposure risks associated with both types of these uncured adhesives and thus mill designers and operators work with resin suppliers to build mills that are safe and minimize exposure risks.

Structural adhesives are generally mixtures of resin with fillers and extenders. Resin companies tailor the adhesives for each different mill based on specific needs. *Extenders* are substances with some adhesive value added to reduce the amount of primary binder needed. One such extender is furfural, a lignocellulose by-product of furfuryl alcohol production. *Fillers* are additives used to prevent adhesive over-penetration; they have little adhesive value. Fine flour produced from wood, bark, or nutshells are used as fillers. Starch and animal blood are sometimes used to modify viscosity, control the penetration of adhesive into veneer, and control tack or stickiness.

Once cured, structural waterproof resins form stable and inert plastics. They are heat and moisture stable. As mentioned previously, these cured resins do not off-gas measurable amounts of formaldehyde (Smulski 1997) or other undesirable compounds into the air. Cured resins are nonleaching should composite boards be exposed to liquid water.

Factors Affecting Production Efficiency

Bark content and strand retention through processing are the primary factors affecting yield in OSB. Compared with veneer production, OSB conversion yields are higher: 85–90% compared with about 60%. Optimizing lathe chargers and high-speed lathes capable of peeling to 90 mm (3.5 in.) diameter or smaller cores make it currently possible to utilize much more of this forest resource. Mills that produce sanded or other high grades of plywood must, however, compete for the high-quality logs. There are some areas of the world where it is still possible to obtain logs of relatively large size at a reasonable cost. The supply of such material is limited, though, and the production of plywood from large logs continues to decline.

Figure 13.30 illustrates the uniformity of southern pine plywood peeler blocks. These blocks have been cut to length and debarked, and are in the heated bath on their way to the lathe. The average block diameter of these is ~25 cm (10 in.).

FIGURE 13.30.   Debarked and slashed southern pine peeler blocks in the heated bath on the way to the lathe. Average diameter for these blocks is ~25 cm (10 in.). Source: Photo courtesy of Tor P. Schultz.

Regardless of size, the quality of veneer logs is determined by the absence of knots or other surface defects, straightness, taper, roundness, and the absence of defects on the ends. The quality of incoming logs is important because it controls the quality and yield of veneer and also the number of blocks that prove to be defective once in the lathe. The yield of veneer varies with log diameter, core diameter, and the efficiency with which the veneer is clipped and utilized. Table 13.3 shows how the core and block diameters affect the theoretical percent yield from perfectly cylindrical southern pine veneer blocks. In reality, yields are lower owing to roundup losses. Note that in large-diameter blocks of 45 cm (17.7 in.), the volume increase in veneer obtained by reducing the core diameter from 12 to 8 cm (4.7–3.2 in.) is relatively small (2% increase) but when 20 cm (8 in.) diameter blocks are being peeled, the same reduction in core size creates a substantial yield improvement (20% increase).

Phillips et al. (1980) reported from a study of southern pine that 25 mm (1 in.) of sweep in a 20 cm (8 in.) diameter block reduced the veneer yield by 44%. Although not as serious in larger blocks, a 50 mm (2 in.) sweep in a 45 cm (17.7 in.) diameter block (a large diameter in most US regions) would reduce recovery volume by 23%. Scanners that measure actual block volume make it possible to correlate true log volume input with actual veneer output. This processing technology provides accurate, consistent, and uniform information regarding veneer production efficiency.

TABLE 13.3.   Four main grades of structural softwood veneer.

Grade	Description and characteristics
A	Smooth, paintable; not more than 18 neatly made repairs permitted – synthetic, boat, sled, or router type and parallel to grain; may be used for natural finish in less demanding applications
B	Solid surface; shims, various synthetics of wood patches or plugs and tight knots to 1 in. across grain permitted; some minor splits permitted
C	Tight knots to 1½ in.; knotholes to 1 in. across grain and some to ½ in. if total width of knots and knotholes is within specified limits; synthetic or wood repairs; discoloration and sanding defects that do not impair strength permitted; limited splits allowed; stitching permitted
D	Knots and knotholes to 2½ in. width across grain and ½ in. larger within specified limits; limited splits permitted; stitching permitted; limited to interior grades of plywood

Source: APA (2001).

Design Considerations and Product Limitations

There are limitations to the use of virtually any glued product, wood or nonwood. It is important to mention some of the potential problems associated with strand-based composites and steps that the industry takes to make better products and avoid legal entanglements. Aerospace, automotive, marine, military, and housing are the five biggest consumers of composite materials and those industries take steps to minimize potential liabilities such as delamination, thickness swell, water absorption, decay, linear dimensional change, and creep deflection.

Thickness swell and water absorption are still problematic with OSB. Higher levels of resin solids can largely remediate these problems. Companies that are willing to invest slightly more in manufacturing (by adding more resin) can receive large dividends in the form of increased demand, based on reputation, and fewer claims. Most OSB is rated for limited exposure to liquid water during building construction. Standing water on sheathing during building construction, however, can cause irreversible thickness swell and associated problems with construction (out-of-level floors, out-of-plumb walls). Thickness swell can result from the densified wood elements attempting to return to their initial thicknesses. In addition to the incorporated wax sizing, panels are edge-sealed to retard moisture penetration. As panels are cut in the field, this edge sealing is often removed. Subsequent moisture absorption, such as in an inadequately vented attic or basement, causes the edges of the panels to swell. The swelled panel edges are then readily detected as ridges through malleable roof shingles or floor coverings. Explicit construction guidelines are available from the manufacturer's grading association and from most retailers. Higher-performance OSB is available. With the addition of extra resin, this product is better able to withstand adverse conditions associated with housing construction with minimal thickness swell and other problems. For many contractors, the higher cost is offset by the more forgiving nature of this type of product.

Ill-conceived architectural details or poor construction practices can lead to failure of OSB sheathing – as is the case with most wood-based materials. Proper cladding, flashing, and building maintenance are critical for the successful use of strand-based composites (Wu and Ren 1999). Linear expansion is limited owing to the cross-laminated nature of OSB; it is generally less than 0.40%. Diligent protection from liquid moisture is the best insurance against undesirable linear dimensional change and is necessary to avoid thickness swelling and its associated problems. In some cases, where stucco finishing was applied directly over OSB, problems have developed. These problems trace mainly to the intrusion and entrapment of moisture in the wall cavity. Liquid moisture can come from the exterior (such as from a leaking window) or the interior (such as that driven by natural vapor pressure, followed by condensation) of the house. The results can be swollen sheathing, buckling, mold, and decay.

Excessive deflection or mechanical creep is another potential liability for strand-based composite products. Many structural situations (joists, beams, stringers, rafters) require high bending strength and stiffness. It has been observed that in high-humidity or cyclic-humidity applications (e.g. damp basements, crawlspaces, poorly vented attics, garages, bridge construction) strand-based products are susceptible to creep deflection many times that which would be encountered in dry and stable applications or in solid-sawn wood members. Such deflection can be unsightly, unsafe, and costly to repair. Potentially hazardous conditions include collection or ponding of water on flat roofs or subfloors, long unsupported spans over damp crawlspaces, or heavy out-of-balance equipment such as HVAC on structural members. It is prudent in structural design to recognize the potential for creep in high- or cyclic-humidity flexural applications.

The issue of volatile offgassing, in service, has largely been put to rest. PF adhesives are now formulated such that there is virtually no residual free phenol or formaldehyde in the panels when they leave the factory. Both compounds are completely reacted to form the nonvolatile and inert adhesive system. MDI adhesive systems also completely polymerize into a stable and nonvolatile plastic. As such, the widespread use of these structural panels in residential, commercial, and industrial installations poses no inherent adverse health risks due to chemical offgassing.

References and Supplemental Reading

APA (1994). *Performance Standards and Policies for Structural-use Panels. PRP-108*. Tacoma, WA: APA–The Engineered Wood Association.

APA (1997). *Panel Handbook and Grades Glossary*. Tacoma, WA: APA–The Engineered Wood Association.

APA. 2005. Softwood plywood industry celebrates its 100th anniversary. PR Newswire, 7 March.

APA (2006). *Market Statistics*. Tacoma, WA: APA–The Engineered Wood Association http://www. apawood.org/pdfs/unmanaged/marketstats/2006EngWoodStats.pdf.

APA (2010). *Voluntary Product Standard for Construction and Industrial Plywood. PS1-09*. Tacoma, WA: APA–The Engineered Wood Association.

APA (2011). *Performance Standard for Wood-based Structural-use Panels. PS2-10*. Tacoma, WA: APA–The Engineered Wood Association.

Baldwin, R.F. (1975). *Plywood Manufacturing Practices*. San Francisco: Miller Freeman.

Baldwin, R.F. (1985). Managing the basics in softwood plywood. *For. Ind.* 112 (4): 18–20.

Baldwin, R.F. (2016). *Plywood and Veneer-based Products*. Montgomery, AL: The Donnell Group.

Chui, Y.H. (2000). Strength of OSB scarf joints in tension. *Wood Fiber Sci.* 32 (1): 7–10.

Cohen, D.H. and Sinclair, S. (1990). The adoption of new manufacturing technologies by North American producers of softwood lumber and structural panels. *For. Prod. J.* 40 (11/12): 67–73.

Dagher, H.J. and Bragdon, M. 2001. Advanced FRP–Wood Composites in Bridge Applications Proceedings, SEI/ASCE Structures Congress, Washington, DC.

Hansen, H.J. (ed.) (1948). *Timber Engineers' Handbook*. New York: John Wiley & Sons.

Howard, J.L. 2016. U.S. Forest Products Annual Market Review and Prospects, 2012–2016. USDA For. Serv., Res. Note, FPL-RN-0343.

Howard, J.L. and K.C. Jones. 2016. U.S. timber production, trade consumption, and price statistics 1965–2013. Res. Pap. FPL-RP-679. Madison, WI: USDA For. Serv., For. Prod. Lab.

Jokerst, R.W. and Geimer, R.L. (1994). Steam-assisted hot-pressing of construction plywood. *For. Prod. J.* 44 (11/12): 34–36.

Kollmann, F., Kuenzi, E., and Stamm, A. (1975). *Principles of Wood Science and Technology*, vol. 2. Chap. 5. Heidelberg, Germany: Springer.

Maloney, T. 1994. The development of wood composite materials. Res. Pap. Wood Mtls Eng. Lab. Wash. State Univ.

Marra, G. (1979). Overview of wood as material. *J. Educ. Modules Mater. Sci. Eng.* 1 (4): 699–710.

McAlister, R.H. (1967). Jet veneer dryers: theory and operation. *Woodwork. Dig.* 69 (6): 32.

Parker, D.J. 1987. Parallam parallel strand lumber: Evolution. Proc. International Particleboard/Composite Materials Symp. Wash. State Univ. Pullman, WA.

Phillips, D.R., Schroeder, J.G., and Clark, A.I.I.I. (1980). Reduce pine veneer losses by selecting blocks properly. *For. Ind.* 107 (4): 40–42.

SBA. 2005. OSB performance by design. Structural Board Association. Toronto, Ontario, Canada.

Schuler, A. and Adair, C. 2001. Continued growth for glulam, I-beams, LVL. Panel World. Hatton Brown. Montgomery AL. 9/01, p. 6.

Schuler, A. and Adair, C. 2003. Engineered wood products – an opportunity to "grow the pie" and benefit from globalization! Raleigh, North Carolina: Fall Meeting, Forest Products Society, Carolina Chesapeake Section, 11 September.

Sellers, T. Jr. (1985). *Plywood and Adhesives Technology*. New York: Marcel Dekker.

Sellers, T., Jr. 1990. Use of lignin as partial substitute for phenol. Panel World. 26–30 September.

Sellers, T. Jr. (2001). Wood adhesive innovations and applications in North America. *For. Prod. J.* 51 (6): 12–22.

Shmulsky, R. (2000). Influence of lumber dimension on VOC emissions from kiln-drying loblolly pine lumber. *For. Prod. J.* 50 (3): 63–66.

Smulski, S. (ed.) (1997). *Engineered Wood Products: A Guide for Specifiers, Designers, and Users*. Madison, WI: PFS Research Foundation.

Thompson, R., England, G., Hansell, D., and Warkentin, K. (1999). Air emissions assessment for a new OSB plant. *Pulp Pap. Can.* 100 (11): 20–23.

Tice, B. (2001). World's first continuous 12 ft. OSB plant is off to good start. *Panel World* 42 (5): 8–11.

Tracy, J.M. (2000). SIPS overcoming the elements. *For. Prod. J.* 50 (3): 12–18.

Watkins, E. (1980). *Principles of Plywood Production*. White Plains, NY: Reichold Chemicals.

Williams, D. and Hopkins, W. 1968. Conversion factors for southern pine products. LA State Univ., Agric. Exp. Stn Bull. 626.

Wu, Q.L. and Ren, Y.K. (1999). Characterization of sorption behavior of OSB under long-term cyclic humidity exposure condition. *Wood Fiber Sci.* 32 (4): 404–418.

14

Nonstructural Composites

Nonstructural panels make up a significant portion of wood composites production in the United States. The estimated 2016 United States' combined production volume for hardboard, hardwood plywood, insulation board, medium-density fiberboard, and particleboard was ~12.9 million m³ (461 million ft³). Net imported volume added another 4.69 million m³ (169 million ft³) to this total (Howard and McKeever 2016). Nonstructural panels are used in many applications such as paneling, cabinetry, countertop stock, underlayment, furniture, and siding. As core stock for upholstered furniture, panels have revolutionized manufacture. Owing to their uniform size, the remanufacture of panels into component parts is easily automated and creates little waste as compared with solid lumber.

Nonstructural panel products are highly varied in their manufacture, properties, and applications. In North America, most of these products are panelized to a standard 1.22 × 2.44 m (4 × 8 ft) size. Nonstructural wood-based veneer and particle products include decorative plywood, particleboard, fiberboards, and cement-bonded panels. These products are made from wood ranging in form from large sheets of veneer to long strands, to relatively uniformly manufactured particles, to individual fibers or fiber bundles.

Decorative Plywood

The art of decorative veneering dates back thousands of years. The Egyptians, c. 1500 BCE, are credited with producing veneer to decorate furniture, tools, weapons, and other possessions. The value of wood furniture was enhanced by applying decorative veneer to its faces. Egypt's arid climate has preserved many such specimens for millenia. These early pieces illustrate that wood composites can last indefinitely, even when only nonwaterproof plant- and animal-based adhesives are used. Plywood did not, however, become

Forest Products and Wood Science: An Introduction, Seventh Edition. Rubin Shmulsky and P. David Jones.
© 2019 John Wiley & Sons Ltd. Published 2019 by John Wiley & Sons Ltd.
Companion website: www.wiley.com/go/shmulsky

a major industry until the 1930s. The adoption of the hot press from Europe and the development of synthetic waterproof adhesives during World War II provided the technological advances that made mass-produced plywood possible.

Decorative plywood, like structural plywood, is a panel product of wood veneers glued together such that the grain direction of some veneer runs at right angles and others run parallel to the long axis of the panel. As with structural plywood, balanced construction, from face to core, is used. The perpendicular arrangement of panel layers yields good strength and dimensional stability both parallel and perpendicular to the long axis of a panel.

Decorative plywood uses appearance-grade hardwood face veneers and interior resin types. Core plies are from lower-value wood species and lower-grade veneers. These panels are intended for applications such as wall paneling, furniture, and other interior use products. A small amount of hardwood plywood is used for nondecorative functions such as container- and pallet-decking. Decorative hardwood plywood production in the United States is concentrated in the South and scattered throughout the eastern half of the country. Worldwide, there is significant production in various regions, including East Asia and the Pacific Rim. Some hardwood veneer is imported from South America, Asia, and Africa, but most hardwood plywood produced in the United States is made using domestic veneer. The production processes of structural and decorative plywood differ significantly. Although the processes of *rotary cutting* the veneer and hot pressing the panels are similar in principle for both hardwood and softwood plywood, the equipment to produce decorative plywood tends to be less automated with more labor input per unit of output, the emphasis being on superior quality production. The decorative plywood industry is diverse, ranging from small mills that produce custom-ordered plywood for specific architectural applications to large, high-capacity mills producing thin panels on production lines similar to those in the structural plywood industry.

Basic Steps in Manufacture

In general, the processes for the production of hardwood veneer for decorative panels and veneers for structural panels are similar and are covered more fully in Chapter 13. However, there are some differences inherent in production of decorative panels as described in this section.

Almost all hardwood blocks are heated to high temperatures prior to cutting veneer. Dense hardwoods are usually heated by soaking at temperatures of up to 93 °C (200 °F). Heating improves both the yield and quality of the veneer. Fleischer (1965) showed that higher wood densities need higher temperatures to obtain the surface quality expected. For example, basswood may cut well at 16 °C (60 °F) but white oak requires about 93 °F (200 °F). Both steam chests and hot water baths similar to that shown for softwood blocks (Figure 13.30) are used to heat the wood prior to veneer production.

Hardwood veneer is both peeled and sliced (Figure 13.9). A great deal of decorative veneer is rotary peeled, similar to the production of structural veneer. Traditionally hardwood veneer lathes have been slower and less automated than structural softwood veneer lathes. The increased raw material costs, however, have made it imperative to maximize both production yield and volume. Therefore, it is now common for hardwood veneer mills to utilize high-speed scanning and lathe charging equipment in order to remain competitive.

FIGURE 14.1.   Production of rotary peeled hardwood veneer. This continuous ribbon of 3 mm (0.125 in.) thick oak veneer was crosslaminated into three-layer plywood, then ripped into 64 mm (2.5 in.) wide strips, profiled, and used as solid laminate flooring.

Figure 14.1 shows a continuous ribbon of oak veneer as it comes off a lathe. This veneer is used for high-value hardwood plywood flooring. In decorative veneer mills the green veneer from the lathe is often wound onto a roll and then moved to clippers that chop out defects. Other hardwood mills use directly coupled conveyors between the lathe and the clipper, but these systems require that the clipper speed be mated to the lathe. Veneer thickness ranges from ~0.3 mm (0.01 in.) for high-grade face material to 3.2 mm (0.125 in.) for lower value core stock.

Slicing involves greater raw material loss and is a relatively labor-intensive process. In preparation for slicing, a square, rectangular or wedge-shaped block of wood, called a *flitch*, is sawn from a log, heated, and rigidly mounted on a slicer. The slicer blade moves up and down or from side to side, or the knife is fixed and the flitch moves up and down. In any case, one piece of veneer is cut with each stroke, a slow process as compared with rotary cutting. Slicing is used for producing decorative veneers from high-quality hardwoods where particular grain patterns are desired that are not achievable with rotary peeling.

Decorative veneer drying occurs at lower temperatures than structural veneer drying in order to preserve wood quality. Dryer temperatures generally remain in the 70–85 °C (160–185 °F) range. Emissions from hardwood veneer dryers are substantially less than those from softwood dryers and generally do not require the use of pollution control technologies.

This difference is due to differences in wood extractives, lower dryer temperatures, and lower production volumes.

Throughout manufacturing, sheets of veneer are graded and sorted. Because decorative plywood and veneer are sold by appearance, sorts are based on defect types and sizes, sheet size, color differences, species, and other factors. Most plywood plants operate veneer upgrading stations. There, veneer sheets can be patched, sewn, taped, filled, sanded, or otherwise doctored to improve veneer grade. Graded veneer is then sold in packages for applications such as architectural doors, millwork, paneling, and specialty products or it is pressed into decorative plywood. The sliced veneer from a flitch may be kept together in order to allow subsequent matching of the grain, a costly process. Figure 14.2 illustrates a musical instrument with a thick layer of high-value high-density rosewood used as the fingerboard laminated onto a lower value and lighter weight base wood stock for the neck.

FIGURE 14.2.   High-value, high-density rosewood fingerboard laminated onto base-material wood neck stock.

The processes of applying adhesives to veneers, assembling veneers into a panel, and moving panels in and out of a press are often the most labor-intensive steps in manufacture. Because appearance is critical and product value is relatively high, most of the layup is done manually. Production is mostly in the form of standardized flat panels but a variety of specialized curved presses are used to make specialty products. Curved furniture parts such as chair arms and rockers, millwork such as window and door arches, and other products are frequently made from hardwood veneer. In those cases, parallel lamination, analogous to structural laminated veneer lumber, is commonly used. For thick or curved stock, high-frequency pressing is usually used such that press time can be reduced.

Regardless of size, the quality of veneer logs is determined by the absence of knots or other surface defects, by straightness and roundness, and by the absence of defects on the ends. High-value veneer logs are often graded individually, sold by log grade, and volume estimated by one of the log rules. Log grades are usually based on straightness, freedom from rot or soft centers, and the proportion of log face that is free from knots or other defects. Rarely are logs purchased by weight. In recent years, the expanded use of scanners that measure actual block diameter, and thus volume, makes it possible to closely monitor production yield.

Decorative Plywood Selection and Use

Decorative plywood is usually selected based upon the appearance of the face veneer. These panels are produced primarily, but not exclusively, from hardwoods. Grain patterns are determined by the species, the log characteristics, and the way the veneer was cut. Flat- or quarter-sawn grain patterns are available in the more expensive plywoods made using sliced veneer. In less expensive plywood with rotary-cut faces, the grain pattern is highly variable, but different species provide a variety of grain effects. Veneer from ring-porous species provides distinct patterns when rotary (tangential about the circumference of the log), rift (between ~30 and 60° to the radial plane), or quarter (parallel to the radial plane) cut. The desired color of the finished panel can also be important when selecting species. Although it is possible to produce a light finish on the darker woods, e.g. walnut, cherry, mahogany, it is easier to start with lighter-colored species such as maple or birch when seeking a light color in the finished product.

Because most decorative plywood is made with nonwaterproof adhesive, panels should only be used where they will not be subjected to liquid water. Decorative plywood with a waterproof glueline can be obtained for situations where the additional cost is justified and the quantities involved are adequate. Decorative plywood is designated *Type 1* or *Technical* if it has waterproof gluelines, *Type 2* if it is moisture resistant, and *Type 3* is not resistant to water (ANSI 2009).

Particleboard

Particleboard is a panel product made by compressing small particles of wood while simultaneously bonding them with an adhesive. The many types of particleboard differ greatly with regard to the size and geometry of particles, the amount of adhesive used, and the density to which panels are pressed. The properties and potential uses of any board depend on these factors.

The major types of particles used for particleboard are:

1. *Shaving* – a small wood particle of indefinite dimensions produced when planing or jointing wood. It is variable in thickness and often curled.
2. *Flake* – a small particle of predetermined dimension produced by specialized equipment. It is uniform in thickness, with fiber orientation parallel to the faces.
3. *Chip* – a piece of wood cut or split from a larger piece of wood by a knife or hammer, as in a hammermill.
4. *Sawdust* – produced by sawing, in a wide range of sizes. It is usually further refined.
5. *Sliver* – nearly square cross-section, with length at least four times the thickness.
6. *Excelsior* – long, curly, slender slivers.
7. *Strand* – a long and narrow flake, wafer, or veneer strip, 0.4–0.8 mm (0.015–0.030 in.) thick, with parallel surfaces.
8. *Wafer* – similar to a flake but larger. It is wider than a strand and nearly square. Usually over 0.5 mm (0.020 in.) thick and over 25 mm (1 in.) long. It may have tapered ends.

A range of particleboard types is shown in Figure 14.3. These boards are visibly different because of the size of particles on the faces. However, boards that appear similar may be quite different in strength and dimensional stability. Boards from small particles have better surface properties and can compete better with medium-density fiberboard (described later in this chapter) for furniture core markets. One of the advantages of particleboard as an industrial material is that it can be tailored to meet a variety of use requirements.

The particleboard industry in Europe and North America has shifted during its development back and forth between an emphasis on use of roundwood to the use of mill *residues* as a source of raw material. In Europe the first plants used wood chips and plywood plant residues. As technology improved it was shown that flakes of uniform thickness, cut from bolts or sawmill lumber scraps, produced a superior product. Many plants were then built that used roundwood as the raw material. In the United States most of the earliest plants relied upon sawmill residues, such as planer shavings and sawdust, and this practice has continued.

The development of particleboard in the United States has been beneficial to the forest products industry in that it provides a strong market for mill residues. When particleboard emerged as a major segment of the industry in the late 1950s and early 1960s, it was based upon the use of mill residues, then available at $1.80–2.70 t^{-1} ($2–3 short ton^{-1}), about the cost of hauling them away. The situation has since changed dramatically. In most parts of the country, few clean wood residues remain unutilized. Because dry residue is also valuable as an energy source, the particleboard industry faces stiff competition for this raw material.

Presently more than one-half of the raw wood material used for particleboard in the United States is planer shavings. The balance is in the form of other mill residues, logging residues, or whole tree chips. Roundwood is a nonsignificant raw material for standard particleboard because it has many competing markets and is generally too valuable to be a major raw material for this product.

FIGURE 14.3. Typical particleboard types. Boards appear different because of color, density, and particle size.

In the United States the cost of the wood raw material currently makes up one-fourth to one-third of the total production cost, a dramatic increase over the situation in the late 1980s. In some cases whole-tree chips and logging residues are used for particleboard. Currently there is much effort to incorporate clean recycled or urban wood waste from sources such as construction sites and manufactured housing operations; the costs of collecting, transporting, and sorting have been major deterrents to the use of this material.

Variables in Particleboard Production

Particleboards differ by particle size and geometry, particle differentiation between face and core, board density, resin type, or method of manufacture. The American standards (ANSI 2016) for particleboard categorize such products using three variables: board density, quality class, and resin type. Canadian standards recognize 11 grades, differentiated principally by physical properties.

Particle Size and Geometry.   Many types of particles are used, singly or in combination, for particleboard production. Figure 14.4 shows a number of particle types used as furnish. At the lower end of the size range are particles so small that they approach the wood fiber bundles used to produce fiberboards. At the upper end are flakes over 12.5 mm (0.5 in.) long. Longer strands are used for oriented strandboard.

FIGURE 14.4.   Some of the many types of particles used as furnish for particleboard.

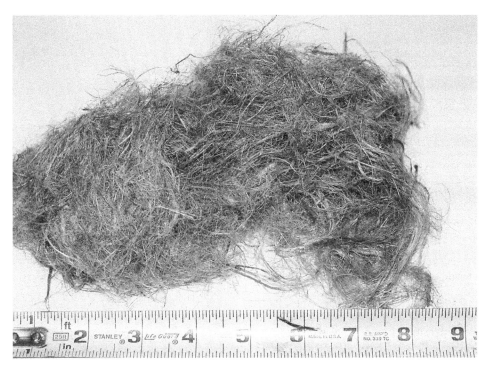

FIGURE 14.5. Residual fiber from kenaf processing. This and other agricultural residues have a promising future as a raw material source for nonstructural bio-based composite products.

The ideal particle for strength and dimensional stability is a thin flake of uniform thickness with a high length-to-thickness ratio; such elements are, however, difficult to produce from residue material. Generally the strength performance and economics of particleboard limit its raw material streams to those associated with wood waste or co-products of other primary production, such as sawdust and planer shavings. Technically, any raw material from pulpwood to wood dust may be used in making particleboard.

Residual agricultural fiber has been viewed and used as a potential fiber source for particleboard. Wheat straw has been investigated heavily as a raw material and a small number of commercial facilities have attempted operation. Other agricultural fibers such as that from corn stalks or kenaf, as shown in Figure 14.5, show promise for similar international applications. In general, the commercial use of agricultural residual fiber as a commecial-scale raw material in the United States has not worked out owing to storage issues, resource availability, and ultimately economics. Trees, which can be stored on the stump and harvested when convenient, hold a major advantage over the seasonal availability of agricultural residues.

Particle Distribution Between Face and Core. Surface properties are of concern when the panels are used in furniture manufacture, are to be overlaid, or will be painted. Small, fiber-like particles are best for the surface layer if a smooth face is needed. For this reason, multilayered boards made from layers of particles of various coarseness are often produced (Figure 14.6). Multilayer boards are superior for many applications because

Single-layer (homogeneous)

Three-layer

FIGURE 14.6.   Particle size gradations in a particle mat.

Five-layer

Graduated

layering makes it possible to increase the bending strength and stiffness of panels by altering the relative properties of face and core without increasing overall panel density.

Panel or Board Density.   The ANSI standard for particleboard (2016) includes three board-density classifications: high density, $800\,kg\,m^{-3}$ ($50\,lb\,ft^{-3}$) or greater; intermediate density, $640–800\,kg\,m^{-3}$ ($40–50\,lb\,ft^{-3}$); and low density, less than $640\,kg\,m^{-3}$ ($40\,lb\,ft^{-3}$). Most particleboard produced in the low-density category is used for door core, which has only minimal strength requirements. Most conventional particleboards range in density from 670 to $800\,kg\,m^{-3}$ ($42–50\,lb\,ft^{-3}$). In all of these products, board density is higher than that of the constituent raw material.

Resin.   Urea–formaldehyde (UF, amino-based) adhesives are the most commonly used resins for particleboard manufacture in Europe and the United States. In the United States, favorable economics help amino-based adhesives control about 60% of the wood adhesives market (Sellers 2001). Their relatively short curing cycle time is another advantage of UF. Boards produced with UF are intended for interior use only. In northern Europe, particleboards produced with UF resins have been used in home construction for floors and interior walls. In North America such boards are not deemed suitable for structural subflooring or sheathing in residential construction. In the United States, UF-bonded boards are, however, used as mobile home decking (a combination of subfloor and underlayment).

It was UF resin that caused much concern regarding formaldehyde offgassing of many products in the 1980s and again in the mid-2000s with the manufactured housing associated with Hurricane Katrina relief along the US Gulf Coast. With this resin, excess formaldehyde is often put into the mixture to accelerate the curing. As a result some free formaldehyde may still exist after it is cured. Advances in processing technology and resin chemistry, in particular the use of formaldehyde scavangers, have greatly limited this offgassing. Additionally in the United States and Europe there are now formaldehyde and other air quality standards with which particleboard must now comply. In general, all of the alternative adhesive technologies add raw material costs or slow production.

Other resin systems are in limited use for particleboard. Combinations of UF resin and melamine resin are used in Europe to produce boards with greater water resistance than those produced using only urea resin. Isocyanate binder, MDI (diphenylmethane-diisocyanate), is used in several European mills and it has been used to some degree in the United States, although currently it comprises only a small percentage of US particle-board binders. The advantages of this resin are fast press cycles and high dryer throughput. MDI is reported to be tolerant of higher moisture content in the furnish, up to 22% MC, than urea or phenolic binders (Youngquist et al. 1982). Also, in the case of wheat straw boards, isocyanate seems to be the only commercially available resin that is compatible with the waxy outer surface of straw. The largest impediment to growth of MDI's current market share is cost.

The quantity of resin used to manufacture a board is the major factor determining strength and dimensional properties. Increased resin solids yield stronger and more dimensionally stable boards. For economic reasons, however, it is undesirable to use a greater amount of resin than necessary to obtain the minimum desired properties. While resin solids comprise only a small portion of finished product mass, resins often account for one-fourth or more of the total production cost for particleboard, and in some cases may cost as much as the wood in a panel.

Commonly, the resin content of urea-bonded boards varies from 6 to 10% based upon the weight of resin solids. Most strength properties are improved by resin and increase at a decreasing rate, i.e. the more resin that is added, the less the incremental improvement. Once enough resin is added that particle strength becomes the factor limit-ing board strength, there is no reason to add more. Figure 14.7 shows the relationship of resin to strength in pine and spruce boards produced from planer shavings.

Basic Steps in Particleboard Manufacturing

Raw material is brought to a plant in the form of shavings, chips, mixed mill residue, or sawdust. It is best to maintain a uniform raw material mix (species and moisture content) because this minimizes process adjustments and simplifies quality control. If large stor-age facilities are available and an adequate inventory can be maintained, material types can be stored and constantly mixed as they enter the process. Most raw material for particleboard is purchased on the basis of weight. Dry planer shavings are usually the most highly valued type of residue and sawdust is valued the least. The alternative value of sawdust or chips is often dictated by the price set by pulp and paper mills in the geographic area.

Some type of milling is required for almost all raw material types. Chips must be cut into flakes or smaller and finer particles. Lumber trim and other types of solid wood resi-dues are often chipped prior to final breakdown into particles. Even planer shavings and sawdust may be further milled to obtain the average particle size and distribution desired. A variety of machines including refiners, hammermills, and flakers are used to produce the type of furnish desired. These machines grind, cut, tear, or otherwise reduce the wood into a range of particle sizes. Figure 14.8 illustrates the operation of a *ring flaker*. This machine has knives that cut thin particles from chips. A *disk refiner* of the general type shown in Figure 14.9 is sometimes used to prepare small fiber-like particles. The rotating disk plates are grooved and as particles move from the center to the outer edges

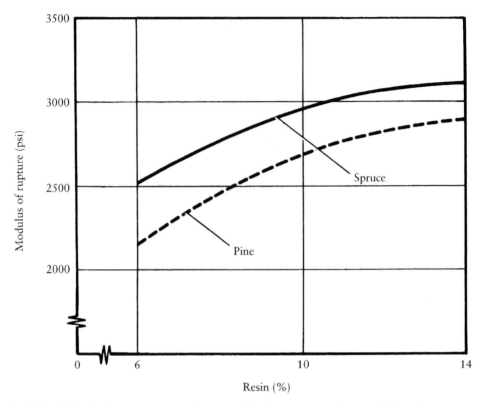

FIGURE 14.7.   Resin level and strength relationship for planer-shavings particleboard.
Source: Maloney (1993).

of the disk, they are ground to fiber bundles. The closer the disks are set together, the finer
the particles produced are. Disk refiners are also used to produce fiber for the fiberboard
and paper industries.

   Particle drying was discussed briefly in Chapter 7. Most types of particleboard are
made with resins that contain water and are applied in liquid form. In this case the
particles are dried to 2–5% MC prior to blending. Then with the addition of 4–6% MC
from the resin, the panels have a final moisture content near 10%.

   Once particles are dried, they are often screened to remove fine dust-like material
(fines). If fines are not removed, they tend to absorb much of the resin, thus lowering the
strength of the panels. Incorporation of fines helps to achieve smooth panel surfaces but
otherwise fines contribute little but weight to the properties of a product. If not incorporated into the surface of the panels, the screened fines are generally burned for heat.

   As with oriented strandboard, the process of adding resin and *wax* to particles is
called *blending*. The wax emulsions (ranging from about 0.25 to 2% by weight) are added
to provide some water repellency or *sizing* to panels. The amount of resin solids (ranging
from 6 to 10% by weight) is low considering the large surface area of the small particles.

   The process of depositing furnish into a mat is termed *forming*. Proper formation
provides as uniform a mat as possible, i.e. uniform mat weight across the area of the
board. For particleboard, formation is dry, using air as the conveying medium. Particles
are dropped or thrown into an air chamber above moving cauls or belts and floated

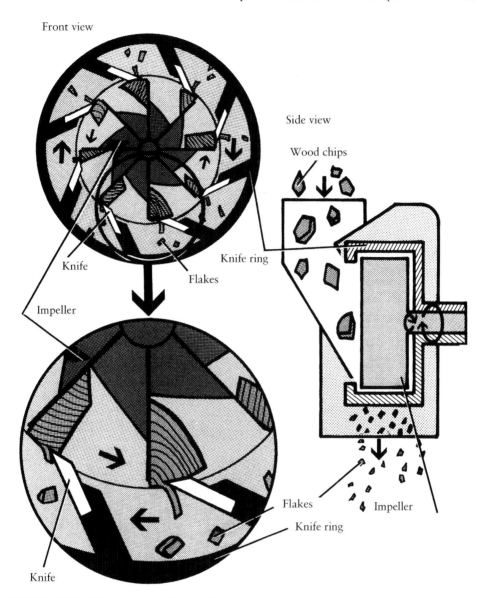

Front view

Side view

Wood chips

Knife

Knife ring

Flakes

Impeller

Flakes

Knife ring

Impeller

Knife

FIGURE 14.8. Elements of a knife-ring flaker.

down to their position. Three to five forming stations are used to produce panels with graduated face-to-core particle sizes.

Once mats are formed, they are conveyed into a press loader, which simultaneously loads all openings in the press. The particle mats are formed either on metal cauls or on belts. In the belt (caulless) system, mats must be strong enough to be moved into the press without a caul plate below it. To accomplish this, mats are prepressed prior to moving them into the hot press. Multiopening presses (Figure 14.10) found throughout North America have platen sizes from 1.22×2.44 to 2.44×9.75 m (4×8 to 8×32 ft). Large single-opening presses are used in some plants.

FIGURE 14.9.   Main components and plates of a disk refiner.

Board thickness is controlled in several ways. Metal stops are used along the edges of the platens in some presses. These presses close until the platens come to rest on the stops. Multiopening presses may also control the distance from the bottom to the top of the press when in the closed position. In essence, this controls the thickness of all the panels in the press. Modern equipment relies on *linear position sensors* to control panel thickness. These positioners digitally monitor and control the press position. This type of process control adds flexibility to the pressing operation and eliminates the time-consuming process of changing press stops each time a panel thickness is changed. Modern presses are also designed such that all of the platens close simultaneously, maintain the same spacing, and provide uniform cure time, a further step toward the production of consistent panels (Figure 14.11). The pressure inside the press decreases throughout the curing cycle. Essentially, only enough pressure need be applied to the panels in order to keep them at the desired thickness and as the resin cures the thickness of the panels becomes fixed.

The rate of cure of particleboard resins increases continually. In the past few decades, the press time for a 12.7 mm (0.5 in.) thick UF bonded board has been reduced from around 10 min to less than 3 min. Maximum press temperatures are on the order of 150 °C (300 °F).

Factors Affecting Particleboard Properties

Raw Materials. The most important characteristic of a species for particleboard manufacture is specific gravity (SG). As a general rule, lower-density species are preferred, the medium-density woods are used if readily available at a good price, and the highest-density woods are avoided.

It might seem that high-density woods should produce the strongest particleboard. In fact, the lower the wood density, the higher the board strength at any given final panel density. This is because lighter weight species have more particles per kilogram of furnish, are easier to densify, and achieve better glueline contact. This also indicates that the strength of particleboard is largely determined by glue bond quality and not by wood strength. For example, if particleboard is produced from aspen, with wood SG of about 0.38, an adequately strong board can be produced at a board SG of about 0.62. However, if paper birch with an SG of 0.55 is used, the board SG must be raised to about 0.75 to obtain a board equally as strong as the aspen board.

FIGURE 14.10. Multiopening particleboard press. Particleboard presses are typically much larger and more complex than plywood presses because of the need to load the mats carefully into the press and maintain a uniform pressure.

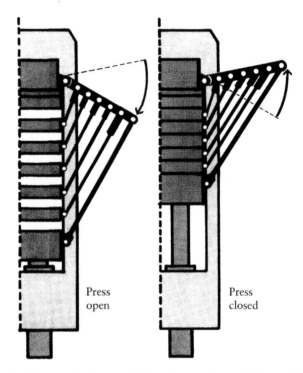

Press
open

Press
closed

FIGURE 14.11.   Simultaneous-closing particleboard press. Source: Maloney (1993).

Low-density woods are also preferable because they reduce density variation within a mat. Suchsland (1959) developed a statistical model explaining this behavior. He demonstrated that because of the variation in density across the face of a board, it is necessary to compress wood furnish beyond the average density of the species to obtain adequate interparticle contact. If a mat containing furnish with an SG of 0.45 is pressed so the total SG of the resulting board is 0.45, only about 50% of the mat particles will be under pressure and have intimate contact. To produce satisfactory contact between particles, it is usually necessary to compress the board to a density 1.2–1.6 times that of the furnish. This ratio between board density and wood density is called the *compression ratio*.

Maloney (1993) pointed out that a compression ratio of 1.3 (or 30%) is a good guideline for determining the minimum board density for a medium-density board. Using this guideline, it is expected that satisfactory particleboard could be produced as shown in Table 14.1. Most commercial boards being produced from Douglas fir and ponderosa pine exceed the density indicated in Table 15.1 by 50–80 kg m^{-3} (3–5 lb ft^{-3}). Means to produce medium-density particleboards from high-density species have been demonstrated experimentally. Furnish from high-density tree species yields composite panels of very high density. In many cases these can become too heavy for cost-effective shipping and handling.

Other species characteristics that affect particleboard manufacture include extractives, pH, and buffering capacity. Extractives are reported to cause problems in some woods with high extractive content, such as western red cedar and some tropical species.

TABLE 14.1. Guidelines for particleboard specific gravity as related to raw material specific gravity.

Raw material	Wood SG	Particleboard SG[a]	Particleboard density[a] (lb ft^{-3})
Douglas fir	0.48	0.69	43
Hickory	0.72	1.03	64
Oak	0.63	0.90	56
Ponderosa pine	0.40	0.57	36
Southern pine	0.53	0.76	47
Yellow poplar	0.42	0.60	37

[a] Including 10% resin and wax at 0% MC.

In these species, difficulties sometimes arise in developing proper cure of resin and blows in panels often occur. A *blow* is an internal rupture of the board resulting from internal gas pressure arising from volatilized extractives or excessive moisture content as the press opens. Moisture control and relatively slow decompression cycles are the two best ways to control and minimize panel blows.

Fibrous cellulosic nonwood crop residues can also be used as raw materials for particleboard (Bowyer and Stockmann 2001). Wheat straw, flax shives, sunflower stalks, bagasse (sugarcane residue), and kenaf are among materials that can be used to produce commercial boards. The quality of such boards can be comparable with that of wood-based boards and these materials can be used in combination with wood with little quality loss (Youngquist and Chow 1993). Bagasse and flax shives are used in several parts of the world for board products. Commercially bonded (MDI) wheatstraw boards have comparable properties to wood particleboard. There are pros and cons associated with using agricultural residues or wood/residue blends for panel products. Less waste, increased carbon sequestration, and decreased wood harvest are among the benefits (Douglas 2000). Wider use of nonwood fiber sources could reduce wood consumption but environmental benefits are not likely because the production of agricultural-based fibers is environmentally taxing. Farming row-crops for the sole purpose of providing fiber makes little sense from an environmental perspective because the environmental impact of growing such crops exceeds that associated with growing trees. Thus, the use of nonwood raw materials is limited to the waste streams of those respective agricultural crops. Furthermore the seasonally limited availability of raw materials is a major detractor from the viability of these feedstocks. Further, the history of economic failures associated with agriculture fiberboard mills limits the attractiveness of large commercial capital investment in such ventures (Lengel 2001).

Board Density and Density Profile.   The higher the overall density of particleboard from a given raw material is, the greater the strength. However, other properties such as dimensional stability may be adversely affected by increased density. The variation in density throughout the thickness of the particleboard is called the *density profile*. This vertical density variation (profile) should not be confused with the horizontal variation of density in the plane of the board illustrated in Figure 14.12.

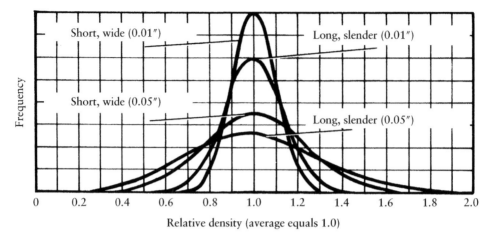

FIGURE 14.12.   Theoretical distribution of density in a compressed particleboard mat for four types of particles.

It is difficult to produce boards with truly uniform density profiles, i.e. equal face and core density, because as the press is closed the surface layers of the mat heat first and to a higher temperature than the core. This softens the surface particles and allows them to become more densified than the core. The development of steam-injected presses, now in common use, makes it possible to heat the furnish more rapidly and to approach the goal of uniform density.

Other factors also affect the density profile. A higher moisture content in the face layers at the time of pressing results in greater densification. The effect of a moisture gradient on density profile within a mat when producing a 12.7 mm (0.5 in.) thick particleboard is shown in Figure 14.13. The rate at which the press is closed also affects the density profile. The more rapidly a press is closed, the greater the relative compression of the face as compared with the core.

Differences in particle geometry between face and core also affect the density profile. More pressure is required to compress a mat of long, slender, thick particles than short, wide, thin particles. Therefore, if the core is made from coarse particles and the surfaces of fine shavings, more densification occurs in the face layers. Experimental boards with densities as low as 735 kg m⁻³ (46 lb ft⁻³) have been successfully made from high-density hardwoods using this principle.

Properties and Use of Particleboard

About one-half of the production of industrial particleboard goes into home or office furniture manufacture and one-third into kitchen and stereo/TV cabinets in the form of core stock. Collectively these products are termed "casegoods." Construction applications (classified as nonindustrial) such as mobile home decks and underlay for carpeted floors provide a significant market.

The most important strength properties of particleboard are modulus of elasticity (MOE), modulus of rupture (MOR), and internal bond. Screw-holding strength is also critical for uses in furniture and cabinets. Screw-holding strength is determined largely by

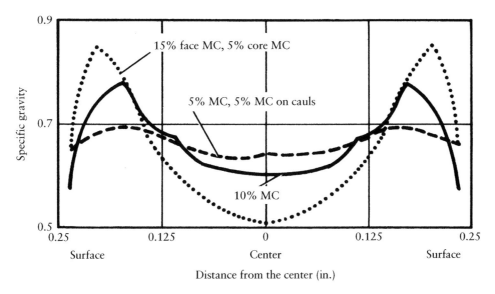

FIGURE 14.13. Effect of moisture distribution on the density profile.

board density, although resin content has an effect. It is advisable to design particleboard components such that screws and fasteners are not required on panel edges. Where components are to be loaded in bending, MOR and MOE are important. The tendency of particleboard to creep under long-term load, such as when used as shelving, often needs to be considered. The ANSI standard for the top-quality UF-bonded medium-density mat-formed particleboard (grade M-3) requires a minimum MOR of 16.5 kPa (2400 psi) and an MOE of 2.76 MPa (400 000 psi).

Internal bond (IB) is the tension strength perpendicular to the plane of the panel. IB is the best single measure of the quality of manufacture because it indicates the strength of the bonds between particles. It is an important test for quality control because it indicates the adequacy of the blending, forming, and pressing processes. Figure 14.14 shows a specimen and blocks for conducting an IB test.

Dimensional change of particleboard, both in thickness and in the panel plane (known as *linear change*) can be important. Generally, particleboard is not as stable in the linear direction as plywood. *Medium-density particleboard* is allowed a *linear swelling* up to 0.35% owing to atmospheric changes from 50 to 90% relative humidity (ANSI 2016). For particleboard made from flakes, linear expansion is limited to 0.20%. Although these changes seem small, they can cause problems if panels are installed with no provision for swelling. When particleboard is used as underlay, it is recommended that it be glued to the subfloor to reduce its linear change and caution is advised in areas that could be subject to water leaks.

Thickness swell of particleboard exceeds the normal swelling of solid wood and can be quite significant, ranging from about 10 to 25% when subjected to dry and then wet conditions. Thus UF-bonded particleboard should not be used where it may be subjected to wetting. Moisture-induced thickness swell is only partially reversible so if particleboard is repeatedly wetted and redried, its thickness will often continually increase. The permanent, nonrecoverable component of the swelling is called irreversible swelling.

FIGURE 14.14.   The setup for testing the internal bond of particleboard. The specimen is glued between two steel blocks.

Wood-based Fiber Products

Wood-based fiber products include hardboard, insulation board, medium-density fiber-board (MDF), and a rapidly growing number of wood fiber–plastic composites. These products are manufactured from wood that has been reduced to individual fibers or fiber bundles that are formed into a mat, compressed, and dried. Hardboard, insulation board, and MDF are products generally 3.0 mm and thicker used in building, furniture, and cabinet construction. For the most part these products are produced in panelized form. Wood plastic composites typically consist of wood fibers or fiber bundles blended with high levels of thermoplastic polymer.

Medium-density Fiberboard

MDF is a panel product made primarily from wood fiber and bonded with synthetic resins to a density of 500–800 kg m^{-3} (31–50 lb ft^{-3}). After particleboard, it is the most prominent nonstructural composite. Although it can be formed using a wet or a dry process, most of the production is dry-formed. Like dry-process hardboard, MDF is made from wood that has been reduced to individual fibers and fiber bundles with a binder added to accomplish board strength. One of the keys to quality production of this product is the use of pressurized refiners which produce pulp of a very low bulk density.

MDF has been commercially produced for ~50 years. Worldwide production is on the order of 30–40 million m^3 (1.06–1.41 billion ft^3) with the United States producing about 10%. MDF consumption continues to grow rapidly both in North America and worldwide owing to its versatility for a range of products from core stock to a substitute for clear lumber.

Manufacturing Process

The first steps in making MDF are similar to those employed in manufacturing hardboard. Logs and other raw materials such as plywood and furniture trim or sawmill cut-off blocks are initially reduced to chips. The chips are then refined using thermomechanical pulping (see Chapter 16). In many cases, wax is added during pulping. Thereafter, the process closely resembles particleboard manufacturing. Special blending and forming machines are necessary for MDF because of the bulky low-density fiber. Most MDF processes use *blowline* blending. There, the resin is injected into a pressurized pipe that contains the fiber from the refiner. UF resin is typically used as the binder with resin solids on the order of 8–10% by weight. For some speciality products, the resin solids content can be 15% or greater. Then, the furnish is dried, which reduces subsequent press time and potential for blow-type delamination. Dried furnish is then dry formed and pressed into panels.

A variety of raw material types can be used for MDF. Wood residue such as planer shavings, sawdust, and plywood trim are regularly used but the input should include at least 25% pulp chips to produce the desired quality of furnish (Maloney 1993). There is considerable potential for use of agriculturally derived fiber in MDF production and a number of mills now make use of such materials. The fiber from oil palm stalks, for example, is widely used in Malaysia and Indonesia for MDF. Many fiber sources can be utilized or mixed as long as the interaction between the chemistry of the raw materials and the resin is properly controlled.

Uses

The most-significant use (70% in the United States) of MDF is in furniture and kitchen cabinets where it is used in much the same way as particleboard or lumber-core plywood. The edges of particleboard are too porous to be shaped or finished directly and thus are commonly edge-banded for furniture applications (Figure 14.15). MDF, in contrast, has more highly refined raw material and a more uniform density, which results in smooth, tight edges that can be machined almost like solid wood. MDF can also be finished to a smooth surface and grain-printed, thus eliminating the need for surface veneers or laminates. For these reasons there is a well-defined market for MDF furniture panels. Another

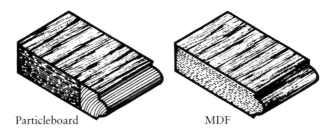

Particleboard   MDF

FIGURE 14.15.   In contrast to particleboard, MDF requires no edge-banding prior to shaping.

application is wall paneling, where the strength of lumber is not required but a smooth, printable, or paintable surface is needed. MDF panels can also be profiled and finished to be used as high-value wainscot.

Hardboard

Hardboard is a high-density wood fiber product that is most commonly manufactured to an SG near 1.0. The product is generally in the form of flat sheets ranging from 1.6 to 12.7 mm (0.06–0.5 in.) in thickness but it can also be molded to a variety of shapes. Hardboard was developed inadvertently in 1924 by William H. Mason who had invented a quick explosion process for transforming chips to pulp and was attempting to make a low-density insulation-type product from such fibers. The invention led to the immediate formation of the Mason Fiber Company, a name later changed to Masonite Corporation. The name Masonite is still sometimes applied in a generic sense to all hardboard. Over the last 30-plus years, US hardboard production has decreased by about 67% from its peak in of 2.15 million m³ (7.3 billion ft² on a 0.125 in. basis) in 1983 to ~705 000 m³ (2.4 billion ft²) in 2015 (Howard and McKeever 2016).

Manufacturing

An important distinction between hardboard and other fiber products is that in hardboard (of the wet-process type), lignin plays a role in fiber-to-fiber bonding. Lignin should be retained during pulping to achieve this benefit. Wood elements are then highly refined by machines such as the Asplund defibrator, invented in 1931 by a Swedish engineer of that name. This invention played a major role in the development of the hardboard industry (Suchsland and Woodson 1986).

   The basic procedure employed in hardboard manufacture is shown in Figure 14.16. Note that resin and wax are added after the pulping process or during drying of the fiber. Water-soluble resins such as phenol formaldehyde are used. Resin–solids concentration varies from 0 to 2% of dry board weight depending on board type. These small amounts of adhesive improve board strength and both the resin and the wax increas water resistance.

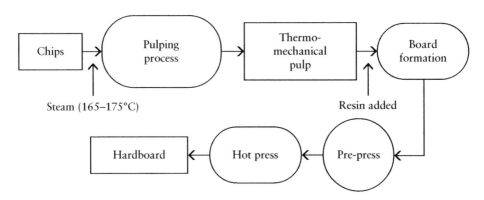

FIGURE 14.16.   Basic steps in hardboard manufacture.

After pulping, fibers are formed into a mat and prepressed. Mat formation is accomplished using either water or air as a forming medium. The difference in these techniques, known as the wet and the dry processes, is explained below. The manufacturing sequence concludes with hot pressing in which high temperatures (190–235 °C) and pressures (3.5–10.5 kPa) are used to bring the lignin to a thermoplastic state and thus to densify the fiber mat.

Wet Process

As the name suggests, in the wet process, pulp is mixed with water as in the making of paper. The mat is formed using a Fourdrinier machine, on which the water–fiber mixture is metered onto a wire screen. Water is drained with the aid of suction applied to the underside of the wire, and the fiber mat along with the supporting wire is moved to a prepress where excess water is squeezed out. The prepress operation is important, because the subsequent step involves pressing at high temperatures. After prepressing, the compressed mat is moved into the hot press along with the wire screen on which it was formed. High pressures and heat serve to form ligneous bonds, squeeze out additional water, and dry the mat. The screen is retained in the hot-pressing operation to allow the escape of water vapor. *Wet-process hardboard* is typified by evenly distributed density (because water is an efficient forming medium) and one rough side caused by the screen. Consequently, the panel is classified as "smooth on one side" (S-1-S). Several firms produce a wet-process board with two smooth sides by drying the fiber mat prior to pressing.

Dry Process

Dry-process hardboard is made using air rather than water as a forming medium. Fiber bonding is accomplished with a synthetic resin as it is in the manufacture of particleboard. Following pulping, the fiber is dried, resin is added, and the furnish is introduced into a forming device that creates a "snowstorm" of the dry, fluffy fiber. The fiber blanket formed in this way is thick (100–125 mm for what will eventually be a 6.5 mm thick panel), so a press roll is placed downstream of the former to compress the loosely piled fibers. Hot pressing completes the sequence. The fiber mat is relatively dry when it enters the hot press and therefore no screen is needed beneath the mat, resulting in panels smooth on both sides (S-2-S). Dry-process hardboard tends to have more density variation, greater linear expansion, and lower bending strength than the wet-process variety. There is substantial less lignin bonding as compared with the wet process, and therefore more resin (about 2% resin solids, dry basis) is used.

Tempering

Hardboard is sensitive to water unless specially treated so it should be used only as an interior product. Moisture causes linear expansion of panels, thickness swelling, and the formation of surface blisters. Hardboard intended for exterior use, such as siding, is treated and finished for resistance to moisture and decay. For other uses, hardboard is often tempered. Traditionally *tempering* was achieved by soaking finished panels in various oils, followed by baking at high temperature to flash off the volatile fractions.

Another process for tempering involves high-temperature treatment without a preliminary oil soak; the purpose of exposure to heat, which may be as great as 200 °C (392 °F), is to increase crosslinking between cellulose and other polymers. Performance under wet conditions can also be improved by using more resin in board manufacture. The latter methods for tempering have become popular because of the air pollution problems associated with baking oil-soaked panels.

Applications

Hardboard in the form of flat panels is used in case-good furniture, cabinets for electronic components, drawer bottoms, dust stops, sliding doors, general purpose backing, tabletops, etc. It is also commonly used for wall paneling, cabinet doors and tops, interior door faces, garage door panels, and store fixtures. Unfinished panels are perforated with holes and used as pegboard for workshop, laundry room, and garage walls. Smooth-faced hardboard sheets are also painted or covered with vinyl or other material for use as exterior siding or as decorative paneling for interior use (Figure 14.17). Appearance-grade panels can be made with contoured or sculptured surfaces by using sculptured platens (or dies) in the hot press for applications such as interior paneling or garage door panels. This technique allows the reproduction of rough-sawn surfaces, simulated brick, or other textures. Interior door skins, made to look like raised panel doors, are one of the

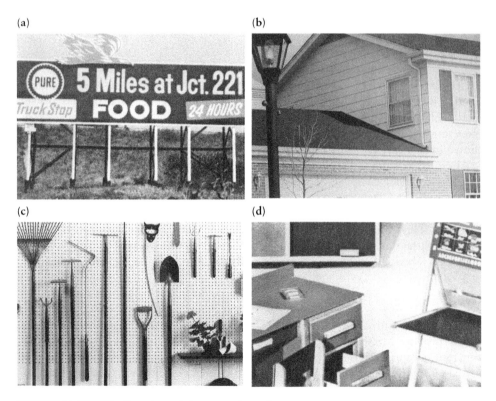

FIGURE 14.17.   Hardboard panels have a variety of uses.

FIGURE 14.18. Hardboard can be molded to different shapes.

most significant markets for hardboard. Other hardboard products are produced by steaming and molding finished panels to various shapes (Figure 14.18). Molded hardboard products are common in the auto industry, where they are used as door and roof panels, back window decks, and dashboards.

Insulation/Acoustical Board

A group of fiber panel products manufactured to specific gravities ranging from about 0.16 to 0.50 are referred to as *insulation board*. These range from low-density acoustic ceiling tiles to higher-density wall-sheathing board used under siding in light-frame construction (Figure 14.19). Insulation board has been produced in the United States since 1916, when a plant using softwood groundwood pulp was built in Minnesota. Currently, most production facilities are in the South. Competition from both wood and nonwood products, primarily fiberglass, plastic foam, and plywood, has greatly reduced the market for wood-based insulation board, although the popularity of this product has increased in recent years.

Manufacturing Processes

Insulation board production is similar to that of wet process hardboard except for the pressing operation. A hot press is not generally used in making insulation board. Instead, the mat is brought to desired thickness using a press roll and then dried. The omission of

FIGURE 14.19.   Structural insulation board is applied prior to siding.

hot pressing means that ligneous bonding is not achieved in insulation board. Fiber-to-fiber linkages are provided primarily by hydrogen bonding, although additives such as starch or asphalt are often used for bond enhancement, similar to paper manufacture. By avoiding a hot press, significant capital and operational costs are achieved.

Formation of the insulation board mat is done with a single- or double-cylinder machine or on a Fourdrinier. As compared with the Fourdrinier, the *cylinder machines* are simple and rugged. The pulp slurry is pumped into a trough that surrounds a large cylinder faced with a wire screen. The cylinder rotates slowly and a vacuum inside the cylinder draws the slurry and fiber against the wire. The fiber mat is thus built up on the wire screen and is transferred to a conveyor as it emerges from the trough. On a double-cylinder machine the two formed mats meet to form one thick mat. This double machine results in a mat that is symmetrical with regard to the fibers on both surfaces. Single-cylinder machines produce mats with the coarser fibers on the cylinder side.

At the upper end of the density range of insulating panels is Homasote. It is manufactured from recycled newspaper fiber and is widely used as wall paneling.

Uses

Traditionally, wall sheathing and ceiling tiles were the major products made from insulation board. Insulation board sheathing provides thermal protection to the wall construction and serves to reduce noise transmission through the exterior walls but only a small portion of the wall sheathing currently used in the United States is insulation board. Also, as a result of competition from more fire-resistant products, little insulation board currently goes into ceiling tile. Other uses of insulation board include backers for aluminum siding and shingles, panels for built-up roof construction, and stock for a variety of industrial purposes such as sound-deadening board. In some cases PF resin is added and several 12.7 mm thick panels are face-laminated to form a 25–76 mm thick structural decking. Fiberglass is also added to some panels to increase structural performance. Such components are manufactured in the 450–500 kg m^{-3} density range. Bagasse fiber, a residue of sugar cane processing, is a successful example of a nonwood raw material that has been used for composite insulation panels.

Other Wood-fiber Products

Several products that use inorganic binders (gypsum, portland cement, or magnesite) and wood fiber have been commercialized. The fiber provides bending strength and flexibility while the binders provide desireable properties such as fire and termite resistance. Some cementitious wood fiber products are said to be immune to water damage, termites, and decay. The decay and insect immunity are attributed to the high pH. Some of these products contain ground sand as well as the binder and wood fiber. Cement-bonded products are generally characterized by high density, up to 1600 kg m^{-3} (up to 100 lb ft^{-3}), and tend to be less flexible than an all-wood-fiber product but of comparable or higher bending strength. These products appear to be well suited for use as lap siding, vertical siding, roof shingles, shakes, tile backer, and tile underlayment. Of note in exterior products, if liquid water is absorbed and allowed to freeze (causing expansion), cracking and splitting can result so priming and sealing are recommended. Sales of wood/cement siding in the United States increased from ~110 000 m^2 in 1997 to 610 000 m^2 in 2001 (Kurpiel 2002), and by 2005 this product commanded 13% of the US residential siding market, up from only 2% in 2000.

Another type of product is a cement/wood fiber shingle that has an appearance similar to a wood shake (Youngquist 1995). Another product, long-established in the United States, is one in which wood excelsior (long strands of wood) is combined with portland cement to make acoustic ceiling tiles for commercial and industrial buildings. In international trade this latter product is termed *wood wool*. Wood wool board is about one-fourth to one-third wood by weight, the remainder being portland cement or other mineral binder. The product is usually produced in densities from 320 to 400 kg m^{-3} (20–25 lb ft^{-3}). It is well suited to developing areas of the world because it can be produced by simple hand-forming methods using locally available mineral binders. Figure 14.20 shows a wood wool board and a prototype home in the Philippines with exterior walls of this product.

Species selection is important in the production of wood wool because many species contain wood sugars and extractives that retard the cure of the cement. This problem can be reduced by long-term storage of wood bolts prior to shredding and by the addition of chemicals that accelerate the cure rate of the cement. The density of the wood is not critical except as it affects shredding on the excelsior machines.

A gypsum-fiber product utilizing gypsum as the binder and recycled paper has become widely used in Europe and is also produced in North America (Maloney 1994, 1996). This product is an alternative to conventional gypsum board used for walls and ceilings. Conventional gypsum board has gypsum sandwiched between two thick sheets

(a)

(b)

FIGURE 14.20.   Board produced from wood excelsior bonded with portland cement and prototype home. The left-hand side of the wall shows the individual panels in place. The finished wall on the right is produced by application of a cement stucco.

of paperboard. This product has the wood fiber distributed uniformly throughout the thickness of the panel.

Recent years have marked the development of a number of thermoplastic composites made using waste paper or waste wood fiber as a reinforcing filler. The wood and plastic combination yields a moderately strong, stable, light-weight, and workable material. In some cases the wood component contributes to the strength of the products while in other cases its main contribution is to reduce density (weight).

Wood plastic composites are commonly formed by mixing finely ground wood fiber (which can be derived from wood waste streams) with thermoplastic polymers. Clean urban construction and remodeling wood waste can also be incorporated. The clean wood is then chipped, hogged, or ground to an elemental form such as particles or fiber bundles. For some processes, it is refined to the consistency of flour. The elemental wood is then dried and screened for size. Because much of the wood has been previously dried as lumber, plywood, or other material, minimal energy is needed for drying. The thermoplastic polymers can be from virgin or recycled sources. The major types are polyethylene (used in bags and soda and milk bottles), polypropylene (used in plastic cups and storage containers), and polyvinyl chloride (used for pipes and as a coating for paper and cloth). Wood plastic composite blends generally contain 50–60% wood and the balance as plastic. Because the wood component is cheaper, products are continually refined which contain higher proportions of wood. In Europe, there is at least one manufacturer that utilizes a 4:1 wood-to-plastic ratio. Additional additives are often combined to enhance production or product performance. These include coupling agents (to assist in bonding), waxes, lubricants, ultraviolet light stabilizers, pigments and dyes, and antioxidants. Nano fibers and particles can be added to enhance properties.

The properties and durability of wood–plastic composites vary widely depending largely on the wood–plastic ratio, the type of plastic used, the characteristics of the wood fiber, and other processing parameters. During processing, wood and plastic components are metered by weight into a forming machine. The machine heats the plastic above its melting point to a molten state under high pressure. As it is mixed with the wood the plastic flows into the wood elements. Other additives such as wax, biocides, or pigment, can be added while the mixture is in this molten state. The mixing action of the manufacturing equipment assures that the wood, plastic, and any additives are uniformly distributed, thereby creating a homogeneous product.

Manufacture is largely based on traditional linear extruding and pultruding equipment that is common to the plastics industry. For production, the finely ground wood fiber is mixed with molten plastic along with other chemicals such as pigments and release agents. Once adequately mixed the material is extruded through a profiled die, pultruded, or injection molded. Because machine dies are interchangeable one machine can produce a variety of profiled products. After forming, the material is rapidly cooled, often with a fine water spray, to facilitate solidification of the plastic resin. The wood–plastic products can then be embossed, printed, textured (to increase slip-resistance), or otherwise worked, crosscut to length and packaged. Successful products include molding, millwork, flooring, curved furniture, and automobile parts.

One of the most commercially successful of these products is decking. Decking is produced with both solid and hollow cross-sections (Figure 14.21). Virtually any type of profiled product can be produced. Other products that have been produced successfully include shelving, trim for windows and doors, and window and door core stock.

FIGURE 14.21.   Hollow-section wood–plastic composite decking material made from highly refined wood waste from a commercial window manufacturer.

Under sustained bending loading, especially in severe exterior environments, some of these products can creep significantly. Over the last two decades the long-term exterior durability (e.g., resistance to decay and ultraviolet breakdown) of many wood–plastic composites has improved and can rival treated wood.

Since the development of such decking in the 1990s and its growing acceptance in the market place, numerous other products have been developed. These include molding and millwork, trim, flooring, and others. In general, these products are costly. However their uniformity in color, size, and performance is an attractive attribute.

Another potential production method is sheet molding. In that case, wood–plastic composite formulations, which may contain nano particles and fibers, are manufactured into sheets, which may be layered. Curved products such as automobile shelves, seat parts, and trunk liners, and other compound molded pieces, can be manufactured from the sheets. This technology is widely used in the plastics field but has yet to develop in the wood–plastic arena.

References and Supplemental Reading

American National Standards Institute (ANSI). 2009. American national standard for hardwood and decorative plywood. ANSI/HPVA HP-1-2010.

American National Standards Institute (ANSI) (2016). *American National Standard for Particleboard. ANSI/A208.1-2016*. Gaithersburg, MD: Composite Panel Association.

American Society for Testing and Materials. 2000. Standard methods of evaluating woodbase fiber and panel materials. ASTM D 1037-99.

Baldwin, R.F. (1995). *Plywood and Veneer-based Products Manufacturing Practices*. San Francisco: Miller Freeman.

Baldwin, R.F. (2016). *Plywood and Veneer-based Products*. Montgomery, AL: The Donnell Group.

Bowyer, J.L. and Stockmann, V.E. (2001). Agricultural residues: an exiciting bio-based raw material for the global panels industry. *For. Prod. J.* 51 (1): 10–21.

Douglas, T. (2000). American agri-fiber industry continues to evolve, mature. *Panel World* 41 (6): feature.

Fischer, K. 1972. Modern flaking and particle reduction. Proc. Wash. State Particleboard Symp. 6:195–213.

Fleischer, H. (1965). *Use of small logs for veneer*. Madison, WI: USDA For. Serv., For. Prod. Lab. Res. Note, FPL-101.

Griffin, G. (1983). Optimizers boost lathe efficiency. *Plywood Panel World* 24 (10): 6–9.

Heebink, B., Lehmann, W., and Hefty, F. 1972. Reducing particleboard pressing time. USDA For. Serv., For. Prod. Lab. FPL-RP-180.

Howard, J.L. 2016. U.S. timber production, trade consumption, and price statistics 1965-2012. Res. Pap. FPL-RP-679. Madison, WI: USDA For. Serv., For. Prod. Lab.

Howard, J. L., and McKeever, D.B.. 2016. U.S. forest products annual market review and prospects, 2012–2016. Research Note FPL-RN-0343. Madison, WI: USDA For. Serv., For. Prod. Lab.

Jokerst, R.W. and Geimer, R.L. (1994). Steam-assisted hot-pressing of construction plywood. *For. Prod. J.* 44 (11/12): 34–36.

Koch, P. (1964). *Wood Machining Processes*. New York: Ronald Press Chap. 12.

Koch, P. 1972. Utilization of Southern Pines. USDA For. Ser. Agric. Handb. 420. Chap. 19, 23.

Kollmann, F., Kuenzi, E., and Stamm, A. (1975). *Principles of Wood Science and Technology*, vol. 2. Chap. 5. Heidelberg: Springer.

Krzysik, A.M., Younquist, J.A., Rowell, R.M. et al. (1993). Feasibility of recycled newspapers for dry-process hardboard. *For. Prod. J.* 43 (7/8): 53–58.

Kurpiel, F. (2002). Fastest growing building material in North America? *Panel World* 43 (2): 56–57.

Lengel, D. (2001). Ag-fiber dot gone: a litany of failure. *Panel World* 42 (4): feature.

Maloney, T.M. (1993). *Modern Particleboard and Dry Process Fiberboard Manufacturing*. San Francisco: Miller Freeman.

Maloney, T.M. 1994. The development of wood composite materials. Res. Pap. Wood Mtls Eng. Lab., Wash. State Univ. Pullman, WA.

Maloney, T.M. (1996). The family of wood composite materials. *For. Prod. J.* 46 (2): 19–26.

Miller, D.P. and Moslemi, A. (1991). Wood-cement composites: effects on hydration and tensile strength. *For. Prod. J.* 41 (3): 9–14.

Sellers, T. Jr. (1985). *Plywood and Adhesives Technology*. New York: Marcel Dekker.

Sellers, T. Jr. (1990). Use of lignin as partial substitute for phenol. *Panel World*. Sept. 31 (5): 26–30.

Sellers, T. Jr. (2001). Wood adhesive innovations and applications in North America. *For. Prod. J.* 51 (6): 12–22.

Suchsland, O. (1959). An analysis of the particleboard process. *Mich. State Univ. Agric. Exp. Stn Bull.* 42 (2): 350–372.

Suchsland, O. (1962). Density distribution in flakeboards. *Mich. State Univ. Agric. Exp. Stn Bull.* 45 (1): 104–121.

Suchsland, O. (1967). Behavior of a particleboard mat during the pressing cycle. *For. Prod. J.* 17 (2): 51–57.

Suchsland, O. and Woodson, G.E. 1986. Fiberboard Manufacturing Practices in the United States. USDA For. Ser. Agric. Handb. 640.

Watkins, E. (1980). *Principles of Plywood Production*. White Plains, NY: Reichold Chemicals.

Youngquist, J.A. (1995). The marriage of wood and nonwood materials. *For. Prod. J.* 45 (10): 25–30.

Youngquist, J., Carll, C., and Dickerhoff, H. (1982). U.S. wood-based panel industry: research and technological innovations. *For. Prod. J.* 32 (8): 14–24.

Youngquist, J., and Chow, P. 1993. Agricultural Fibers in Composition Panels. Proc. International Particleboard Symp. Wash State Univ. Pullman, Wash.

15

Pulp and Paper

Paper and paperboard are manufactured from wood that has been reduced to individual fibers or fiber bundles, which are then formed into a mat and dried. The distinction between paper and paperboard is primarily based upon product thickness, with sheets over 0.3 mm thick classified as *paperboard*. The linerboard and corrugating medium used to make corrugated cartons are examples of paperboard. *Tissues* include a wide range of products with low-weight sheets such as facial and bathroom tissues, napkins, and paper toweling. Other products that are manufactured from wood that has been reduced to individual fibers or fiber bundles include the previously discussed hardboard, insulation board, medium-density fiberboard, and a rapidly growing number of wood fiber–plastic composites.

Paper

It was not until the late 1800s that wood became an important source of papermaking fiber. Prior to that time cotton and linen rags were the major source. In 1840 the ground-wood pulping method was developed in Germany. In 1856 the soda chemical pulping process was developed in England, and in 1884 the kraft pulping process was developed in Germany. These processes were then adopted in the United States. Currently, wood is the dominant raw material for paper manufacture.

Worldwide, 2016 production of wood (mechanical, chemical, and semichemical) pulp for paper and paperboard was on the order of 172×10^6 metric tons, with the quantity of nonwood fiber used in paper and board production reported at $11.4.4 \times 10^6$ metric tons (FAO 2017). Nonwood fibers in use include primarily straw, bagasse, and bamboo. Other sources include reeds, abaca, esparto and Sabai grasses, oil palm, cotton linters, rags, sisal, and kenaf. Wood is clearly the dominant worldwide raw material for pulp, with ~93% of production. Recovered fiber provided another 228×10^6 metric tons of raw material for production of paper and paperboard. Together, virgin and recycled production is on the order of 400×10^6 metric tons.

Forest Products and Wood Science: An Introduction, Seventh Edition. Rubin Shmulsky and P. David Jones.
© 2019 John Wiley & Sons Ltd. Published 2019 by John Wiley & Sons Ltd.
Companion website: www.wiley.com/go/shmulsky

Paper (including paperboard and tissues) has become so ubiquitous in people's daily lives that most rarely consider its source or its importance. It serves as a primary packaging product, a communications medium, a base for sanitary and disposable products, and an industrial sheet material. For 2014, annual production capacity for paper and paperboard in the USA was on the order of 78.9 million metric tons (FAO 2015). This value was down from 92.5 million metric tons just 10 years before.

Commercial operations are highly capital intensive. To achieve economies of scale, pulp and paper mills are generally large capacity and operate around the clock. Bark and lignin are commonly burned for energy, which is used to heat the dryer rollers and in some cases to generate electricity. In the future, as energy costs and globalization increase, it is reasonable to assume that some pulp and paper mills will operate in a dual manner, that is, depending on the prevalent market cost structures (for wood, paper, and energy), wood may be converted to pulp and paper or it may be converted to electricity. Another developing concept is that of the bio-refinery. A bio-refinery is essentially a mill that converts bio-based raw materials into chemicals, energy, and/or reconstituted products (Sims 2002). Another developing concept is that of the bio-refinery. A bio-refinery is envisioned as a mill that will convert bio-based raw materials such as wood into chemicals, energy, and/or reconstituted products (Sims 2002). Whereas kraft papermills of today subject wood to chemicals that remove lignin and the hemicelluloses in the pulping step, with these compounds then burned for power, the chemical paper mill of the future may extract hemicelluloses from pulp chips prior to subsequent processing, with these used to produce a variety of chemicals and polymers. Then, in the bio-refinery, the potential will exist for conversion of wood to *syngas*, liquid fuels, pulp, and a wide array of industrial chemicals and feedstocks. Syngas consists primarily of carbon monoxide and hydrogen, is combustible, and has about half the energy density of natural gas. To date, this potential transition has not occurred because, given a fixed amount of wood or other fiber as raw material, the value of paper manufactured therefrom has remained much higher than the value of energy derived therefrom. In the future, the black liquor that results from pulping of the extracted chips will be used to generate power, as today, except that much more efficient gasification systems will be used. The net effect may be that pulp and paper mills will become net providers of energy to the national energy grid, with pulp (and paper) but one of many product options for the bio-refinery manager. In the process of conversion from paper mills to full bio-refineries, wood-processing facilities may switch from net emitters of carbon dioxide (traceable to the use of fossil-based fuels for a portion of the process energy), to facilities that sequester substantial quantities of carbon through avoided emissions. Such facilities may become more flexible, more responsive to societal needs, and potentially more profitable.

The Manufacturing Process

In simple terms, the process of paper manufacture involves (i) reduction of wood to constituent fibers (pulping), (ii) suspension of the fibers in water, (iii) beating or refining of the pulp, (iv) blending of additives (fillers, sizing materials, wet-strength binders, etc.), (v) formation of a fiber mat, (vi) drainage of water, and (vii) drying of the sheet. Surface treatment to improve printing qualities follows in many types of paper.

Pulp Production

A primary difference among various paper manufacturing processes is the method used to break the wood into fibers, i.e., *pulp*. Mechanical, chemical, or heat energy, or combinations of these processes are employed to accomplish pulping. The forms of energy used to make the pulp determine to a large extent both yield and properties.

Mechanical Pulping.   There are two methods of producing mechanical or groundwood pulp. These are the stone-groundwood and the refiner mechanical pulping processes. To produce *stone-groundwood pulp* a large abrasive stone, or a wheel made up of stone segments, is rotated while the tangential surface of wood bolts is pressed against the stone (Figure. 15.1). As the abrasive stone grinds against the wood, the fibers are compressed, loosened, and separated. Frictional heat softens the lignin, helping to separate the fibers from the wood mass. Groundwood pulp made from spruce is pictured in Figure 15.2. Note the pieces of fiber and bundles of unseparated fibers in the mixture. There are few stone groundwood processes in operation; almost all mechanical pulp is produced using refiners.

   Another predominant method of mechanical pulping involves the use of a machine with opposing serrated steel disks. One type, called a double-disk refiner, is composed of two fluted metal disks that can be closely spaced and driven in opposite directions.

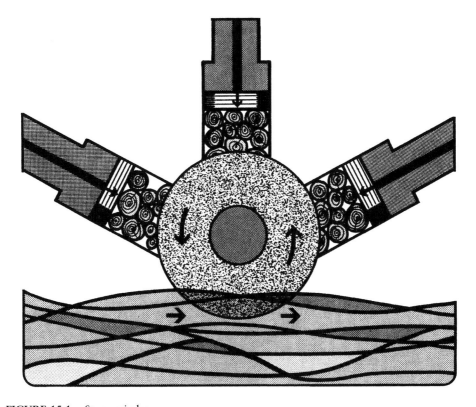

FIGURE 15.1.   Stone grinder.

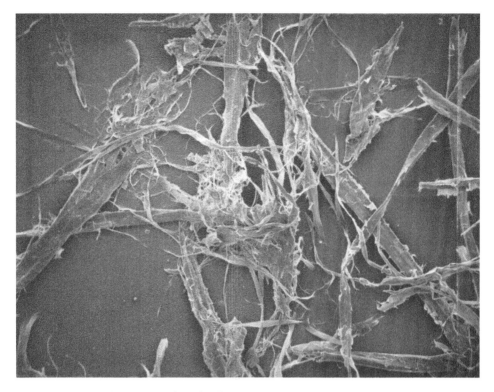

FIGURE 15.2.   Unbeaten groundwood pulp (spruce) ×125.

A variation of this machine, called a single-disk refiner, has one fixed disk (stator) and one disk that rotates (rotor). In both types of refiners, wood chips are moved by a screw-feed mechanism into the center of the machine where they pass between the two closely positioned disks wherein mechanical action reduces the chips to fiber. Pulp produced in this way is called *refiner mechanical pulp*.

Because the wood is physically torn apart, little material is lost in mechanical pulping as long as the fibers are flexible enough to avoid the production of fines. Dense species are generally not pulped by mechanical processes because the wood cells frequently shatter. The short, thick-walled fibers and large proportion of vessels and parenchyma in some hardwoods reduce their usefulness for mechanical pulp. Resinous softwoods like southern pine can also present problems but techniques to overcome such difficulties can be successfully employed.

In mechanical pulping, the proportion of wood that becomes usable fiber is commonly on the order of 95–99%, a fact translating to relatively low-cost pulp. Unfortunately, high yield also results in pulp unsuitable for many uses because the cellulose, hemicellulose, and lignin that make up the roundwood are still present. The lignin, which serves to strengthen solid wood through stiffening of fibers, gives rigidity to the fibers of mechanical pulp. These rigid fibers have little fiber-to-fiber bond potential and form a coarse, bulky mat. The paper thus formed has low strength and relatively poor surface quality. The presence of lignin in mechanical pulp contributes to yet another problem: long-term durability. Lignin, and to a limited extent certain carbohydrates, yellow with age, particularly when exposed to ultraviolet light. This is the reason for the yellowing commonly seen in old newspapers.

Semimechanical Pulping.   Several pulping methods are combinations of mechanical, thermal, and/or chemical processes. One is the *thermomechanical process* (TMP), in which chips are subjected to steam as they pass through refiners, generally in two passes. In most TMP processes, chips are presteamed in a vessel under pressure (superheated), refined under pressure, and then refined again at atmospheric pressure. In some systems steaming is done at atmospheric pressure. The heat serves to soften lignin, allowing fiber separation with less fiber damage than can be obtained by the stone-groundwood or refiner mechanical pulp process. This improves both strength and absorbency.

Chemi-thermomechanical pulping (CTMP) is a process similar to TMP but with chemicals added to the chips during the process to assist in softening and separating the wood fibers. Alkaline sulfite liquor, kraft black liquor, and sodium sulfite are among the chemicals used. CTMP makes possible the use of denser hardwoods such as oak, birch, and eucalyptus for magazine paper and newsprint.

The *semichemical-mechanical pulping* (SCMP) process is similar to CTMP with the exception that the chips are refined at atmospheric pressure, and thus at a lower temperature. These two processes provide the highest-quality mechanical pulp and have been the most widely used in recent pulp mill construction of any of the mechanical pulping processes. Semimechanical processes typically have pulp yields from 80 to 90%.

Semichemical Pulping.   Semichemical processes consist of a chemical treatment more severe than that used in the semimechanical processes but still requiring some degree of mechanical treatment to complete fiberization. Wood pulped in this way combines the high-yield advantages of mechanical processing with some of the high-quality fiber characteristics of chemical processing. The *cooking liquor* (chemical solution) causes partial degradation of the ligneous bonds and serves basically the same function as heat in the thermomechanical process. The mechanical energy needed for fiber separation is reduced and the damage to fibers is decreased. The most widely used process in this category is the *neutral sulfite semichemical process*. In this case, sodium sulfite buffered with sodium carbonate is used to pretreat the chips. Semichemical pulp yields of 65–75% are common and may occasionally be higher. The neutral sulfite semichemical process utilizes a wide range of hardwoods for products ranging from corrugating medium to newsprint.

Chemical Pulping.   The use of chemicals to achieve fiber separation is the most widely used pulping method worldwide. The majority of the pulp produced in North America uses processes of this type. Wood chips are placed in a chemical solution (cooking liquor) and heated in a pressurized vat (*digester*). Fiber separation occurs as the lignin that cements the cells together is dissolved. These processes result in low yields, as compared with other pulping methods, varying from 40 to 55% depending upon the severity of the pulping and bleaching processes.

There are two major chemical pulping processes now in use, the sulfite and kraft (sulfate) processes, which differ in the chemicals comprising the cooking liquor. The *sulfite process* uses a mixture of sulfurous acid and ammonium, magnesium, calcium, or sodium bisulfites. The sulfurous acid (H_2SO_3) reacts with lignin to form lignosulfonic acid. This relatively insoluble compound is, in turn, reduced to soluble lignosulfonic salts in the presence of the basic bisulfites. The sulfite process yields high-quality pulp of the type desired for fine writing papers.

Formerly, calcium-based bisulfite was the most commonly used chemical with sulfurous acid. The calcium compound was cheap and worked quite well in the pulping of long-fibered species such as spruce, hemlock, and true fir. There were, however, several problems associated with its use. The most serious was that recovery of cooking chemicals and process heat was technically difficult and economically unfavorable. As a result, sulfite mills constantly had a used cooking liquor disposal problem, on the order of ~6250 l t^{-1} (1500 gal short ton^{-1}) of pulp produced. Another problem was that the process did not work well in pulping highly resinous softwoods such as pine. There, the strongly acidic (pH < 2) conditions needed for pulping caused the lignin and the resins to bond together into larger molecules which counteracted the pulping process. As a result, new sulfite installations were designed to use ammonium or magnesium bisulfites. Development of chemical recovery technology makes it possible to achieve complete chemical recovery of magnesium-based cooking liquors through a relatively simple process. In recent decades sulfite pulp mill capacity has declined steadily while the kraft process has grown rapidly. Year 2005 production statistics by pulp type are shown in Table 15.1. This table illustrates that chemical processing of wood-based fiber dominates the market.

The *kraft* (or sulfate) process is based on the use of a cooking liquor made primarily from sodium hydroxide and sodium sulfide. The sodium hydroxide attacks lignin, breaking it down into segments whose sodium salts are soluble in the cooking liquor. The sodium sulfide, when exposed to water, breaks down to sodium hydroxide (increasing the amount of that compound available for pulping) and sodium hydrosulfide, which serves to increase the solubility of lignin.

An interesting piece of science history is the explanation of how this process became known as the kraft process. In the course of operating a Swedish soda process mill, a digester full of partially cooked pulp was accidentally blown. The material was about to be thrown away when the mill manager decided to use it in making some low-quality paper; the surprising result was that the paper produced was far stronger than any previously made. The Swedish (and German) word *kraft*, meaning "strength," soon became the commonly used name for what is also called the sulfate process.

High recoverability of cooking liquors and process heat makes the kraft process comparatively free from residue disposal problems. The kraft process is effective in pulping

TABLE 15.1.   Estimated year 2004 pulp for paper and paperboard production in the United States in thousands of metric tons (percentage of total).

Mechanical	4.14×10^6	7.7%
Chemical	46.2×10^6	84.5%
Semichemical	3.2×10^6	5.7%
Dissolving wood	900×10^3	1.7%
Total wood pulp	54.5×10^6	99.6%
Other fiber pulp	245×10^6	0.4%

Source: FAO (2006).

any species, including those that are highly resinous. These factors, plus the fact that high-strength pulp is produced, explain the overwhelming popularity today of the kraft process. One negative aspect is a characteristic rotten cabbage smell caused by volatile sulfur compounds. The costs of eliminating this smell are high. Because the human olfactory system can detect even minute concentrations of sulfur compounds, they must be removed almost entirely from stack gases to completely solve the odor problem.

Because no mechanical action is needed to achieve cell separation, chemically produced pulp is composed of smooth and virtually undamaged fibers (compare Figure 15.2 with the top frame in Figure 15.3). Moreover, when a high proportion of the lignin is removed in the process, an important component of age-induced yellowing in bleached paper is eliminated. The penalty paid for the high quality is low yield (and therefore high pulp cost). The yield, expressed as dry weight of usable fiber divided by the dry weight of chips placed in the digester, is lower (40–55%) than the lignin content might suggest. The reason for this is that the conditions that solubilize lignin also degrade both cellulose and the low-molecular-weight hemicellulose.

The advantages of the kraft and sulfite processes can be summarized as follows. The kraft process produces the highest strength pulp, allows efficient recovery of pulping chemicals, utilizes a wide range of species, and tolerates bark in the process. The sulfite process provides a slightly higher yield of bleached pulp and provides pulp that is easy to bleach and easier to refine for fine papers.

Fiber Recycling. Wastepaper recovery and recycling has long been practiced in the United States. As long ago as 1965 the US wastepaper recovery and reuse rates were about 23%, and these figures remained nearly constant through the early 1980s.

In the mid-1980s the American paper industry initiated a program to increase wastepaper recovery and reuse. Thereafter, the recovery rate began to rise steadily; by 1993 the waste paper recovery rate passed 39%, and by 2005 it reached 51.5%. In 2016, in the United States, the production of recovered fiber pulp and paper was 130×10^3 tons (143×10^3 short tons) and 47.6×10^6 tons (52.4×10^6 short tons), respectively (FAO 2017). This production level remained steady during the following decade.

Much of the recycled paper used in the United States goes into the manufacture of corrugating medium, the paper used in the fluted plies of corrugated boxes. Other uses are for newsprint, other printing grades, and structural wood fiber products.

Recycled fiber mills face several significant problems, including removal of contaminants (adhesives, plastics, waxes, latex, asphalt, etc.) and elimination of inks from fiber to be used in making printing papers. In addition, there is a limit to the number of times that fiber can be recycled. With present technology it is considered that 70% represents a practical maximum recycling rate. A relevant question when evaluating recycling performance is what reuse rate is possible. Put more simply, why is the recycling rate not 100%? For any one country the reasons for low reuse rates can include high paper production capacity in relation to population and net export of waste paper. More fundamentally, however, there are at least five reasons why total reuse of paper has not been achieved. These reasons are that:

- Some paper is not collected for reuse. Paper that is discarded or burned is effectively removed from the recyclable waste stream.
- Some used paper not compatible with reuse (i.e. tissues, food papers).

(a)

(b)

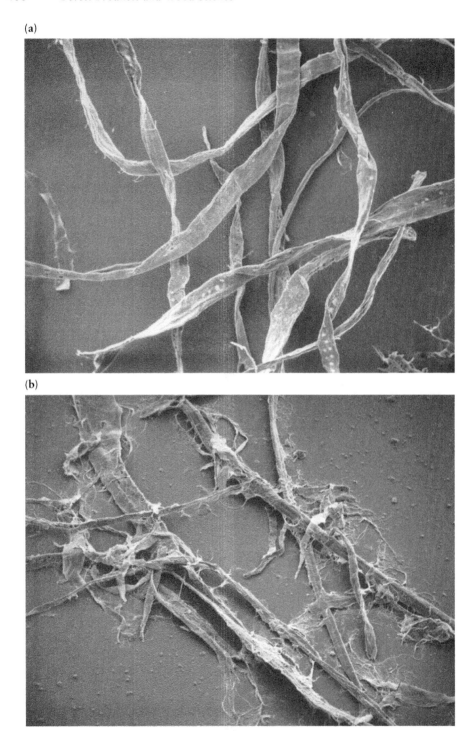

FIGURE 15.3. Beating flattens and partially unravels fiber walls (chemically pulped southern yellow pine).

- Some paper is put into long-term use (i.e. library books, photographs, and papers used to coat gypsum board and as a part of other construction products).
- A significant portion of the weight of some papers is composed of nonrecyclable inorganic materials (i.e. binders and fillers). In some coated, glossy papers, such as those used to produce calendars, posters, brochures, etc., the clay filler used to produce a smooth and glossy surface may comprise as much as 40% of the paper's weight. In recycling, this becomes sludge that cannot be reused in the coating process.
- Fiber is degraded each time it is recycled. This action is termed "attrition loss." Eventually each fiber is broken down into tiny molecular-level microfibrils that are lost in the manufacturing process. Depending upon the grade of paper being made, a fiber can be reused about three to nine times. This reality means that from 11 to 33% of fiber is lost with each recycling process, requiring a continual source of virgin fiber in paper manufacture.

Nanotechnology applications have emerged from the laboratories of scientists studying paper production and recycling. Recent results suggest that it may be possible to repair recovered fibers through the addition of nanoparticles to damaged surfaces of the fiber (Eadula et al. 2006). This raises the possibility that it may someday be possible to reuse fibers indefinitely.

Washing and Bleaching

It is necessary to clean pulp after it is formed to remove cooking liquor and/or impurities. After chemical pulping, the wood fiber and cooking liquor are blown from the digester into what is known as a *blow pit*. There fiber is collected and initially separated from the spent cooking liquor and gases that have been produced. Fiber is next cleaned in a multistage washing process to remove residual liquor.

Untreated wood pulp is brown to tan in color, owing mainly to the presence of lignin. Unbleached fiber is used for bag paper, linerboard, and other uses where strength, but not color, is important. For writing or book papers or other products where whiteness is important, fiber must be bleached. This is usually done using strong chlorine-based compounds. The technology of bleaching chemical pulps, however, went through rapid technological change in the 1990s and chlorine bleaching processes began to fall out of favor. New processes based upon oxygen-bleaching techniques are gaining commercial acceptance. Pulp referred to as TCF (*totally chlorine-free*) is being produced in the United States and Europe (Singh 1993).

Bleaching attacks residual lignin and can be carried to the point where lignin is either totally removed, as with the highest-quality writing and printing papers, or simply lightened in color, as in the manufacture of newspaper or catalog-quality stock. The latter degree of treatment is less expensive, having little effect upon yield, but it results in only temporary whiteness. Bleaching to achieve lignin removal gives virtually permanent whiteness but is expensive. In this case, water use is high and pulp yield is significantly reduced.

Beating and Refining

Much of the strength of paper results from the hydrogen bonding of cellulose molecules between adjacent fibers. To provide the maximum potential for bonding, fibers are pounded or ground to flatten them and to partially unravel microfibrils from the cell walls; the surface area of fibers (and thus the area available for bonding) is greatly increased by even a small degree of such flattening and unraveling (Figure 15.3).

The mechanical flattening and unraveling of fibers, called *beating*, is accomplished in various types of refiner machines. Disk refiners are commonly used for this purpose (see discussion under mechanical pulping). Another common type of machine is the conical refiner (known as a Jordan refiner). In this type of machine (Figure 15.4) a conically shaped, longitudinally fluted plug rotates inside a similarly shaped and ribbed housing. The location of the plug inside the housing controls the spacing between the flutes and ribs. Fibers are subjected to a mechanical rubbing action as they pass through the conical refiner.

Because fiber-to-fiber bonding has a great impact on paper properties, it is desirable to have a quantitative measure of the pulp's bond potential. In North America, bond potential is usually expressed in terms of Canadian Standard freeness. This is measured by suspending a given amount of fiber in water and then determining the rate at which water drains through a wire mesh onto which the fiber has been allowed to settle. Because the rate of drainage is inversely related to the surface area of the fiber, a mat of well-beaten fiber is quite resistant to drainage of water and the Canadian Standard freeness of well-beaten fiber is thus low.

The relationship between beating time, freeness, and various strength properties is illustrated in Figure 15.5. Note that freeness is decreased by extended beating. Burst and tensile strengths tend to be higher with longer beating times. Burst and tensile properties are closely related to interfiber bonding and are thus directly affected by any treatment that increases bond potential (Figure 15.5a, b). Figure 15.5c shows that tear strength is significantly increased as beating starts but is reduced rapidly thereafter. The explanation for this is that tear strength is somewhat influenced by interfiber bonding but is influenced much more by the integrity and length of individual fibers. Within the first few minutes of beating, flattening of cells and some unraveling of microfibrils occurs, which greatly increases surface area while causing little reduction in either the length or the strength of fibers. The increased surface area resulting from further beating is offset by the damage to the individual fibers.

Sheet Formation

Following beating, and in some cases secondary refining, fiber is mixed with water at a ratio of about 1% fiber by weight. It is quite common to mix different types of pulp (i.e., mechanical and chemical) at this stage, with the proportion of each dependent on the type of paper to be manufactured. Additives such as starch (for increased bond strength) or wet-strength resin are often added to the mixture at this point as well, as are clays (for brightness and opacity) and rosin size (for decreased liquid absorption). This mixture is then formed into a thin mat.

The most commonly used machine to form the fiber mat is called a *Fourdrinier* (Figure 15.6). It is basically a rapidly moving horizontal screen fitted with a device called a head box that meters the pulp slurry onto the screen. As pulp flows onto the screen, water drains away with the aid of vacuum boxes or other drainage-enhancing devices mounted under the wire, leaving a mat of fibers. Two other types of machines use two wires or a rotating wire cylinders. After the fiber mat is formed using one of these

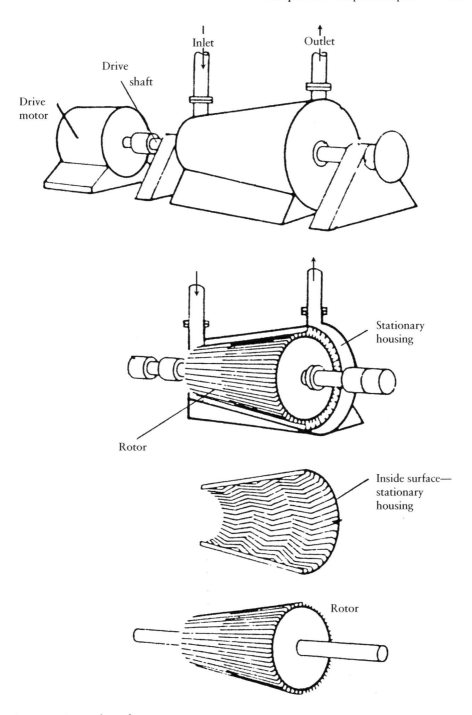

FIGURE 15.4.   Jordan refiner.

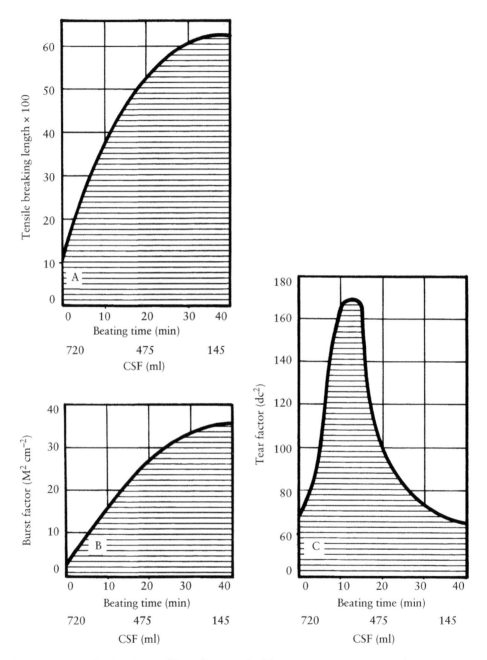

FIGURE 15.5. Beating time and Canadian Standard freeness versus paper strength.

machines, it is wet pressed, dried over a series of steam-heated drums, pressed again to desired thickness, and wound into large rolls. The application of coating, sheet polishing operations (known as calendering), the winding of the sheet onto a reel, and the cutting of large rolls into smaller ones are operations that can follow. The entire process is summarized in Figure 15.7.

FIGURE 15.6.   Fourdrinier paper machine.

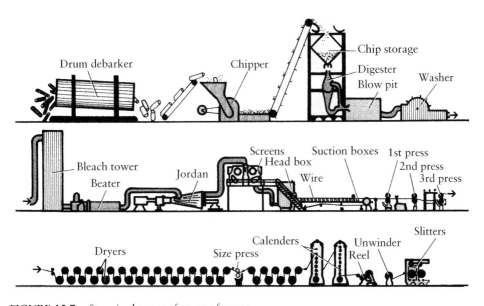

FIGURE 15.7.   Steps in the manufacture of paper.

Paper Consumption

Certain types of paper are almost invariably associated with specific types of pulps. Brown wrapping paper and grocery bags, for example, are almost always made from high-strength, unbleached kraft pulp, whereas gift wrap, including hard tissue paper, is typically produced from fine sulfite pulp. Although other examples of this kind can be given, most paper products are made from a blend of pulps. Wallpaper, for instance, often uses a blend of sulfite, sulfate, and mechanical pulps in order to incorporate the advantages of tear strength (sulfate), printability (sulfite), and low cost (mechanical) into the product. Similarly, low-cost newsprint, which is made primarily from mechanical or semimechanical pulp, often contains a certain amount of sulfite pulp to improve sheet quality.

The blending of 50–80% bleached hardwood kraft pulp with that from softwoods is common when making most types of printable papers for periodicals, catalogs, and containers. The hardwood pulp provides a smooth, opaque paper, while the softwood fibers add strength.

Paper Properties

Common Measures of Quality

There are many ways to define paper quality. When making grocery bags, for instance, strength is quite important. A bag filled with heavy canned goods is often picked up by its sides, so it must have high tensile strength. A bag that contains an exceptionally heavy item such as a large soft drink bottle should be able to resist this kind of concentrated load, a property measured by burst strength. High resistance to tear is another property obviously needed in an all-purpose bag. Moreover, the bag should retain its strength when wet.

When book paper is manufactured, tear strength is critical but other factors are also significant. The sheet must accept ink, but it must have low absorbency to prevent ink diffusion and the development of fuzziness around printed characters. High *opacity* is necessary so that printing does not show through to the other side. Other important properties might be brightness, permanent whiteness, and surface smoothness. If paper is to be used in making a product like a restaurant menu, all of the properties outlined above are needed as well as folding endurance.

Paper used for toweling needs an entirely different set of properties. Strength, particularly wet strength, is important in a towel, and such paper should also be highly absorbent. For each of the thousands of paper products, a similar list of important properties can be enumerated. There are dozens of tests to evaluate both pulps and paper. Many of these tests are spelled out in TAPPI (Technical Association of the Pulp and Paper Industry) and CPPA (Canadian Pulp and Paper Association) standards.

Sheet Properties and Fiber Characteristics

Knowledge of only one characteristic of wood, i.e., density, allows prediction of the yield of pulp per unit volume of wood as well as a number of paper properties. Density is directly related to cell wall thickness. The general rule is that the lower the density, and thus the lower the proportion of thick-walled latewood cells, the better the wood is as a papermaking material. It should be noted that this rule does not hold if high tear strength is desired.

Thick-walled fibers result in paper with low burst and tensile strengths but with high tear resistance. Paper made primarily from thick-walled cells also tends to have very low folding endurance. The relationship of burst and tensile strengths to cell wall thickness is explained by the fact that these properties are very dependent upon a high degree of fiber-to-fiber bonding. The primary reason for apparently low bond potential of thick fibers is that paper is manufactured on a weight basis, meaning that the number of fibers in a sheet is inversely related to the density of fiber walls. Second, thick-walled fibers have less surface area per unit weight than thin-walled fibers. Also they do not collapse as readily to form ribbons as thin-walled fibers. These two factors lessen opportunities for interfiber bonding.

Tear strength, like burst and tensile strengths, is influenced by the extent of interfiber bonding. More important, however, is the effect that individual fiber strength has upon tear resistance. It has been found that thick-walled cells that provide high tear strength are usually composed of a high proportion of hemicellulose. High levels of hemicellulose usually result in rapid hydration of pulp, the formation of more and better interfiber bonds, and the development of dense mats.

A second characteristic of wood that impacts paper properties is fiber length. Tear strength is the property most affected and the relationship is direct up to a fiber length of 4–5 mm. Generally, the greater the fiber length is, the higher the tear resistance.

The proportion of various cell types in wood affects the quality and quantity of pulp. This relationship is particularly true of the portion of hardwood volume accounted for by vessels. Because of their shape, vessels do not bond readily to fibers, thereby they contribute little to strength. Vessels may separate from the surface of the finished sheet in printing and are more likely to break up during processing. Thus woods containing a high proportion of these cells, such as oak, are likely to give lower pulp yields than those with a higher fiber content.

Important Types of Paper

Linerboard

This is a relatively lightweight board typically produced from unbleached kraft fiber. It is used as the outer surface on corrugated containers. The production of linerboard in the United States is large, exceeding even that of newsprint in tonnage. Linerboard generally is made from softwood fiber and is produced on a Fourdrinier machine. Its most important properties are stiffness and burst resistance, although some degree of printability is also desired on the outer surface.

Corrugating Medium

This lightweight paperboard is used for the fluted inner plies of corrugated boxes. Linerboard is glued to both sides of corrugating medium to produce what the public calls "cardboard." Because corrugating medium provides much of the rigidity to corrugated containers, it must have good stiffness properties and good resistance to crushing. Some is produced from 100% recycled fiber, but most is produced from a mixture of semimechanical and recycled pulp.

Newsprint

The primary requirements of newsprint are that it can be run through modern high-speed printing presses, provide a reasonably good printing surface, and be low in cost. Newsprint is generally produced from a mixture of mechanical, semimechanical, and chemical pulps. In order to develop adequate strength for printing at high speeds, some unbleached sulfite or bleached kraft is usually added to the groundwood pulp. Because mechanical pulp is cheaper than chemical pulp, only enough chemical pulp is used to meet the print-speed requirements.

Publication Grades

Papers for high-quality printing purposes must be coated to improve the gloss, detail, and brilliance that can be obtained in printing. The addition of fillers and coatings at various stages of manufacture can greatly alter the properties of the paper. Coatings can make up over 30% of total sheet weight in some lightweight grades. A higher proportion of chemical pulp is usually required as paper weight is reduced. TMP and other semimechanical pulps are used in these papers, as are bleached kraft pulp from both hardwoods and softwoods.

Fine Paper

This is the classification for white, uncoated printing, and writing paper containing only a small amount of mechanical pulp. Sulfite and highly refined bleached kraft pulp are used in cases where wood furnish is incorporated. Nonwood fibers such as cotton and linen are also used.

Tissue

This category covers a wide variety of facial and bathroom tissues, paper napkins, and toweling. Because these are lightweight papers that must have a loose structure, they cannot be produced on conventional paper machines. Key to success in this business is an effective proprietary system of producing a low-density sheet. These products generally require a high-quality furnish with long, highly refined fibers because softness is a function of fiber properties and bulk.

Paperboard

This category of thick paper includes the linerboard and corrugating medium described above. However, there are a number of other important types of paperboard that usually have a multi-ply construction. Folding boxboard is made with virgin pulp in the outer ply and secondary fiber for the inner plies. Foodboard, with 100% bleached virgin pulp, is used for food packaging. The paperboard for the outer layer of gypsum board (sheetrock) is usually made from 100% recycled fiber.

Kraft Sack Paper

This paper is produced from unbleached softwood kraft pulp. Tear strength and tensile energy absorption are two of the most important properties. Sizing is often added to the well-refined fibers in the papermaking process to provide additional internal and wet strengths.

Measurement and Sources of Raw Material

Wood

As noted earlier, about 90% of paper worldwide is made from wood. In the United States this figure is virtually 100%. About half of the wood used for the manufacture of paper in the United States is in the form of small-diameter pulpwood bolts (Figure. 15.8). Because measuring individual pieces is not practical, loads of pulpwood bolts can be scaled to determine the number of cords, measured with some other quasi-volumetric units, or simply weighed. Most pulpwood is currently purchased by weight in order to save the time and expense of volume scaling. Payment is then made either directly by weight ($ t^{-1}), or the weight is converted to cord equivalents, with payment then made on a cord basis. Elsewhere in the world, pulpwood volume is expressed in cubic meters. In the United States pulp chips are often traded on a weight basis (by the green or estimated dry ton), volume basis, or a combination weight/volume basis (by converting weight measurements to cords).

FIGURE 15.8.   A load of pulpwood begins the trip to the paper mill.

A common unit of measure for pulpwood in the United States is the cord, which is defined in Chapter 13. It is important to remember that a cord does not contain 3.62 m³ (128 ft³) of wood but 3.62 m³ (128 ft³) of space. A cord of 178–254 mm (7–10 in.) diameter undebarked bolts, 1.2 m (4 ft) long, for example, contains only about 2.3 m³ (80 ft³) of wood, about another 0.3 m³ (10 ft³) of bark, and the remainder as air. A greater amount of wood is contained in cords composed of larger-diameter and/or shorter bolts.

Agricultural Fiber

About 10% of fiber used in papermaking worldwide comes from fiber crops (crops planted specifically as a fiber source), agricultural crop residues, and a few other sources such as oil palm stalks. In North America, nonwood fiber is often promoted as an alternative raw material in papermaking.

Understanding the possibilities for nonwood substitution in papermaking requires knowledge of the alternatives. A discussion of two major wood-fiber alternatives, that is, wheat stalks (an agricultural residue), and hemp and kenaf (two fiber crops) follows.

The Wood Cell Wall in Comparison with that of Wheat Straw.   The chemical composition of wheat stalks (wheat straw) is very similar to that of the other cereal straws such as barley and flax. Note in Table 15.2 that the cellulose content is very similar to that of wood, that the hemicellulose content tends to be higher, but that the lignin content is considerably lower. As will be discussed later, the lower lignin content is an advantage in papermaking.

The silica and extractives content of wheat straw are considerably higher than that of either softwoods or hardwoods (Table 15.3). In both cases this is undesirable from the standpoint of papermaking and in the production of energy. Extractives include waxes, oils, resins, fats, gums, tannins, and aromatic and coloring materials. These compounds are so named because they are not incorporated into the cell structure.

Comparing cell lengths and diameters, the fibers of wheat straw are about the same length as hardwoods, but generally shorter than softwoods (Table 15.4). The fibers are smaller in diameter (i.e. finer), and the cell walls are thinner, as evidenced by the greater percentage of volume contained within cell lumens and by the lower tissue density. These differences can be viewed as minimal or significant depending upon the intended application.

TABLE 15.2.   Chemical composition of wheat straw versus softwoods (SW) and hardwoods (HW).

Cell wall composition (percentage of dry weight)	Wheat straw			SW	HW
	Stalk	Node	Leaf	SW	HW
Cellulose	40	40	35	42	42
Hemicellulose	45	40	55	28	35
Lignin	15	20	10	30	23
Total	100	100	100	100	100

TABLE 15.3.   Cell wall, silica, and extractive content of wheat straw versus SW and HW.

Plant composition (percentage of dry weight)	Wheat Straw				
	Stalk	Node	Leaf	SW	HW
Cell wall	90	79	76	9.5	93
Silica	3	8	11	<0.5	<0.5
Extractive	7	13	13	4.5	6.5
Total	100	100	100	100	100

TABLE 15.4.   Cell characteristics of wheat straw versus wood.

Cell characteristics	Wheat straw				
	Stalk	Node	Leaf	SW	HW
Length (mm)	1.3	0.5	1.5	2.0	1.0
Diameter (mm)	0.015	0.015	0.015	0.030	0.020
Lumen (%)[a]	75	50	80	65	55
Tissue density (g m^{-2})	0.34	0.68	0.27	0.49	0.63

[a] Determines density of plant tissue with cell wall for wood at $1.40\,g\,cm^{-3}$ (9% lumen) and for wheat straw at $1.35\,g\,cm^{-3}$ (0% lumen). For example: multiply $1.35\,g\,cm^{-3} \times 0.75 = 0.34\,g\,cm^{-3}$.

A comparison of cell structure within the stems of softwoods, hardwoods, and wheat straw shows great differences. Whereas cells in a woody stem occur in a rather organized fashion, the cells of a wheat stalk are rather dramatically divided into regions in which cell walls are thick and quite thin. Only the thick-walled fibers are useful in making paper or other fiber products. When manufacturing paper, thin-walled parenchyma cells and the waxy cells found at the very outer layers of the straw must be removed prior to sheet formation.

The Wood Cell Wall in Comparison with that of Hemp.   Hemp is sometimes suggested as an alternative papermaking fiber to wood. It is a herbaceous annual plant with a single, straight, unbranched hollow stem that grows over a 4–5 month growing season to a height of about 1–5 m (3–16 ft) and a diameter of 10–60 mm (0.4–2.3 in.). The stem is characterized by a relatively thin outer layer (referred to as bark or bast), and a wood-like core that surrounds a hollow center. The bast constitutes, on average, about 30–35% of the dry weight of the stem, with the proportion of bark variously reported from 12 to 48%. Primary bast fibers are highly variable in length, ranging from 10 to 100 m (0.4–4 in.), with an average length of 20–40 mm. These fibers are thick-walled and rigid. Secondary bast fibers are reported as comparatively short: about 2 mm or about 0.55 in. in length. The woody core makes up the remaining 65–70% of stem weight, and consists of short fibers that are reportedly a rather constant 0.50–0.55 mm in length (Table 15.5). Recall that these fibers are significantly shorter than even the juvenile fibers of most hardwood and softwood species.

TABLE 15.5.   Physical characteristics of hemp and wood.

| Characteristic | Unit | Hemp bark | | Hemp core | Softwood | Hardwood |
		Primary	Secondary			
Fiber length	mm	$10\text{--}100^a$ (20)	2^a	0.55	$2.5\text{--}5.5^b$	$0.8\text{--}1.9^{b,c}$
Juvenile fiber length	mm				$1.3\text{--}3.0^d$	$0.8\text{--}1.3^e$
Alpha-cellulose[g,h,i]	%[f]	$67 \pm 5^{a,g,h}$	$38 \pm 2^{a,g,h}$	40.6	42 ± 2^i	45 ± 2^i
Holocellulose[g,h,i]	%[f]	$80 \pm 1^{a,g,h}$	$69 \pm 3^{a,g,h}$	74.9	69 ± 4^i	75 ± 7^i
Lignin[g,h,i]	%[f]	$4 \pm 2^{a,g,h}$	$20 \pm 2^{a,g,h}$	16.0	28 ± 3^i	20 ± 4^i
Extractives[g,h,i]	%[f]				3 ± 2^i	5 ± 3^i
Ash content[g,h,i]	%[f]				$<0.5^i$	$< 0.5^i$

[a] DeMeijer (1994).
[b] Panshin and de Zeeuw (1980).
[c] Manwiller (1974).
[d] See Chapter 6, Table 6.1.
[e] Koch (1985).
[f] Expressed as a percentage of the dry weight.
[g] Ranalli (1999).
[h] Kirby (1963).
[i] Thomas (1977).

Chemically, the bark fibers of the hemp stalk contain considerably more cellulose and holocellulose and significantly less lignin than either hardwoods or softwoods. Hemp core contains less cellulose than wood, about the same holocellulose fraction, and generally the same lignin content as hardwood species.

The Wood Cell Wall in Comparison with that of Kenaf.   In the mid-1950s, the US Department of Agriculture initiated a crops-screening program. The intent was to identify new crops containing major plant constituents different from those in the crops then available, and to evaluate and promote their potential for industrial use.

One plant species that was identified early on as having industrial potential was kenaf (*Hibiscus cannabinus*), a fast growing annual plant native to east-central Africa, that has long been cultivated in Africa, India, and other parts of Asia for use in making such products as twine, rope, and carpet backing. Kenaf is also reportedly grown extensively in Central America. Early identification of the commercial potential of kenaf is not surprising in view of the fact that it had been under examination as a possible substitute for jute, a cordage-making fiber needed by the military, for some 15 years prior to initiation of the crops-screening program. In any event, upon completion of screening of more than 500 crops for potential use as papermaking fiber, kenaf emerged as the top candidate for further research into utilization options and technologies.

Since the early 1960s, a substantial research effort in the United States has focused on the technological aspects of using kenaf as a papermaking fiber. As a result, scores of research papers are now in the literature which generally indicate excellent properties of kenaf pulp and a number of kenaf paper products, with only minor process modifications. Limited production of kenaf paper is currently reported in several countries, including Mexico, Thailand, and China.

TABLE 15.6. Physical characteristics of kenaf and wood.

Characteristic	Unit	Kenaf bark	Kenaf core	Kenaf whole	Softwood	Hardwood
Fiber length	mm	2.5–4[a,b]	0.5–0.6[a,b]	–	2.5–5.5[c]	0.8–1.9[c,d]
Juvenile fiber length	mm	–	–	–	1.3–3.0[e]	0.8–1.3[f]
Alpha-cellulose[g,h,i]	%	51.0	34.9	40.6	42 ± 2	45 ± 2
Holocellulose[g,h,i]	%	81.1	71.6	74.9	69 ± 4	75 ± 7
Lignin[g,h,i]	%	11.8	18.3	16.0	28 ± 3	20 ± 4
Extractives[g,h,i]	%	2.8	4.8	4.1	3 ± 2	5 ± 3
Ash content[g,h,i]	%	2.8	1.8	2.1	<0.5	< 0.5

[a] Kaldor et al. (1990).
[b] Clark and Wolff (1965).
[c] Panshin and de Zeeuw (1980).
[d] Manwiller (1974).
[e] See Chapter 6, Table 6.1.
[f] Koch (1985).
[g] Expressed as a percentage of the dry weight.
[h] Data for kenaf from Pande and Roy (1997).
[i] Data for softwood and hardwood from Thomas (1977).

Kenaf is a herbaceous annual plant with a single, straight, unbranched stem that grows to a height of about 4–6 m, and a diameter of 25–50 mm (1–2 in.) over a 5–7 month growing season in a variety of climatic conditions. Similar to hemp, the stem is characterized by a relatively thin outer layer or bark, and a wood-like core. The bark constitutes 35–40% of the dry weight of the stem, which contains long, slender fibers some 2.5–4 mm in length – about the same length as the fibers of southern pine and other softwood species. The woody core makes up the remaining 60–65% of stem weight, and consists of short fibers that average only 0.5–0.6 mm in length. Again as with hemp, these core fibers are shorter than the fiber tracheids of most hardwoods, and generally shorter than even the juvenile fibers of most wood species (Table 15.6).

Chemically, the kenaf stalk is similar to wood. The alpha-cellulose and extractive yields, as percentages of dry weight, are essentially the same. The holocellulose fraction of kenaf is 0–8% higher than that of commonly used woods of the United States, while the proportion of lignin in kenaf is only about 60% of that found in wood – a very desirable characteristic when making almost any kind of paper or fiber product. An undesirable characteristic of kenaf is that its ash content (its mineral content) is about four times that of wood.

Papermaking with Nonwood Fibers. The nonwood fibers discussed briefly above have some features that are superior to those of wood fiber and others that represent clear disadvantages. Nonetheless, it is technically possible to manufacture high-quality paper from all of these fibers. A major issue with agriculturally derived fiber is that in most regions of the world this tends to be harvested within a narrow window of time, requiring storage of vast quantities of biodegradable material for periods approaching a year. Inventory, storage, and fiber loss costs can be quite significant.

Trees have a significant advantage over agricultural crops as a source of fiber in that they can be harvested at almost any time of year and can be left growing for longer periods if markets are soft. Logs or bolts can also be stored in mill yards for longer periods than crop materials.

Because almost any fiber can be used to make paper, the question of which types of fiber to use hinges mostly on availability and overall cost. Considering the comparative costs and problems associated with storage of agriculturally derived fiber, wood tends to be the most attractive fiber source.

From an environmental point of view the use of agricultural crop residues makes a great deal of sense, assuming that volumes removed from the land do not exceed residuals needed for conservation tillage and assuming that fiber storage problems can be resolved. The same cannot be said, however, for fiber crops. These crops not only have problems of degradation in storage following harvest, but they also carry all of the environmental burdens of intensive agriculture. With regard to hemp and kenaf, it has been observed that while there may be valid reasons for promoting these crops as sources of papermaking fiber, environmental protection is not one of them (Bowyer 2001).

References and Supplemental Reading

American Paper Institute (API) (1984). *Wood Pulp and Fiber Statistics*, 47e. New York: American Paper Institute.

Biermann, C. (1996). *Handbook of Pulping and Papermaking*. Amsterdam: Elsevier Science and Techology.

Bowyer, J. (1999). Economic and environmental comparisons of Kenaf growth versus plantation grown softwood and hardwood for pulp and paper. In: *Kenaf Properties, Processing, and Products* (ed. T. Seller, N. Reichert, E. Columbus, et al.), 323–346. Starkville, MS: Mississippi State University.

Bowyer, J. 2001. Industrial hemp (Cannabis sativa L.) as a papermaking raw material in Minnesota: technical, economic, and environmental considerations. Department of Wood & Paper Science Report Series –2001, funded by the Minnesota Environment and Natural Resources Trust Fund (May).

Clark, T. and Wolff, I. (1965). A search for new crops – VIII. Sulfate pulping of kenaf (*Hibiscus cannabinus*). *TAPPI* 48 (6): 381–384.

DeMeijer, E.P.M. (1994). Hemp variations as pulp source researched in the Netherlands. *Pulp Pap.* 67 (7): 41–43.

Eadula, S., Lu, Z., Grozdits, G., and Gibson, M. 2006. Paper properties that can benefit from LbL nano-assemblies. Newport Beach, CA: Forest Products Society International Meeting, 28 June.

FAO. 2006. Forestry production and trade. http://www.fao.org/faostat/en/#data/FO

FAO. 2015. Food and Agriculture Organization of the United Nations. Pulp and Paper Capacities Survey. 2014–2019. Rome.

FAO. 2017. Food and Agriculture Organization of the United Nations. Forest Products Statistics. http://www.fao.org/forestry/statistics/en (accessed 3 October 2017).

Iyengar, S. and Ackley, K. (1993). Paper producers should consider options for secondary fiber use. *Pulp Pap.* 67 (8): 45–47.

Kaldor, A., Karlgren, C., and Verwest, H. (1990). Kenaf – a fast growing fiber source for papermaking. *TAPPI* 73 (11): 205–208.

Kirby, R. (1963). *Vegetable Fibers: Botany, Cultivation, and Utilization*. New York: Interscience Publishers.

Koch, P. 1985. Utilization of hardwoods growing on southern pine sites. USDA For. Serv., Agric. Handb. no. 605, pp. 315–336.

Krzysik, A.M., Younquist, J.A., Rowell, R.M. et al. (1993). Feasibility of recycled newspapers for dry-process hardboard. *For. Prod. J.* 43 (7/8): 53–58.

Manwiller, F. (1974). Fiber lengths in stems and branches of small hardwoods on southern pine sites. *Wood Sci.* 7: 130–132.

Pande, H. and Roy, D. 1997. Delignification kinetics of soda pulping of kenaf. University of Toronto, Faculty of Forestry, Toronto, Canada (unpublished).

Panshin, A. and de Zeeuw, C. (1980). *Textbook of Wood Technology*, 4e, 134–170–171. New York: McGraw-Hill.

Ranalli, P. (1999). Agronomic and physiological advances in hemp crops. In: *Advances in Hemp Research* (ed. P. Ranalli), 61–84. New York: Food Products Press.

Saltman, D. (1978). *Paper Basics*. New York: Van Nostrand Reinhold.

Sims, R.E.H. (2002). *The Brilliance of Bioenergy in Business and in Practice*. London: James and James.

Singh, R.P. (1993). Technological approaches to TCF pulp. *Pulp Pap.* 67 (10): 35–36.

Sixta, H. (2006). *Handbook of Pulp*. New York: Wiley.

Thomas, R. (1977). Wood: structure and composition. In: *Wood Technology: Chemical Aspects*, ACS Symposium Series 43 (ed. I. Goldstein), 1–23. Washington, DC: American Chemical Society.

16

Energy and Chemical Products

The use of wood for energy is not a new idea. The use of fire for heating and later for cooking, pottery making, and to provide light goes back at least 200 000 years. In the United States where both wood and fossil fuels are available, burning wood for energy is generally the lowest common denominator. If one can make lumber, paper, veneer, or other solid/composite products from wood, then those invariably have higher economic value. The woody and other bio-based materials that get directed to energy are those which cannot be used for any other higher-value option.

Worldwide, the primary use of wood today is still as a fuel for heating and cooking. In the United States, wood was a principal fuel for heating and cooking until about 1900. Thereafter, coal, oil, and natural gas became dominant fuels. The oil embargo of the early 1970s resulted in an abrupt increase in the cost of imported crude oil – from ~$3 to 30 per US barrel (one barrel equals ~159 l or 42 gal) – and marked growth in wood consumption for fuel, with the volumes consumed for that purpose briefly rising to close to the levels of 1900. During the past decade, oil has ranged from ~$35 to 135 per barrel. As this text is being written in 2018, crude oil is ~$60 per barrel.

After decades of debate about how long the age of petroleum abundance might last, it remains inconclusive. During the past decade, alternative fuels and energy sources have been researched heavily and to a limited extent adopted. The last decade has demonstrated that the development and adoption of these are directly related to the price of of crude oil. Beyond simple supply and demand economics, biofuels are tied to complex policies and legislation.

Wood and other biomaterials are a major potential source of energy and chemicals. However, biomass as produced by trees and other plants through solar-energy-driven photosynthesis has the potential to provide significant quantities of energy, as well as a wide array of chemical compounds that are useful to industry.

Biomass currently provides less than 3% of US energy needs, but about one-half of energy from renewable energy sources. Today in the United States, some 190 million tons

Forest Products and Wood Science: An Introduction, Seventh Edition. Rubin Shmulsky and P. David Jones.
© 2019 John Wiley & Sons Ltd. Published 2019 by John Wiley & Sons Ltd.
Companion website: www.wiley.com/go/shmulsky

of biomass is used annually for the production of energy or bioproducts that directly displace petroleum-based feedstocks.

Over the past few decades the vast majority of wood used to generate industrial energy has been consumed by the forest products industry itself. As a result, this industry has a high degree of self-sufficiency, as over half of all energy used in the primary forest products industry is self-generated. Wood energy provides process and space heat from combustion of mill residues, with residues also burned to create direct and indirect heat for wood drying. Some plants also generate electricity in cogeneration facilities. The importance of woody biomass for industrial energy will continue and will likely expand to other industries unless other more economical alternatives to petroleum and natural gas are developed. A national goal of the United States, which is discussed in more detail later in this chapter, is for the production of 30% of domestic energy from biomass by 2030 (Perlack et al. 2005); future uses of wood for energy are likely to include widespread use for electricity generation and for production of liquid fuels.

The heating value of wood fuel compared with alternative industrial fuels is shown in Table 16.1. On a volumetric basis, wood has significantly lower embodied energy than fossilized fuels. On a cost basis, however, wood often competes favorably in situations where it is technically feasible to utilize it as fuel.

The amount of useable heat, and thus value, from wood is inversely related to the moisture content. The energy value of wood fuel is generally sufficient to justify the costs associated with converting it to energy. The value of wood as a fuel can present problems for mills that are dependent on low-cost wood residue for particles or fiber when there is a competing interest in that raw material for fuel.

Traditionally, the portion of the forest-related biomass not used in the manufacture of wood products has been referred to as residue. *Mill residues* consist of planer shavings, bark, slabs, cut-off blocks, plywood trim, and saw- and sander-dust. Pulp chips are considered a primary product and not a residue. *Woods* or *logging residues* include tops, limbs, thinnings, and cull trees normally left in the woods.

Worldwide the use of wood as a home heating and cooking fuel remains of great significance. Globally, fuel (combustion) has always been the single largest use of wood. As the price of crude oil continues to display wide variability, the United States, Europe,

TABLE 16.1.   Heat energy comparison values of wood and alternative fuels for the United States in 2004–2005.

Fuel type	Approximate bulk fuel cost (US\$ t^{-1})	Approximate energy content (GJ t^{-1})	Approximate cost, US\$ GJ^{-1}
Wood (planer shavings, 15% MC)	\$22	17	1.29
Wood (sawdust, 90% MC)	\$12	10	1.20
Coal (industrial, nonelectric)	\$43	23	1.88
No. 2 fuel oil	\$464	42	11.05
Natural gas	\$444	47	9.44
Electricity (all sectors)	\$0.08 (\$ kWh^{-1})	—	21.17

Source: Department of Energy (2006).

and other industrialized areas are seeking to curb their petroleum demand by researching, developing, and installing wood-based energy solutions. That said, wood and other bio-based raw materials are likely best suited to being a minor but important part of the overall US energy portfolio. The primary limitation to their use is raw material availability. History has shown that energy-hungry societies that rely heavily on virgin wood energy can strip their forests in relatively short order. The use of woody residuals and by-products as a means to help stabilize energy supplies perhaps makes more sense in the name of sustainability given the immense size of the energy market.

Materials Available for Energy

There are five major potential sources of wood-related biomass for energy generation. These are roundwood from growing stock, mill residues, logging or woods residues, urban and industrial waste streams (paper, paperboard, wood pallets, railroad ties, demolition wood, urban tree removals, etc.), and dedicated short-rotation woody crops called *energy plantations*. Roundwood and mill residues are, of course, also in demand for production of fiber and particle products. Logging residues could technically be used for particle and fiber products, but because of a high content of bark, soil, and other contaminants, this material is generally better suited to energy production. Recovered paper is in demand for many types of recycled paper products. Marks (1992) points out the advantages of wood powder from residues as an upgraded wood fuel.

The use of short-rotation intensive forest management to produce biomass in energy plantations caught public attention in the 1970s and has generated considerable study and controversy since then. Such plantations have existed in a number of countries for some years. In the United States, bio-based energy for rural development is receiving increased attention. Currently, in many parts of the Southeastern United States, plantation pine thinnings are being used for energy. These are being converted to pellets, microchips, and other fuels. Two decades ago, these small-diameter trees went into pulp production but as domestic pulping has declined, energy has developed as a market for some of these trees.

The primary potential for biomass production is from energy plantations, from agricultural crop residues, and from forest thinnings. Regarding plantations, some 55 million acres within the continental United States have been identified as available and having high potential for the production of energy crops such as switchgrass, reed canary grass, poplar, eucalyptus, and other species. An estimated 377 million dry tons of biomass crops could be produced annually from these 55 million acres. In addition, an estimated 428 million dry tons of agricultural residues in excess of that needed for conservation tillage could be removed annually from US farmland for the production of biofuels. Another 368 million dry tons of woody biomass could be sustainably removed annually from the nation's forest lands and gleaned from current waste streams. Part of the woody biomass would come from noncommercial forest thinnings conducted for the purpose of reducing wildfire danger; the 368 million dry ton number is viewed as being extremely conservative. Commercial thinnings are now entering the market as a result of the decline in the domestic pulp and paper market. Their high quality, ready availability, and infrastructure for production make them a feasible option in areas where the forests must be thinned to produce subsequent sawlogs, but there is no other market for the pine thinnings.

Not only is the energy production potential from such material substantial, but combustion of such material is carbon neutral. The growth of replacement crops and reforestation after harvest and increased forest growth rates following thinning operations sequester atmospheric carbon in a quantity equivalent to that released when the harvested crop is burned. This represents a substantial advantage over the combustion of fossil fuels.

In the southern United States sweetgum, sycamore, and cottonwood fiber have been used for energy for decades. Hybrid poplar, willow, and eucalypts have been researched as energy plantation species (DOE 2001). Loblolly pine thinnings are now entering the picture on a commercial scale. Whether biomass is obtained from forests or from energy plantations, transportation costs represent a considerable challenge to the widespread adoption of biomass-derived power.

Mill residues are a highly desirable wood fuel because they are available at the mills (no transportation cost) and they are often partially dried. In the United States, most bark and wood residue available in sufficient quantity has a ready market as fuel, landscaping material, mulch, or other uses, although residues at some small plants are still not completely utilized.

The material left in the woods after logging, termed *slash*, represents a significant storehouse of fuel. Because of its dispersion, soil content, and small size, however, this material is expensive to collect and use. Utilization depends upon the development of cost-effective recovery systems that are able to keep the woody materials as clean as possible.

Numerous means have been investigated with respect to harvesting, collecting, and transporting woody biomass. Because it is relatively bulky and contains much water it is expensive to transport over long distances. To be economically viable as an energy source, the biomass should be as compact as possible. One of the most efficient means of facilitating handling, storage, and transport is via chipping. Chipping is a means of reducing residual stems and branches into a relatively uniform material that is easily handled, loaded, and shipped. Whole-tree chippers used to harvest stands for pulp chips can be used to produce chips for energy. A number of firms have used whole-tree chippers for this purpose. Figure 16.1 illustrates a high-capacity mobile chipper that is moved to various logging sites. When actively fed, the machine is capable of reducing ~73 t (80 short tons) of wood per hour.

A major deterrent to the use of the total forest biomass for fuel is harvesting cost. Specialized equipment to harvest small-diameter stems and collect and transport the material is needed. Several machines to do this job have been developed and tested by equipment manufacturers and forest products firms. Figure 16.2 illustrates some specialized logging equipment that is designed to quickly process small-diameter thinnings in the wood. The tracked cutter harvests, cuts to precision length, and delimbs small-diameter stems. The articulated forwarder carries the cut stems to the landing for transport. Each of these is suited for operation in wet or environmentally sensitive areas. This kind of automated processing increases production rates, keeps the stems clean and off the ground as compared with traditional skidding, and eliminates the need for the highly hazardous chainsaw felling, bucking, and delimbing.

Another means of handling biomass is baling. Much like the agricultural baling of hay or straw, it is possible to orient woody biomass, bundle it together, tie it off, and trim the ends. Figure 16.3 shows bales of loblolly pine biomass. These bales contain logging

FIGURE 16.1.   Mobile tracked chipper capable of reducing ~73 t of woody biomass to chips each hour.

residue, from needle size material up to ~15 cm (6 in.) in diameter. The bales are ~3 m (10 ft) long and have a diameter of ~0.75 m (30 in.). Biomass for the bales is consolidated such that less volume is required for transport. An advantage of shipping bales versus chips is that the bales can be comminuted at a mill site where better control over chipping or fiberizing can be achieved. This technology, now more than a decade old, has not yet shown major economic feasibility or commercial adoption in the United States. Another disadvantage of including bark, needles, small-diameter stems, and other underutilized biomass is that these typically contain high proportions of minerals or ash, which is undesirable in virtually every type of fuel technology.

Wood construction and demolition residues are a relatively untapped source of energy. Clean new construction and remodeling wood debris as well as spent pallets and crates are highly valuable as a fuel. These dry materials have a high heat value and are clean burning (so long as they are not preservative treated or otherwise contaminated). Urban wood from landscape tree, stump, and leaf removal are additional potential energy sources. The major obstacles preventing further use continue to be sorting, collection, and consolidation. Logistically, it remains unfeasible to use urban wood waste for any major and scalable energy project.

Nature of Wood as Fuel

The most common, simplest, and least expensive method of converting wood into energy is by *combustion*. The first stage of combustion is the evaporation of the water. Next, the volatile components of wood, both combustible and noncombustible, are driven off at temperatures from 100 to 600 °C (212–1110 °F). Some 75–85% of the wood can be volatilized. In the last stage of combustion, the carbon is oxidized and the combustible volatile gases are burned. A standard test method to evaluate solid fuels, termed *proximate analysis*, provides

(a)

(b)

FIGURE 16.2.   The upper frame is that of a tracked cutter. Its head works to fell, delimb, and cut to length small-diameter logs. The lower frame is that of a self-loading mobile forwarder that carries the stems up off the ground to keep them clean and free from inorganic debris. Both highly mobile machines produce minimal ground pressure and thus minimal soil disturbance.

FIGURE 16.3. Bales of woody biomass that contain small stems, branches, and needles. The cylindrical bales are ~3 m long and have a diameter of about 0.75 m.

the percentages of carbon and other volatile compounds. The proximate analysis of several fuels is shown in Table 16.2.

Combustion involves combining carbon from the wood with oxygen to form carbon dioxide and combining hydrogen from the wood with oxygen to form water vapor. The oxygen in these reactions comes partly from the wood but mostly from the air. Wood contains about 6% hydrogen, 49% carbon, and 44% oxygen by weight. The amount of oxygen (and thus air) required for the burning process can be theoretically calculated based upon the chemical analysis (termed *ultimate analysis*) of the species. In practice *excess* (or more) *air* is required for complete combustion. In modern wood furnaces, air supply is regulated to assure efficient burning; too little oxygen (rich) produces excess hydrocarbons, carbon monoxide, and particulates while too much (lean) favors excess nitrogen oxides (smog precursors).

TABLE 16.2.   Proximate analysis of various fuels.

Fuel type	Other volatile matter (%)	Carbon (%)	Ash (%)
Douglas fir			
Wood	86.2	13.7	0.1
Bark	70.6	27.2	2.2
Western hemlock			
Wood	84.8	15.0	0.2
Bark	74.3	24.0	1.7
Hardwoods (avg.)			
Wood	77.3	19.4	3.4
Bark	76.7	18.6	4.6
Western coal	43.4	51.7	4.9

Source: Corder (1975), Arola (1976), Pingrey (1976).

TABLE 16.3.   Average heating values of wood and bark.

Wood type	Ovendry higher heating values ($MJ\,kg^{-1}$)	
	Wood	Bark
Nonresinous	19–20	17–23
Resinous	20–23	21–25

Source: Corder (1975).

Table 16.3 gives average heating values for wood and bark. Resin in wood has a heating value almost twice as high as that of wood; therefore, resinous woods have a somewhat higher heat value (HHV) than those without resin. Bark and wood from softwood species tend to be somewhat higher in heat value than from hardwoods. Heating values also vary by species because of the varying proportions of carbon, oxygen, and hydrogen. In engineering practice however, average heats of $21\,MJ\,kg^{-1}$ ($9050\,BTU\,lb^{-1}$) for dry resinous woods and $20\,MJ\,kg^{-1}$ ($8600\,BTU\,lb^{-1}$) for other dry woods, including hardwoods, are often used.

The total heat generated by complete combustion under controlled conditions is termed the HHV. The actual heat that can be recovered in conventional burners is considerably less than the HHV owing to the losses from vaporizing the water in the fuel and other process losses. Process losses include energy to heat the excess combustion air, drive the water off from the wood, and loss of flue-gas heat. Figure 16.4 (Ince 1979) shows the relationship between the potential and *recoverable heat energy* per kilogram and the moisture content of the fuel.

The available potential heat at any moisture content, as illustrated in Figure 16.4, is sometimes called the *gross heating value* (GHV). The example shown in the figure is for a fuel with a HHV of $19.7\,MJ\,kg^{-1}$ ($8500\,BTU\,lb^{-1}$) in a combustion system with a $260\,°C$ ($500\,°F$) stack gas (i.e. exhaust gas) temperature. Note that the GHV represents the HHV in the portion of fuel composed of dry wood. The GHV of wood can be calculated from the HHV as follows:

$$GHV = HHV \times \left[1 - \text{percentage MC}\left(\text{wet basis} \right) / 100 \right]$$

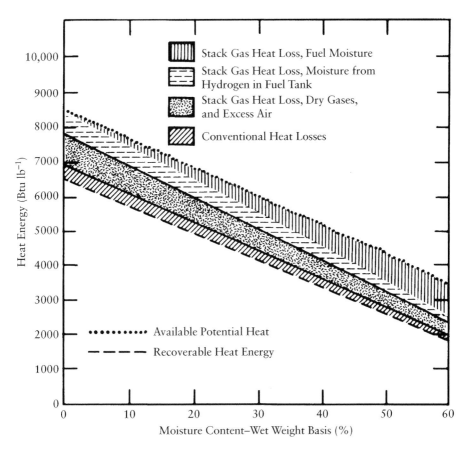

FIGURE 16.4. Recoverable heat, available potential heat, and heat losses for a pound of typical wood fuel burned in a boiler.

The ratio of the recoverable heat to the available potential heat (see Figure 16.4) is called the *combustion efficiency*. With wood fuels and current combustion equipment, combustion efficiencies range from about 80% for dry fuels to 60% for wet fuels. Unfortunately, green rather than dry fuels are most commonly available. When comparing the costs of alternate fuels, the recoverable heat value is most important. This is the energy that produces the steam for industrial processing, powering turbines, and space heating.

The terms *net heating value* (NHV) and *lower heating value* (LHV) are sometimes encountered in combustion engineering. NHV and LHV are used to describe the net heat released by a fuel after reducing the HHV by the heat of vaporization of the water generated by combustion of the hydrogen in the fuel. LHVs are customarily used in Europe for purchases of fuel, while in America HHV is used as the basis on which fuel is bought and sold (Georgia Institute of Technology 1984).

As discussed in Chapter 7, the moisture content of forest products is normally calculated on a dry weight basis; if moisture content is based upon the wet weight, this should be indicated. Caution is needed when reading wood fuel literature as the wet basis is often used without reference to its basis. Moisture content on a dry weight basis can be converted to moisture content on a wet basis by the following relationship:

$$\text{percentage MC(wet basis)} = \frac{\text{percentage MC(dry basis)}}{100 + \text{percentage MC(dry basis)}}$$

An example may aid in understanding the relationship between the moisture content and the recoverable heat. Assume it is known that a species of wood is at 75% MC (dry basis) and has an HHV of $20\,400\,\text{kJ}\,\text{kg}^{-1}$ ($8800\,\text{BTU}\,\text{lb}^{-1}$).

1. What is the gross heating value?

   $$75\% \,(\text{dry basis})\,MC = 75 / (100 + 75) = 43\%MC\,(\text{wet basis})$$

Thus the gross heating value is:

$$20\,400\,\text{kJ}\,\text{kg}^{-1}\,(1 - 0.43) = 11\,600\,\text{kJ}\,\text{kg}^{-1}\,(5000\,\text{BTU}\,\text{lb}^{-1})(\text{wet})$$

2. If the combustion efficiency of the boiler is 70%, how much recoverable heat will be obtained per kilogram of wood fuel at 75% MC?

   $$11\,600\,\text{kJ}\,\text{kg}^{-1}\,(\text{wet}) \times 0.70 = 8150\,\text{kJ}\,\text{kg}^{-1}\,(3500\,\text{BTU}\,\text{lb}^{-1})(\text{wet})$$

3. If recoverable heat is valued at \$4.20 per million kJ, how much more could be paid for this fuel without increasing the cost of energy if it were purchased at 15% MC rather than at 75% MC? (At 15% MC the combustion efficiency of the boiler is 78%.)

   $$15\%MC = 15 / (100 + 15) = 13\%MC\,(\text{wet basis})$$

   $$20\,400\,\text{kJ}\,\text{kg}^{-1}\,(1^{-}0.13) = 17\,800\,\text{kJ}\,\text{kg}^{-1}\,(\text{gross heat})$$

   $$17\,800\,\text{kJ}\,\text{kg}^{-1} \times 0.78 = 13\,900\,\text{kJ}\,\text{kg}^{-1}\,(\text{recoverable heat})$$

Therefore, the usable heat is increased by $13\,900 - 8150 = 5710\,\text{kJ}\,\text{kg}^{-1}$ ($2450\,\text{BTU}\,\text{lb}^{-1}$) when using wood at 15% MC rather than 75%. The difference in value, i.e. the added amount which could be paid without increasing the energy cost, would be:

$$(5710\,\text{kJ}\,\text{kg}^{-1} \times \$4.20/\text{million kJ})/1\,000\,000 = \$0.024\,\text{kg}^{-1} : \$24\,\text{t}^{-1}\,(\$21.8\,\text{per short ton}).$$

A forest products engineer should know how to estimate the amount of wood fuel required to supply an industrial boiler or an electrical generation plant. Such estimates require consideration of the operating characteristics of the boiler system as well as the nature of the fuel. The energy output of a boiler is commonly expressed in mass of steam per hour (kg water evaporated h^{-1}). The heat required to produce a kg of steam varies depending upon the pressure and temperature, but it is generally in the range of

2550–3000 kJ kg⁻¹ (1100–1300 BTU lb⁻¹). Typical boiler sizes in the forest products industry run from 750 to 180000 kg h⁻¹ (that is, 500–120000 kW or 50–12000 hp).

The following example explains the general approach to estimating fuel requirements. Assume that a 50000 kg h⁻¹ boiler requires 2900 kJ kg⁻¹ steam to heat the feed water and generate the steam. This boiler is to be fired with fuel at 75% MC having a gross heat value of 11600 kJ kg⁻¹ (5000 BTU lb⁻¹) (see 1 in the example above). The efficiency of the boiler is 67%. The weight of fuel required per hour is estimated as:

$$\frac{50000\,\text{kg steam h}^{-1} \times 2900\,\text{kJ kg}^{-1}\text{steam}}{0.67(\text{boiler efficiency}) \times 11600\,\text{kJ kg}^{-1}\text{fuel}} = 18600\,\text{kg fuel h}^{-1}$$

Calculations for estimating the wood requirements of a wood-fired electric generating plant are discussed by Garrett (1981). The process can be outlined by the following example: 1 kWh is the equivalent to 3600 kJ h⁻¹. Thus, if an electric-generating plant operates with a typical overall efficiency of 25%, it requires a 14400 kJ input (3600/0.25) to produce 1 kWh. The daily energy requirement for a 20 MW plant is thus:

$$14000\,\text{kJ kWh}^{-1} \times 20000\,\text{kWh} \times 24\,\text{h} = 6.91 \times 10^{9}\,\text{kJ day}^{-1}$$

If the wood fuel going into the plant has an HHV of 19300 kJ kg⁻¹ and 45% MC (wet basis), then the GHV is 19300 (1×0.45) = 10600 kJ wet kg⁻¹. Thus, the daily wood fuel requirement is:

$$\frac{6.91 \times 10^{9}\,\text{kJ day}^{-1}}{10600\,\text{kJ wet kg}^{-1}} = 650 \times 10^{3}\,\text{kg fuel day}^{-1}$$

Therefore, a moderate-scale 20 MW power plant requires about 650 t (715 short tons) of green chips per day, about as much wood fiber as a 180 t (198 short tons) per day pulp mill consumes.

The gross heat and the usable heat, assuming typical combustion efficiencies, of hardwood at different moisture contents are shown in Table 16.4. Note the significance of the combined effect of gross heat per ton and combustion efficiency on the amount of usable heat which can be obtained.

There are two main approaches to reducing moisture in wood fuels. The most common means of drying is by supplying heat energy to vaporize the moisture, i.e., by thermal means. The second is by applying mechanical energy to squeeze the free water

TABLE 16.4. The gross and usable heat of hardwood at different moisture contents and assumed combustion efficiencies.

Percentage moisture content (ovendry basis)	Gross heat (GJ t⁻¹)	Assumed combustion efficiency (%)	Usable or recoverable heat (GJ t⁻¹)
0	19.2	80	15.4
15	16.7	78	13.0
30	14.8	76	11.3
60	12.0	72	8.6
100	9.6	67	6.4

from the wood or bark. This is referred to as *dewatering* or *compression drying* and is applicable only to very wet fuels.

Thermal drying is used most often to dry wood fuel. The installation of thermal fuel-drying equipment can often be economically justified, particularly if increased boiler capacity is needed. The heat for these fuel dryers is typically supplied by burning fines from the fuel being processed. Cascade dryers, which utilize waste heat from stack gases, are also available. Such equipment is widely used in Scandinavia and there are several installations in the United States.

Drying wood chips or particles by moisture evaporation in a tube, drum, cascade, or screen dryer makes it possible to attain any wood moisture content desired, but it has the disadvantage that it requires high energy input, generally greater than 4200 kJ for each kilogram (1800 BTU lb^{-1}) of water removed. Thermal drying thus requires an energy input that is greater than the increase in heating value resulting from the drying. Nonetheless, drying wood fuels by evaporative means may prove advantageous in those cases where the energy consumed is from flue gas or another low-cost and otherwise wasted source.

Another approach to the drying of wet wood fuels is *mechanical compression* or *dewatering*. Although equipment for mechanically dewatering bark and very wet residue has been available to the forest products industry for many years, these existing processes are limited in application because they can reduce the moisture content to only about 50% (wet basis). Mechanical dewatering systems are thus best suited for removing water added that has been added in processing, not the water naturally occurring in green biomass. For example, the wet wood dust that is collected in electrostatic precipitators (a type of pollution control device) can be inexpensively dewatered in order to burn and recover the energy from what would otherwise be waterlogged waste.

Researchers have investigated lower-cost means of reducing moisture in wood chip fuels. Methods investigated include drying in the forest after felling but prior to chipping, air drying forest residue in ventilated bins, and high-pressure compression drying of chips. The latter method can reduce the moisture content to about 40%, which is lower than commercial equipment can accomplish (Haygreen 1981, 1982). This method shows economic potential if chips from energy plantations, typically with a high moisture content, are used as fuel.

The use of dry wood fuel has advantages in addition to the increased heat value described above. Boiler capacity is increased when dry fuel is used. The greater efficiency of the furnace combined with lesser quantities of water vapor generated as a result of burning lower MC chips results in lower flue gas volume. This situation permits an increase in the amount of wood that can be burned, which raises heat production. If fuel is burned at 40% (wet basis) rather than 50%, the boiler's steam output typically increases by about 10%. Likewise, a smaller boiler can be installed if drier fuels are to be burned. Newby (1980) cites a 7% increase in steam generation as a result of reducing fuel moisture from 60 to 55% (wet basis).

When drier fuel is burned, the volume of stack gases decreases per kilogram of steam produced. Thus boilers designed for specific steam capacity using dry fuel have lower capital and operating costs for stack gas emission-control equipment than boilers designed for wet fuel. The emissions levels listed on permits required by state and federal agencies are also lower. Because mill production capacity is often limited by total site emissions, maintaining minimum pollution at the boiler allows greater capacity at the productive machine centers (dryers, presses, sanders, etc.). In a test to evaluate the effects of fuel moisture on particulate emissions, Johnson (1975) showed that an increase of fuel moisture from 52 to 63% (wet basis) doubled the rate of particulate emissions. Thus, utilizing drier fuels yields direct dollar savings, lower pollution, and increased production.

Conventional wood residue boilers require constant adjustment of controls as fuel moisture fluctuates. As a result, the maximum efficiency for the boiler may not be obtained. Fuel-drying systems that provide fuel at relatively uniform moisture levels allow boilers to be fine-tuned to the incoming fuel moisture content and eliminate this efficiency problem (Vanelli and Archibald 1976). Dry planer shavings are an example of a highly efficient wood fuel. In high-production southern pine planer mills, about 55 kg of planer shavings are produced per cubic meter of dressed 50 mm (2 in.) thick lumber (~280 lb per thousand board feet). More shavings are produced with thinner or narrower lumber. The available heat energy in these shavings is on the order of 750 MJ (710 thousand BTU). The heat required to dry a cubic meter of green lumber is ~1400 MJ (1.3 million BTU). Thus over half of the fuel needed to heat the dry kilns can come directly from the planer mill. Figure 16.5 illustrates the flaming particles inside of a wood-fired boiler as seen through one of the boiler's fire eyes.

Gasification and pyrolysis processes also realize benefits from using dried wood fuels that are similar to those for combustion systems. The use of fuel at high moisture levels reduces the temperature of combustion products and the efficiency of the system. In low-energy gasifiers, efficiency may be lowered by about 15% when burning green wood. In some gasification systems, only dry wood can be used. During the past decade, gasification systems have been developed to run cleanly and efficiently on a wide range of biofuels and in the future their uasage will likely increase. Pyrolysis systems have not yet become commercially viable. Although liquid fuels are generally more valuable than solid ones, the costs associated with vaporizing solid biomass followed by condensing and upgrading the output to fuel have kept that technology out of the range of commercial viability to date.

Despite the advantages of burning dry fuels, many wood-residue fuels are combusted as received, without the benefit of drying. With wet fuels, the drying is accomplished in the boiler during the first stage of combustion. For very wet fuels, some burning systems include a separate drying step prior to burning in the boiler or reactor.

As an industrial fuel, wood is low in sulfur and nitrogen. This reduces the cost of pollution control equipment compared with fossil fuels that produce significant sulfur dioxide and nitrous oxide emissions. The major air pollution problem with industrial wood fuels is *particulate emissions*, primarily unburned carbon particles. Particulate emissions can be controlled by efficient combustion, mechanical collectors such as cyclones and bag houses, wet scrubbers, and in some cases by electrostatic precipitators.

There is a popular belief that dense species such as oak are better than others for fuel. This is true relative to the volume of wood that must be handled to obtain a given amount of heat. Density also affects the rate at which the wood burns in a stove or fireplace. However, in terms of industrial applications, species is relatively unimportant because the amount of heat per ovendry kilogram varies little among species; the moisture content of the wood is much more important. Difficulties exist, however, in handling and grinding the bark of some species. If extremely low-density woods are burned, the weight of wood in the furnace and corresponding heat output at any time may be reduced.

FIGURE 16.5.   Dry wood particles moving through a boiler as seen through the boiler's fire eye.
Source: Photograph courtesy of Donald L. Buckner.

Wood for Energy in the Forest Products Industries

Wood-based materials range from products that require low-energy inputs for manufacture, e.g. lumber, to those requiring high levels of energy, e.g. pulp and paper. Unfortunately, the low-energy level manufacturing processes generate more residues than do the high-energy-level products. Wood composite industries usually fall somewhere between these two extremes in terms of energy requirements and residues generated. If a typical sawmill were to burn all of its sawdust, bark, and other residues it would produce more energy than it consumes in the manufacturing process. Therefore, it is potentially energy self-sufficient. Pulp and paper mills, in contrast, are about 50–70% energy self-sufficient when using all of their residues. In the future, black liquor gasification may allow net energy production. In fact, as outlined in Chapter 15, the paper mill of the future is likely to be a major producer of both energy and industrial chemicals – becoming, in effect, a fully integrated biorefinery. Instead of removing the hemicelluloses and lignin during the pulping process as is done today in chemical pulping operations, the hemicelluloses will be extracted from pulp chips prior to subsequent processing, with these used to produce a variety of chemicals and polymers. The potential will exist for conversion of wood to energy, liquid fuels, pulp, and a wide array of industrial chemicals and chemicals and feedstocks.

The possible use of internal mill residues to supply energy for manufacturing a variety of forest products is shown in Table 16.5. These data show that there is potentially an excess of residues in some industries but a deficit in most. The data do not include the use of residues from logging or forest thinning. If fiber products and particleboard plants are to generate their energy from wood residue, fuel from other sources must usually be purchased. Newer figures are not readily available; however, the recoverable energy from wood has gone down since this table was first produced

TABLE 16.5.   Energy required to produce wood-based and nonwood materials.

Commodity	Harvesting	Manufacture	Total	Potential residual fuel recovery
Softwood lumber	1.0	5.6	6.6	9.6
Oak flooring	1.3	6.6	7.9	13.2
Laminated veneer lumber	0.8	7.7	8.5	4.1
Softwood plywood	0.8	8.0	8.8	4.3
Structural flakeboard	1.2	8.7	9.9	10.0
Medium density fiberboard	0.9	10.8	11.7	3.1
Insulation board	0.7	12.2	12.9	0.8
Hardwood plywood	1.2	11.8	13.0	12.3
Underlayment particleboard	5.3	9.4	14.7	1.7
Wet-process hardboard	0.8	22.8	23.6	0.9
Gypsum board	0.1	3.1	3.2	0.0
Asphalt shingles	0.0	6.6	6.6	0.0
Concrete	0.6	8.8	9.4	0.0
Concrete block	0.6	8.8	9.4	0.0
Clay brick	0.7	8.9	9.6	0.0
Carpet and pad	7.7	33.3	40.9	0.0
Steel wall studs	2.9	53.6	56.5	0.0
Steel floor joists	2.9	53.6	56.5	0.0
Aluminum siding	31.1	199.4	230.5	0.0

Energy (GJ t^{-1})

Source: Jahn and Preston (1976).

because sawing efficiencies have improved drastically, which has reduced green target sizes and thus reduced planer shavings.

Most forest products industries use wood combustion to generate energy in the form of heat. Their single largest use of heat is for drying; about 70% of the energy used in lumber manufacture and 40% in papermaking is for this purpose. In relatively large mills that operate continuously it is also economically feasible to use wood for electrical power generation. In small plants, generating electricity from residues has not been practical owing to the cost of the steam boilers, turbines, and generators that are required. However, pulp and paper mills have large electrical power needs and continuous operation, which can make such installations cost-effective. Wood composite plants and sawmills generally rely upon purchased electricity but some generate their own power.

Because the energy requirements of the forest products industries are mainly in the forms of heat and electricity, they are well suited to the simultaneous generation of electricity and low-pressure steam, a process called *cogeneration*. Cogeneration can be accomplished with a high-temperature and -pressure boiler and a special turbine that generates electricity from the high-pressure steam and exhausts low-pressure steam for process heat. Figure 16.6 illustrates a cogeneration system for a mill that requires only one pressure of process steam.

Cogeneration is more economical than the electricity generation alone because the heat from the exhausted steam is utilized. Generally, there are economies of scale related to cogeneration, i.e. larger mills achieve greater benefits. Large mill complexes can run wood-fired cogeneration plants to provide their heat and electrical needs and may sell their excess capacity to public utilities.

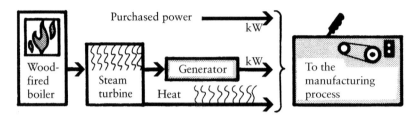

FIGURE 16.6.   Typical cogeneration system for a mill requiring one pressure of process steam and where additional electricity can be purchased if needed.

The Use of Wood for Commercial Electrical Generation

The EIA (2016) reports that US biomass electrical generation for 2015 was 64.2 TWh. This value is up ~20% from 2005, 10 years before. This value is equivalent to nearly 2% of the total US electricity production. Some advantages of biomass to electricity are the minimal refining that is required and its wide availability. Two major disadvantages are high ash content and relatively low energy density. It is possible to burn biomass in a coal fired generator without major retrofitting (relatively speaking). The mixing of biomass with coal reduces overall emissions, particularly of sulfur and mercury. However, biomass is less energy dense than coal. As such a greater volume of biomass is required to produce the same amount of energy. Ongoing efforts to increase the energy density of biomass, such as through pelletization, are used commercially in some coal plants.

The Use of Wood for Production of Liquid Fuels

Ethanol

Ethanol from corn has proven highly controversial as it appears to compete with the production of food. Wood to ethanol conversion technology is not new; wood ethanol was used extensively in the United States and Europe during World War II to fuel vehicles. At the time of this writing, commercially viable wood to ethanol conversion technology remains elusive. What is currently holding back the commercialization of cellulosic ethanol is the cost of production.

Hydrogen

Another possibility for producing fuels from biomass involves the production of hydrogen using gasification or other technologies in combination with steam reforming and what is referred to as the water-shift reaction. Despite low levels of hydrogen within biomass (about 6% by weight), the National Renewable Energy Laboratory estimates that hydrogen could replace a substantial percentage of current gasoline consumption, while also dramatically reducing transportation-related CO^2 emissions. Biomass-derived hydrogen remains not economically feasible. Should it be developed in the near term, on the order of 10 years, then parallel development of hydrogen fueling stations and vehicles to use hydrogen fuel will be needed to achieve commercialization.

Pyrolysis Oil

Another possibility for producing liquid fuel from wood is that of pyrolysis oil. In that case, wood is heated in an anoxic environment. The result is the production of char, condensible gases, and noncondensable gases. The condensable gases can then be burned as a low-value fuel or catalytically upgraded into highly refined gasoline, diesel, jet fuel, or other hydrocarbon-based fuels. This technology is developed, technically, but is not cost-effective while crude oil is available in the $40–60 per barrel range.

Wood Energy Use by Other Industries

Although other industries purchase wood fuels where their local situations make this feasible, they are generally at a disadvantage compared with forest products mills because of the additional transportation and handling costs involved and the uncertainty of a continuing supply. Nonforest products industries consumed only about 15% of the wood energy used in the United States in 2000 according to estimates by the US Department of Energy. That figure was largely unchanged 15 years later.

One of the disadvantages of wood as a solid fuel for small commercial firms, public buildings, and homes is that it is bulky and cumbersome to handle and store. Also, the heating value varies as the moisture content changes. One way to reduce these problems is to densify the wood residue from sawmills into briquettes or pellets. The advantages of wood briquettes and pellets as fuel are that they are dry and therefore of a uniform and high heating value; they have a high bulk density so storage space is minimized; they are clean and easy to handle; and they can be burned in many systems designed for coal. Because of the cost of densification, there will be little use of briquette or pellet fuel within the forest products industry. Pellets are widely used in electrical generation and home heating systems in Europe, providing an attractive fuel where convenience is important and where a reliable supply is available. A significant industry has developed wherein wood pellets are manufactured in the United States and shipped to Europe for energy. Several types of wood fuel briquette logs are shown in Figure 16.7.

Chemical Wood Products

Future developments notwithstanding, a significant chemical wood products industry currently exists. A number of products in use today are made from wood or bark that has been reduced to basic chemical components such as cellulose, hemicellulose, or lignin. The raw material for many of these products is waste liquor, which is a by-product that results from chemical pulping. Also included in this category are products made from the resins of pines and other softwood species. Chemical products are seldom recognizable as wood-based. These include cellulose ethers, lignosulfonates and lignin-based chemicals, modified cellulose, regenerated cellulose, ethyl and methyl alcohol (ethanol and methanol), and naval stores.

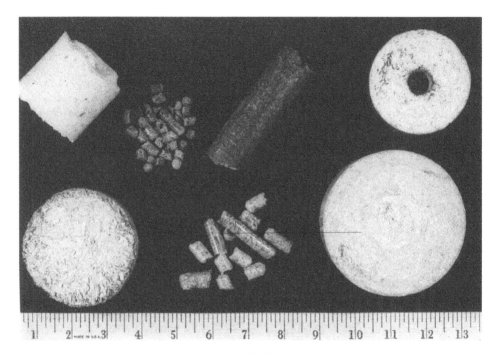

FIGURE 16.7.   Several types and sizes of wood fuel briquettes; scale is in inches.

Cellulose Ethers

These are made by treating alkali cellulose with various reagents. Cellulose ethers include carboxymethyl cellulose, which is used in making products as diverse as laundry detergent additives, toothpaste, adhesives, and strengtheners in unfired ceramics. Other cellulose ethers are used as sizing in papers and textiles and as emulsifying agents in paints and foods.

Lignosulfonates

These versatile compounds originate from spent cooking liquors of chemical pulping. They are used as dispersing and stabilizing agents in oil well-drilling muds, printing inks, dyes, and concrete and as binders in such things as gravel roads, animal feed pellets, wood briquettes, and textiles. Artificial vanilla, used widely in products such as ice cream, cookies, and cakes, is also derived from spent pulping liquors.

Modified Cellulose

This group includes the cellulose acetates and cellulose nitrates. Both of these are important ingredients in adhesives and lacquers. Acetylated cellulose is used in rayon acetate, a material from which textiles are made. In addition, photographic film is made from cellulose acetate, as are a number of extruded and injected molded plastics. Cellulose nitrate is an important source of plastics and was the primary ingredient in celluloid, the first commercial synthetic plastic. Molded plastic articles such as table tennis balls and piano keys are made from this material. Nitrated cellulose is used in guncotton and cordite, both common ingredients in explosives.

Regenerated Cellulose

These products are produced by partially breaking down cellulose through chemical treatment and then recombining components to form a synthesized fiber. Products in this category include cellophane and viscose rayon, a colorfast material used extensively for curtains and drapes, clothing, and bedspreads. Rayon fiber is also used in the inner plies of radial tires and in conveyor belts. Rayon is drawn from a liquefied cellulose fluid. A new product, known as lyocell or tencel, is made by chemically dissolving cellulose and then producing the fiber from a "liquid crystalline" phase. Fabrics from both rayon and lyocell fibers are strong, soft to the touch, and commercially produced as a clothing material.

Ethyl, Furfuryl, and Methyl Alcohol

Each of these is each produced by a different process. Ethanol can be produced from wood by hydrolysis and fermentation of five- and six-carbon sugars. Separation and hydrolysis of crystalline cellulose has been approached in several ways, but the fact that ethanol can be produced much more easily from petroleum feedstocks or grains has limited the use of wood for ethanol production. If the real cost of subsidized commodities (e.g. corn) is considered, biomass-to-ethanol makes sound economic sense. Furfuryl (or wood) alcohol can be made from hardwood waste material. It has been used successfully as a dimensional stabilizer. A closely related wood-derived compound, furfural, is used in the adhesives industry, as an industrial solvent, and as a chemical feedstock. Methanol can be produced by thermal decomposition of wood sometimes referred to as *pyrolysis* or *gasification*. Until the 1920s wood pyrolysis was the only source of methanol but today most is produced from a *synthesis gas* obtained from reformed natural gas.

Naval Stores

This classification includes turpentine and rosin. Both of these materials, along with pine pitch and tar, were once essential to the construction and sea-worthiness of wood sailing ships, explaining the term *naval stores*. Today almost all naval stores are obtained from the nonvolatile (tall oil) and volatile fractions recovered after the chemical pulping of pine wood. Tall oil is an ingredient of some lubricants. Turpentine and its derivatives are used in the manufacture of paints and lacquers and various chemicals including insecticides, perfumes, and artificial flavors (e.g. lemon, orange, cloves). The volatile organic compounds in softwoods are mainly monoterpenes and can be speciated into numerous individual molecules. One derivative, pine oil (α- and β-pinene), is used in making cleaners and disinfectants. *Rosin*, produced when turpentine is distilled from pine gum, is an important industrial chemical. Rosin should not be confused with *resin (oleoresin, pitch, pine tar)*, a common component in epithelial cells and resin canals of some conifers. Used principally in sizing of paper to reduce penetration of liquids, rosin is also used in paints, lacquers, varnishes, hot-melt adhesives, printing inks, plastics, vinyl floor coverings, and as a plasticizing agent in synthetic rubber.

Tannins

Tannic acids, or tannins, are extractive compounds found largely in the bark of certain species (largely acacia, pine, oak, and sumac). Tannins are extracted with water or alcohol, settled, then collected after solvent evaporation. Used primarily in the leather tanning industry, these compounds are becoming important in the wood adhesives industry. Hybrid systems which substitute a portion of tannin adhesive (phenolic in nature) for straight phenol, or resorcinol with phenol, have shown efficacy. In regions of the world where a steady stream of petrochemical-based adhesives is not economically feasible, tannin-based adhesives have great utility value.

Pharmaceuticals

Medicinal compounds from wood have long been recognized. Salicylic acid (aspirin) from willow bark (genus *Salix*) is one of the most famous. Taxol from the yew tree (genus *Taxus*) has received much attention as an anticancer drug in the last 10 years. A significant amount of modern medical research to isolate compounds from domestic and exotic trees and study their respective benefits to health is ongoing and promising. Such research is leading to drugs that combat cancer, aging, arthritis, and an array of other diseases.

References and Supplemental Reading

Arola, R.A. 1976. Wood fuels: How do they stack up? Proceedings of the Conference on the Energy and the Wood Products Industry, Forest Products Research Society.

Corder, S.E. 1975. Fuel characteristics of wood and bark and factors affecting heat recovery. Proceedings of the Conference on Wood Residue as an Energy Source, Forest Products Research Society.

Department of Energy. (2001). U.S. DOE Energy Efficiency and Renewable Energy Network, Biopower Renewable Energy Technical Reports on Wood Residues and Business and Market Opportunities.

Department of Energy. (2006). U.S. DOE Energy Information Administration. (http://www.eia.doe.gov).

EIA. (2016). Southern states lead growth in biomass electricity generation. US Department of Energy, Energy Information Administration. Independent statistics and analysis information sheet. https://www.eia.gov/todayinenergy/detail.php?id=26392

Engelken, L.D., and Farrell, R.S. (1979). Cogeneration in plant operation. Proc. Conf. Hardware for Energy Generation. For. Prod. Res. Soc.

FAO (2005). *State of the World's Forests*. Rome: Food and Agriculture Organization, Forestry Department.

Garrett, L.D. (1981). Evaluating feedstock requirements for a 50-megawatt wood-fired electric generating plant. *For. Prod. J.* 31 (1): 26–30.

Georgia Institute of Technology (1984). *The Industrial Wood Energy Handbook*. New York: Van Nostrand Reinhold.

Haygreen, J.G. (1981). Potential for compression drying solid wood cubes and chip mats. *For. Prod. J.* 31 (8): 43–54.

Haygreen, J.G. (1982). Mechanics of compression drying solid wood cubes and chip mats. *For. Prod. J.* 32 (10): 30–38.

Ince, P.J. (1979). How to estimate recoverable heat energy in wood or bark fuels. USDA For. Ser. Gen. Tech. Rep. FPL-29.

Jahn, E.C. and Preston, S.B. (1976). Timber: More effective utilization. *Science* 191: 757–761.

Johnson, R.C. (1975). Some aspects of wood waste preparation for use as a fuel. *TAPPI* 58 (7): 102–106.

Marks, J. (1992). Wood powder: an upgraded wood fuel. *For. Prod. J.* 42 (9): 52–56.

National Renewable Energy Laboratory. (2005). Minnesota Biomass – Hydrogen and Electricity Generation Potential. Report prepared for Minnesota Department of Commerce and Minnesota Office of Environmental Assistance (February). (https://www.pca.state.mn.us/sites/default/files/MNbiomass-NREL.pdf)

Newby, M.W. (1980). An overview of combustion technology available to the pulp and paper industry. Pap. Trade J. May: 30–34.

Overend, R.P. and Chornet, E. (eds.) (1999). *Biomass: A Growth Opportunity in Green Energy and Value Added Products (Vols I and II)*. Oxford: Pergamon Press.

Perlack, R., L. Wright, A. Turhollow, R. Graham, B. Stokes, and D. Erbach. (2005). Biomass as feedstock for a bioenergy and bioproducts industry: The technical feasibility of a billion-ton annual supply. US Department of Energy, Oak Ridge National Laboratory/US Department of Agriculture. (https://www1.eere.energy.gov/bioenergy/pdfs/final_billionton_vision_report2.pdf)

Pingrey, D.W. 1976. Forest products energy overview. Proceedings of the Conference on the Energy and the Wood Products Industry, Forest Products Research Society.

Rice, R.W. and Willey, R.M. (1995). Higher heating values for pellets made from wood waste and recycled newsprint. *For. Prod. J.* 45 (1): 84–85.

US Department of Energy. (1999). The technology roadmap for plant/crop-based renewable resources 2020. Office of Industrial Technologies, Energy Efficiency and Renewable Energy. (https://www.nrel.gov/docs/fy99osti/25942.pdf)

US House of Representatives. (1993). Global warming (Part 2). Committee on Energy and Commerce, Subcommittee on Energy and Power, Record of Hearings, 29 July, 27 October, and 16 November, Serial no. 103–92.

Vanelli, L.S., and Archibald, W.B. (1976). Economics of hog fuel drying. Proc. For. Prod. Res. Soc. P-76-14.

Walsh, M., Perlack, R., Turhollow, A. et al. (1999). *Biomass Feedstock Availability in the United States: 1999 State Level Analysis*. Oak Ridge, TN: Oak Ridge National Laboratory.

Walsh, M., TorreUgarte, D., Shapouri, H., and Slinsky, S. (2003). Bioenergy crop production in the United States – potential quantities, land use changes, and economic impacts on the agricultural sector. *Environ. Resour. Econ.* 24 (4): 313–333.

Warsco, K. (1994). Conventional fuel displacement by residential wood use. *For. Prod. J.* 44 (1): 68–74.

17

Global Raw Materials

The Importance of Wood

United States

As outlined in the Introduction, wood is one of the most important raw materials in the United States (see Table I.2). For example, the weight of wood used every year in the United States exceeds the weight of all metals and all plastics *combined*! One reason for misperceptions about the magnitude of wood use is that, as the sophistication of wood products increases, many products are becoming scarcely recognizable as wood. Other products are so common that they are often taken for granted. As noted throughout this text, however, of all of the material choices available on the planet, wood is one of the most environmentally sound of all.

An idea of the importance of wood can be gained by considering personal use of this material. A look around the home, for example, can be revealing. As described in previous chapters, the structural shell of the house – walls, roof rafters, floor joists – is probably wood, as are the roof and floor decks. The roofing felt is most likely asphalt-impregnated wood fiber. Even the paint on the house may contain rosins and resins traceable to distillation of softwood chips.

The wall sheathing is likely wood, in the form of plywood, oriented strandboard, or insulation board. The siding may be lumber, hardboard, wood shakes, or wood fiber-reinforced cement. More than likely, the windows have wood frames. The gypsum board that forms the interior walls and ceiling is faced with heavy paper, a wood product. Other common wood features include doors, molding, trim, and kitchen and bathroom cabinets. Countertops of Formica or other high-density laminates cover a wood particleboard core, and the laminate itself is made by impregnating sheets of wood-fiber paper with plastic. The floors may be wood (solid or laminated), but if not, the carpet or vinyl floor covering is almost certainly placed over a wood subfloor.

Forest Products and Wood Science: An Introduction, Seventh Edition. Rubin Shmulsky and P. David Jones.
© 2019 John Wiley & Sons Ltd. Published 2019 by John Wiley & Sons Ltd.
Companion website: www.wiley.com/go/shmulsky

Most of the furniture and furniture framing is likely wood. The curtains, drapes, and bedspreads could be a rayon blend. Rayon is made of regenerated cellulose, usually from wood. The basement may have a wood workbench, shelves, and pegboard. The high-impact plastic handles of your screwdrivers and chisels likely contain finely ground wood fiber. If there is a ping-pong table, it probably has a hardboard top, and the faces of the wood paddles are overlaid with natural rubber, made from latex that is tapped from trees in the tropics. The ping-pong ball itself is made completely of celluloid, derived from wood cellulose.

Next, consider the family car. Wood is difficult to recognize, but it's there. The door liners (the interior part of the door covered with cloth or vinyl) are probably thin sheets of hardboard. So too may be the deck between the back seat and back window. The dashboard may be made from molded hardboard. The seats, trunk liner, and interior exposed surfaces may be plastic that contains up to 50% wood, a filler that greatly reduces cost. The roof insulation and the insulation between the engine compartment and the car interior are usually a fire-resistant wood-fiber mat. The steel-belted radial tires contain a rayon inner ply. Even the oil in the crankcase was probably brought to your service station through the aid of wood-based lignosulfonates, used in controlling the consistency of oil well-drilling mud.

A drive to the supermarket reveals wood almost everywhere along the way. For example, the concrete road surface probably contains about 0.3% lignosulfonates, dispersing agents that help strengthen cured concrete. Wood guardrail posts and timber bridges blend into the surroundings and often go unnoticed. Billboards are readily seen but their wood backing and support posts are also largely unseen.

At the store there are paper boxes, packaging, and labels, almost all made from wood fiber – over one-third of which is recycled. Notebook paper, tablets, pencils, and natural rubber pencil erasers are wood- or forest-based. Photographic film is made from cellulose acetate. The flavoring in vanilla ice cream and cookies may be vanillin, a food additive made as a by-product of wood-pulping operations. The toothpaste and milkshake contain carboxymethylcellulose, ultimately derived from wood. At the checkout stand are wood-fiber books, newspapers, and magazines. Even the check or cash handed to the clerk is a wood product.

Casual observation through the course of a day or two would reveal many more products made from or derived from wood. In any event, wood plays a principal role in sheltering and meeting a variety of needs of US citizens.

Global Trends

As in the United States, wood is important as a raw material globally (see Table I.1). In 2015, the annual global wood harvest approximated 3.75 billion m³ (131 billion ft³), with slightly over half of that used as fuelwood (Table 17.1; FAO 2017). The majority (just under two-thirds) of the total annual harvest was composed of hardwoods. Historically, the vast majority of the hardwood harvest has occurred in developing countries for use for fuel, whereas the primary use of softwood has been in the developed nations where wood has largely been used for industrial purposes (Figure 17.1). There are signs that this traditional pattern may be changing. The economies of several developing nations, and most notably China, have grown at very rapid rates over the past several decades. Accompanying rapid economic growth has been a rapid increase in consumption; one result has been a

TABLE 17.1.   Global production of roundwood in 2015.

	Million cubic meters	Million cubic feet
Annual roundwood harvest	3490	123 560
Fuelwood	1799	63 686
Industrial wood	1691	59 874

Source: FAO (2017).

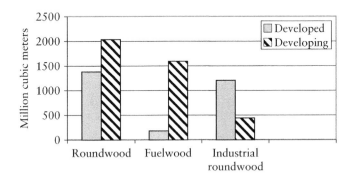

FIGURE 17.1.   Production of industrial roundwood and fuelwood for the United States and the world's least developed countries for 2016. Source: FAO (2017).

steep rise over the past 5–10 years in industrial wood consumption on the part of China and several other nations. Currently, global fuelwood and industrial roundwood production are more or less equal.

Fuelwood comes both from the world's commercial forests (a little less than one-half the total forest area globally) and from woodlands that do not meet the definition of forestland. Industrial wood is harvested largely from commercial forests, including productive plantations. Considering removals from the commercial forests only, some 60% of the harvest is industrial roundwood, with the rest fuelwood.

Global per capita consumption of wood remains steady at ~0.5 m³ (or 18 ft³) in 2015, a figure that is more or less the same as it was a decade earlier. There are however great difficulties in accurately estimating actual consumption in less-developed parts of the world.

Forests and Their Condition

In view of increasing demand on forests, examination of forest conditions and trends is worthwhile. Facts in this regard are often different than widely held perceptions.

Forests and Population Growth

Growth in consumption of wood is closely linked to population growth but also varies widely with the economy. It is important to recognize, however, that population growth is not only directly linked to demand growth but that increasing population also has the

TABLE 17.2.   US historical forest area per capita and year 2100 projection.

Year	Population	Forest area (million acres)	Forest area/ capita (acres)
1785	3 000 000	1044	348
1850	23 300 000	926	40
1910	77 000 000	730	9.5
2005	296 410 000	749	2.5
2100	571 000 000	749	1.3

effect of reducing the area of forestland on a per capita basis. The historical record in this regard is dramatic (Tables 17.2 and 17.3). Taking projected US and global population for the year 2100 into account yields a significant change. By the end of this century it appears that the United States will have only 0.5 ha (1.3 acres) of forestland per capita. Globally, the average may be only about 0.3 ha (0.74 acres). Moreover, these figures include all forestland; the area available for periodic harvest of timber will obviously be even less. That said, various manufacturing technology advances, recycling and reuse, material-efficient multifamily housing, and other ongoing changes that enhance sustainability will help curb overall per capita demand.

Will this kind of per capita reduction in forestland allow wood production to keep pace with increases in population? Without significant changes, the answer would appear to be no.

US Forest Service figures for 2012 show that the forest and woodland acreage has increased to 819 million acres. This land area has increased slightly, about 1%, over the past decade. Softwood growth has increased at a faster rate, overall about 3% during the 2007–2012 interval.

The average US resident consumed the equivalent of 1.25 m³ (44 ft³) of roundwood annually in the form of various products (Howard and Jones 2016). Worldwide, this figure is significantly lower at 0.5 m³. Without question, in the future, there will be increasing demands on forest land to produce the goods and services so desired by the US and global market. These demands present ample career opportunities for forest managers, forest products professionals, and other related disciplines, a trend that will continue into the foreseeable future. Major options available to balance availability with societal demands include: (i) increasing the intensity of management on forested lands generally; (ii) substantially increasing the area of high-yield forest plantations; (iii) significantly improving conversion techniques; (iv) improve durability and longevity; (v) reducing per capita demand through national changes such as shifting toward multifamily housing structures; and/or (vi) incorporating additional nonwood raw materials. The invisible hand of the market pulls on each of these options and in combination they foretell a bright future.

An increase in management intensity in domestic and global forests today is challenging. In some cases, societal pressures are, in fact, leading to increased areas of forest reserves and a lower intensity of management on those lands that are managed for timber production. That said, in much of the US South, intensive and productive management is among the best in the world. Another option for increasing the wood supply that has received a great deal of attention in recent decades is the establishment of vast areas of

TABLE 17.3.   World historical forest area per capita and year 2100 projection.

Year	Population (million)	Forest area million hectares (acres)	Forest area per capita, hectares (acres)
1800	1000	4500 (11 100)	4.5 (11.1)
2000	6000	3900 (9600)0	0.6 (1.6)0
2100	11 000	3900 (9300)0	0.3 (0.7)0

high-yield forest plantations. In parts of the world, this has begun. Because demand is influenced by the economy and there are always global economic swings, it is difficult to predict the potential expansion of new high-yield forest plantations.

In general, the best arable land is reserved for agriculture owing to the great pressure to feed the world's population. Lesser-quality land is useful for forestry. Generally, agricultural demands slowly draw in and convert forested lands; rarely does the conversion go the other way.

The World's Natural Forests

In 2015 forests covered an estimated 40 million km² (15.4 million mi²) or about 31% of the land area globally (World Bank 2017). Trees within urban areas are generally not included in this total. Approximately 59% of the forested area is considered either production or multiple use forest (FAO 2015b). Natural forests make up ~93% of the global forestlands. Plantation forests account for the balance, more or less. Approximately three-fourths of these are classified as productive plantations.

US Forests.   US forests in 2012 covered 309 million ha (766 million acres), or ~33% of the total land area of the country. Of this area, ~204 million ha (504 million acres) were potentially harvestable (that is, sufficiently productive and not in a designated reserve status).

The area covered by forests in the United States in 2012 was greater than it has been in over a century. The forested area is about 75% of the area that was covered by forests in the year 1630. Some 124 million ha (307 million acres) of forestland have been converted to other uses, namely agriculture and the development of cities, towns, and neighborhoods, in the 380-plus year interval between 2012 and 1630. Most of that land conversion was to agricultural uses in the course of the nineteenth century. Observations about US forests provide a useful overview of trends over the past century:

1. During the past 100 years, forestland area has been stable (Figure 17.2).
2. Nationally, the volume of growing stock in United States forests increased by about 58% between 1953 and 2012 (Figure 17.3). The growing stock volume increased in all regions (Oswalt et al. 2014).
3. Populations of whitetail deer, wild turkey, elk, pronghorn antelope, and many other wildlife species increased more than 10-fold between 1950 and the early 1990s (MacCleery 1993).
4. The tens of millions of acres of cutovers or *stumplands* that existed in 1900 have long since been reforested. Many of these areas today are mature forests or have been harvested a second time, and the cycle of regeneration has started again (MacCleery 1993).

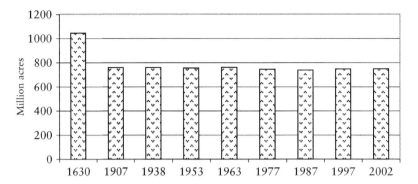

FIGURE 17.2.   Trends in US forestland area (millions of acres), 1907–2012. Source: Oswalt et al. (2014).

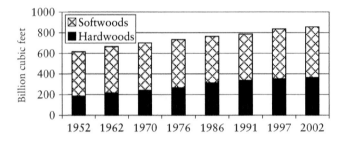

FIGURE 17.3.   Net volume of growing stock, billion cubic feet, on US timberland, 1952–2012. Source: Oswalt et al. (2014).

5. Forest growth nationally has exceeded harvest since the 1940s, with each subsequent decade generally showing increasing margins of growth over harvest (Figure 17.4). In 2012 net forest growth volume (26 billion ft^3) was more than double the harvest volume; 26.4 vs. 12.8 billion ft^3.

6. The efficiency of wood utilization has improved substantially since 1900. Much less material is left in the woods, many sawmills produce more than double the usable lumber and other products per log input, engineering standards and designs have reduced the volume of wood used per square foot of building space, and preservative treatments have substantially extended the service life of wood. These efficiencies have reduced by millions of acres the area of annual harvest that otherwise would have occurred.

7. American society in the twentieth century changed from rural and agrarian to urban and industrialized. This change has been accompanied by a corresponding physical and psychological separation of people from the land and resources. Today's urbanized nation is however no less dependent on the products of its forests and fields than were the subsistence farmers of the past (MacCleery 1993).

Despite healthy growth/harvest ratios in most regions of the United States, pressures to reduce harvesting activity mounted greatly in the 1980 and 1990s. Growing concern about the environment has led to an increase in legislative, legal, and other actions designed to protect and enhance environmental quality; the nation's forests have been a

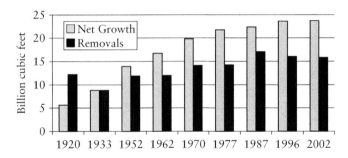

FIGURE 17.4.   US timber growth and removals, 1920–2002.

major focus of such actions. Activity has centered on issues ranging from biodiversity and ecosystem health to water quality and esthetics. Many resultant policies have focused on protectionism which in turn often creates biological stagnation and extreme fire risk (particularly in the arid portions of the Southwest). The realm of forest policy has proven to be one of great emotional polarization.

Global Forests.   Natural softwood forests of the world are found in the Northern Hemisphere. Hardwood forests dominate the tropical and subtropical regions and the Southern Hemisphere, and they occur in extensive regions of the Northern Hemisphere as well. Overall, hardwoods are present in greatest volume worldwide.

Overall, deforestation amounts to ~7.6 million ha annually (3.3 million ha annually when establishment of forest plantations is taken into account), including tropical regions, where most of this decrease occurs, driven primarily by agricultural clearing. Encouragingly, this rate of deforestation has decreased by about 50% over the past decade. Outside of the tropics, forests worldwide are generally increasing in area coverage or are stable in size (FAO 2015b).

Comparison of net annual increment and annual harvest for the world's temperate and boreal forests show the volume of net growth to be almost double the harvest. As a result, ~1.2 billion m^3 are added to standing timber volume annually in the world's temperate and boreal forests at the same time that these forests are providing substantial volumes of industrial wood. Significant concerns about forest health and sustainability of forest use persist in virtually every nation.

Given the large gap between net annual growth and annual harvest in the world's temperate and boreal forests, there would appear to be potential for significantly increasing wood production from natural forests. However, there is significant public opposition to increasing harvest levels in many developed and developing countries, including the United States and Canada. In fact, the potential for increased harvest of natural forests is today constrained, by both limitations of growth and by public perception and politics, to relatively few areas of the world. These areas include Siberia, the far eastern region of Russia, northern Europe, and several European nations. Additionally, the vast (mostly hardwood) forests of Brazil and other parts of South America could conceivably support a larger sustainable harvest. The extent to which economic constraints and environmental concerns limit the potential for increases in harvest in various regions of the world remains to be seen.

Plantation Forests

Plantation forests have been established globally for the purposes of providing industrial wood and fuelwood, and to restore degraded lands for purposes of soil and water conservation. Forest plantations dedicated to production of industrial wood generally produce much more wood per geographic area than natural forests because they are usually established on highly productive sites, intensive silviculture (including fertilization) is practiced, and genetically selected growing stock is used. Plantations will continue to play a significant role, and perhaps even a dominant role, in providing future wood supplies.

The FAO (2015b) showed the area of productive forest plantations worldwide to be 291 million ha (719 millon acres) in 2015, comprising about 7% of the total forest area globally. As noted by the FAO, Asia has by far the largest productive forest plantation estate of any region, accounting for 41% of the world's total plantation area. Productive plantations account for about 8% of all forests in Asia. The 10 countries with the largest reported areas of forest plantation together account for 70% of the global forest plantation area. China, the United States, and the Russian Federation together account for more than half of the world's productive forest plantations.

The area of plantations is increasing steadily, with more than 3.3 million ha (8.2 million acres) added to the global plantation area each year during the period 2010–2015. The reality of expanding plantations is evidenced by the fact that productive forest plantations made up only l.9% of forests globally in 1990, compared with the current 7%.

Industrial forest plantations are typically highly productive compared with natural forests. Plantations often produce $10\,m^3$ of wood ha^{-1} annually, and wood yields of 20–$25\,m^3\,ha^{-1}$ per annum are not uncommon. In addition, annual yields as high as 45–$70\,m^3\,ha^{-1}$ per annum have been attained with some hardwood species. In contrast to plantations, natural tropical moist forest commonly yields 1–$2\,m^3\,ha^{-1}$ per annum, which can be contrasted with perhaps $6\,m^3\,ha^{-1}$ per annum with management; similar yields are attainable for natural forests in temperate regions.

The high growth rates attainable in plantations coupled with the fact that plantations are often established on highly productive sites mean that a great deal of wood can be produced on a relatively small land area. Over one-half of industrial wood is expected to come from plantations within a few decades.

In 2002, nationwide in the United States, managed softwood plantations were projected to rise in acreage from about 6% of commercial timberland area to about 9% by 2050. By 2012, the 9% level was met (Oswalt et al. 2014). The US South has the largest concentration of plantation forest land acres. Coupled with its high proportion of private landownership, the US South is the largest commercial wood provider.

Beyond near-term expectations, there is considerable potential for establishment of additional areas of tree plantations in the United States. It was reported in 1992 that there are about 160 million ha (392 million acres) of US nonfarmed land capable of supporting production of wood-producing crops without irrigation (Wright et al. 1992). More than half of this land (91 million ha or 225 million acres) is believed to be capable of supporting a sustained production of $4.5\,t\,ha^{-1}$ or ~8–$10\,m^3\,ha^{-1}$ per annum (2 short tons ac^{-1} per annum) of wood per year. But there has been considerable discussion of late about the possibility of using this land as a source of biomass for energy production

(Perlack et al. 2005), including energy crops. Others view lands not now in commercial use as opportunities for creating additional wildlife or wilderness reserves. So, although there is considerable potential for expanding the area of tree plantations in the United States, competing interests are likely to substantially limit the amount of land dedicated to this purpose.

It should be noted that despite the great promise of intensively managed plantations for providing needed volumes of wood, there is nonetheless considerable controversy over expansion of plantations around the world. One of the controversial aspects of plantation establishment is the reality that a portion of plantations established each year arise through conversion of natural forests to plantation areas. Controversy is also rooted in perceptions of low biodiversity of plant and animal species in planted forests, long-term forest health risks associated with limited genetic variation in planting stock, and land tenure issues.

Increased Efficiency of Wood Use

The past five decades have brought dramatic technological change to the utilization of timber in the United States. Bingham (1975) provided an example of what was accomplished in utilizing old-growth Douglas fir timber in western Oregon between 1948 and 1973. The logs harvested on an acre of this timberland typically contained about $17\,900\,ft^3$ of wood. He reported that the extent of utilization was as follows:

In 1948, $1250\,m^3$ ($17\,900\,ft^3$) of logs yielded
  $252\,m^3$ ($3600\,ft^3$) of lumber
  $998\,m^3$ ($14\,300\,ft^3$) of residue (fuel and waste)
  Total = $252\,m^3$ ($3600\,ft^3$) of products

In 1963, $1250\,m^3$ ($17\,900\,f^3$) of logs produced
  $321\,m^3$ ($4600\,ft^3$) of lumber
  $265\,m^3$ ($3800\,ft^3$) of paper
  $56\,m^3$ ($800\,ft^3$) of plywood
  $608\,m^3$ ($8700\,ft^3$) of residue (fuel and waste)
  Total = $642\,m^3$ ($9200\,ft^3$) of products

In 1973 $1250\,m^3$ ($17\,900\,ft^3$) of logs produced
  $349\,m^3$ ($5000\,ft^3$) of lumber
  $119\,m^3$ ($1700\,ft^3$) of plywood
  $412\,m^3$ ($5900\,ft^3$) of paper
  $105\,m^3$ ($1500\,ft^3$) of particleboard
  $265\,m^3$ ($3800\,ft^3$) of residue (fuel and waste)
  Total = $985\,m^3$ ($14\,100\,ft^3$) of products

Bingham's figures show that, in 1948, a log sawn to lumber yielded only 20% finished product and 80% waste! Then, as now, the lumber that was produced in North America had many uses, most of which were directly related to homebuilding and remodeling. For example, wood in the form of lumber was almost singularly the material used for the framing of walls, floors, and roofs.

Currently, softwood lumber yield is often measured in the form of (short) tons of logs per thousand board foot conversion, with 4.5 (4090 kg or 9000 lb) being a typical value. A dressed and kiln-dried thousand board foot unit of southern pine lumber weighs ~1000 kg (2200 lb). Approximately 182 kg (400 lb) will be removed as planer shavings (this value fluctuates with the efficiency of the mill and the size of the lumber). In the kiln, ~1800 lb of water are removed from that rough green lumber. About 9% (370 kg or 810 lb) is removed as bark. The remaining 3790 lb is dispersed among chips, sawdust, and trim blocks. Where possible, green trim blocks (short pieces chopped off at the trim saw) are chipped for pulp and paper. Dry trim blocks may be sold for short-length remanufactured products, such as survey stakes, or ground for fuel that can be used to dry lumber. In all cases, pulp chips are more valuable than green sawdust and thus mills seek to optimize their production.

During the last half-century, the amount and importance of new products and processing technologies have increased tremendously. The following information highlights some of the advancements of the last 50 years. By 1963, for example:

- The average large log sawn to lumber yielded 26% lumber, and 51% usable products overall, up from 20% usable product only 15 years earlier.
- Hardboard siding accounted for 9% of the US siding market.
- Vinyl siding was publicly exhibited for the first time at the 1963 World's Fair.
- Steel-reinforced concrete had been developed.
- The wide availability of North American softwood plywood led to large-scale replacement of lumber in home construction for applications such as subflooring, wall sheathing, and roof decking.
- Waferboard had been patented and the first production plant built in Idaho in 1955.

By 1973:

- The average large log sawn to lumber yielded 28% lumber and 78% usable products overall, with almost all of the remaining volume used as fuel to generate process energy. From 1948 to 1973, the usable products obtained from similar acres of Douglas fir increased nearly fourfold.
- The chipper-canter, commercially introduced in the mid-1960s, was widely adopted. This development led to a rapid increase in processing small-diameter logs to lumber.
- Best Opening Face (BOF) technology, which increased lumber yield via computerized evaluation of initial log positioning, had been introduced.
- Log merchandisers, which allowed systematic bucking of long logs and sorting of resulting segments into various use categories, had become common in the West and South United States.
- Development of the retractable chuck lathe made it possible to economically peel small-diameter logs to veneer. This development fostered the rapid expansion of the southern pine plywood industry.
- Hardboard siding had grown to a 20% market share (from 9% 10 years earlier); vinyl had grown to a 5% share, up from 0% 10 years earlier. The remainder of the siding market was shared almost equally by plywood, brick, and aluminum (about 20% of market share each), with small quantities of solid lumber and steel.

- Particleboard was now a standard core stock and underlay material.
- North American softwood plywood held a 99% + share of the US domestic structural panels market. Lumber had been totally displaced as a sheathing material.
- Waferboard was being commercially manufactured in the United States and in Canada. This technology made it possible to produce high-strength panels from small-diameter trees. In 1973, waferboard accounted for 0.05% of the US structural panel market.
- Patents had been issued (in 1968 and 1971, respectively) for wood structural I-beams, and for laminated veneer lumber (LVL).
- Patents had been awarded in Europe and Japan for equipment to produce plastic lumber of large cross-section.
- Recovery of waste paper for reuse in paper and fiber and paper products manufacture approximated 23% of domestic paper consumption.

Advancements in technology did not cease in 1973. By 1983:

- Particleboard was facing a serious challenge from MDF in core markets.
- Waferboard shared the market with oriented strandboard (OSB), and together the two products accounted for 5.5% of the US structural panel market.
- Hardboard siding had increased its market share to 31% of the US siding market. Vinyl siding had increased its market share to 12%.
- A new product, the all-vinyl window, was introduced in 1980. Market share was 3% by 1983.
- Wood structural I-beams and LVL were both being sold on the commercial market. This development permitted the use of small-diameter trees in making large-size structural products.
- Centerless lathe technology for producing veneer had been introduced. This technology allowed the use of logs that previously could not be used in making veneer and allowed the peeling of a log down to a smaller core, thus increasing the volume of veneer recovery.
- Wood scientists turned their attention to juvenile wood as average harvested log diameters declined and as tree plantations began to supply significant quantities of wood.
- Misawa Homes of Japan had begun production of precast autoclaved lightweight cement modular housing units.
- Recovery of waste paper for reuse in paper and fiber and paper products manufacture approximated 26% of domestic paper consumption. The reuse rate remained at the 1973 reuse rate of 22%.
- Technologies for producing lightweight coated papers had been developed in Europe.

By 1993:

- Parallel strand lumber had been developed in Canada and was being sold commercially. Oriented strand lumber, a related product, was also on the commercial market.
- LVL and wood I-beams had achieved wide acceptance in homebuilding applications, such as garage door headers and beams, and in commercial/industrial applications as a substitute for steel.

- Annual North American softwood plywood production had surpassed 19.4 million m^3, 9.5 mm thick basis (22 billion ft^2 3/8 in basis). However, OSB now accounted for one-third of the structural panels market, up from 5.5% only a decade earlier.
- BOF technology, which had been developed in 1971, was now used in conjunction with automated scanners and computer-interfaced production equipment in half of US softwood sawmills, accounting for at least 75% of production.
- The use of fingerjointing to produce softwood studs from small pieces of wood that had been previously wasted or burned for power was common practice.
- Veneer overlay technology allowed the use of thin veneers over complex profiles of substrate materials to produce high-quality moldings, trim, and raised panels.
- All-vinyl windows commanded 7% of the total domestic window market and 13% of the window replacement market.
- Insulating concrete forms had become commercially available before the end of the year (1993). Concrete overall accounted for 3% of the above-grade wall area in US homes.
- Wood fiber–cement siding materials were commercially available.
- Musashi Works of Sekisui Chemical Company Ltd had begun production of precision-made, steel-frame, modular home boxes.
- A wood polymer composite, made from 100% recycled plastic resins (primarily HDPE and LDPE) and wood waste, was commercially available.
- An all-plastic house had been shown four years earlier at the National Association of Home Builders annual convention.
- Plastic lumber had captured some 1% of the US decking market.
- Recovery of waste paper for reuse in paper and fiber and paper products manufacture approximated 39% of domestic paper consumption. The reuse rate had increased to 31%.

By 2000:

- North American production of OSB exceeded the production of softwood plywood.
- Wood I-beams had captured 60% of the North American market for joists and planks and 30% of the market for beams and headers, displacing two large traditional markets for solid sawn lumber.
- Fiberglass and metallic filaments were being used on tension faces of structural beams to enhance strength.
- Insulating concrete forms accounted for almost 3% of above-grade wall area in US homes. Overall, concrete accounted for over 13% of above-grade wall area.
- Plastic lumber made from recycled plastic grocery bags, pallet wrap, and waste wood accounted for over 10% of the United States decking market, up from 1% only seven years earlier.
- Particleboard was being commercially manufactured from agricultural residues (primarily wheat straw) in the United States and Canada.
- An extrudable wood waste/waste plastic composite had been developed and commercialized by the wood-framed window industry.
- Recovery of waste paper for reuse in paper and fiber and paper products manufacture approximated 48% of domestic paper consumption. The reuse rate had grown to 37%.

By 2005:

- Wheat straw particleboard had all but disappeared from commercial markets owing to high production costs and persistent performance issues.
- Waste paper recovery for reuse in paper and fiber products stood at 51.5%. The reuse rate was 37%. The reuse rate declined slightly from 2000 despite steady increases in volumes of paper recovered and reused during the 2000–2005 period, reflecting rising paper consumption and increasing export of recovered paper.
- Wood fiber–cement siding commanded 13% of the United States residential siding market, up from 2% in 2000, and vinyl accounted for 32% of the market, down from 37% in 2002.

Between 2005 and 2015:

- Significant movements of global paper production occurred. Multiple pulp, paper, and chip mills were closed in the United States and in some cases the machinery and equipment were relocated overseas.
- Structural (lumber, plywood, OSB, etc.) and nonstructural (particleboard, MDF) building products dominate the markets for US roundwood. Demand for wood pellets for energy grew. Domestic demand for pulpwood experienced significant decline.
- Housing starts followed a national economic recession between ~2009 and 2012. During that time lumber production dropped by ~60% but had largely recovered by 2015. At the time of writing, timber and wood products manufacturing continues to grow, largely coaxed by increasing housing demand.

Many more examples could be given. The point, however, is that ongoing technology improvements, driven by competition and rising costs of raw materials, are serving to continually increase the quantity of useful products that can be obtained from a given quantity of logs. Furthermore, improvements in forestry practices are increasing the yield of raw materials from a given area of land.

Nonwood Renewable Materials

Research efforts worldwide are beginning to focus on the possibility of utilizing agricultural crops or crop residues as raw materials for the production of paper and various structural and nonstructural composite materials. As an example, promotion of houses constructed largely of straw bales began in the United States in 1994.

Although early efforts to commercialize construction products made from agriculturally derived fiber have failed, this kind of fiber remains potentially useful as a raw material for some of the products now made largely or wholly of wood. Such products include paper and a variety of structural and nonstructural structural panel products. The most interesting potential source of agricultural fiber is agricultural residues. These are by-products of food production that in many areas of the world currently represent a disposal problem. At first glance, the use of these materials is both socially and environmentally attractive, as long as volumes removed from the land do not compromise soil conservation. A closer look at the environmental impacts associated with high-production agriculture, however, diminishes much of this allure. Currently, there is great interest in agricultural residues as raw materials for production of energy. To date however, high amounts of silica and ash in agriculture-related feedstocks limit their viability as various types of bioenergy.

The idea of using agricultural residues as an industrial raw material is not new. Paper was invented in China in CE 105, but it was not until about 1850 that wood began to be used as a principal raw material for papermaking. Early sources of fiber included flax, bamboo, various grasses, cereal straw, cottonseed hair, leaves, and the inner bark of trees (Isenberg 1962; Miller 1965). Wheat straw chemical pulp was first produced in 1827 (Moore 1996). Crop residues, such as bagasse (or sugarcane residue), have long been used in making paper in China, India, Pakistan, Mexico, Brazil, and a number of other countries (Pande 1998). In recent years, a several decades-long trend toward greater production of paper and paperboard from crop residues appears to have plateaued, largely owing to efforts within China to improve papermaking fiber quality through greater use of wood fiber. Estimates of the percentage of nonwood fiber (including crop residues and a small quantity of specially cultivated industrial fiber crops) utilized in world pulp production vary widely. Pande (1998) reported that nonwood fiber in 1994 comprised 11.7% of global pulp production from virgin fiber, a figure almost double the 1994 and 1998 nonwood fiber estimates of 6.2 and 6.5% by Jaakko Poyry Consulting (McNutt and Rennel 1997; Paavilainen 1998). Predictions of nonwood fiber use in 2010 also differ considerably, with the percentage of nonwood fiber estimated to range from 7.5% (Paavilainen 1998) to 12–15% (Pande 1998) of global pulp production in that year. All estimates suggest that the proportion of nonwood fiber in global pulp and paper production will increase.

The use of agriculturally derived fiber for production of structural and nonstructural panels is in its infancy. There are examples of the commercial use of oil palm stalks for production of medium density fiberboard (MDF) in Malaysia and elsewhere, and of use of cereal grain stalks for the production of MDF and other panels on a limited basis in North America. Production volumes, however, remain modest.

Should agricultural crops become a viable source of fiber for paper, energy, or other products, from both technical and economic perspectives, much of the land discussed earlier as available for establishment of tree plantations would be available instead for raising annual crops. However, the raising of intensively cultured agricultural crops for fiber production, instead of trees, makes little sense from an environmental point of view.

Nonrenewable Raw Materials – Trends and Outlook

Demand

Nonrenewable raw materials that can be substituted for wood that is used for structural purposes include steel, aluminum and other metals, portland and masonry cement, and plastics. Because of growing populations and expanding global economies, consumption of all of these materials is increasing significantly. For example, in the 45-year period from 1970 to 2015, the world population grew about 100%, from 3.7 to 7.3 billion people. During that same time span, world production of steel, aluminum, and portland and masonry cement grew by ~200–400%, depending on the product and information source. A similar, though less dramatic, pattern can be seen in the United States over the same 45 year period. Thus, not only is consumption of many materials rising in step with population growth, but per capita consumption is tending to rise as well.

Supply

The life index values for many important minerals and metals like tin, copper, and lead used worldwide, including fuels, are currently estimated at 20–50 years. These values fluctuate each year and are difficult to accurately forecast because both mining and recycling technologies continually improve. Estimates like these are sometimes used to support the contention that the world is on the verge of running out of many of these materials. However, because the life index of a material is based only on the stock of known reserves that are economically available at current technology, it is totally incorrect to forecast resource depletion using the life index value. Low-quality mineral ore is far more abundant than ore of high quality. Therefore, as the ore progressively becomes lower in quality, the quantity of that ore becomes geometrically larger. Thus, for all practical purposes, the Earth will never "run out" of most mineral resources, including the materials used in making cement. Although there is little prospect of running out of various metallic and nonmetallic ores, as supplies of the highest grades and most accessible ores diminish, the environmental impacts and energy consumption associated with mining and metals production can be expected to increase.

As is the case with wood, ongoing technology improvements in growth, extraction, processing, conversion to products, and product use are serving to stretch or extend raw material supplies. Automobiles, for example, are today made from thinner metal skins than formerly, and lightweight plastics have substituted for metals in many of the vehicle parts. Advances in rust protection and the use of noncorroding materials in key locations have, moreover, extended vehicle life. Similar developments are occurring in virtually all industries. When consumption of raw materials is expressed on the basis of consumption per unit of gross national product, the value of all goods and services produced in an economy, a drop in consumption of several key materials can be seen within Western European nations. Similar patterns of raw materials use are in evidence in the United States. Should such trends continue, they will help moderate population-driven increases in global raw material demand.

Environmental Impacts of Forest Harvesting and Wood Use

It is essential that forests be managed in such a way as to ensure the sustainability not only of wood production but also of water, wildlife, and forest biodiversity. It is also critical that actions taken to protect the environment do, in fact, benefit the environment from a global point of view.

Based on environmental concerns, it is often suggested that the harvesting and use of wood in the United States be substantially reduced. The argument is frequently made, for example, that periodic harvesting of forests should be curtailed because harvesting has negative environmental impacts. However, careful consideration of global environmental concerns, given the realities of today's world, leads to a much different conclusion: to protect the environment, forests should be utilized to the maximum extent possible within sustainable limits.

Essential factors to consider when contemplating the proper role of forests include the following:

1. Growing populations worldwide consume vast quantities of raw materials, and those raw materials must come from somewhere.
2. Despite ongoing advances in technology and an increasing focus on recycling, global raw material demand is increasing rapidly.
3. Although the production of wood and wood fiber does have environmental impacts, so too – and often to a greater extent – does the production of potential substitutes for wood – metals, plastics, concrete, and agriculturally derived materials.
4. Environmental impacts associated with production of wood products are less and, in many cases, substantially less, than those associated with production using other materials.
5. Wood is a dominant raw material in the United States and worldwide. In the United States its use exceeds that of all metals and all plastics combined.
6. The United States is a net importer of most categories of raw materials – most metals, petroleum (the basis for plastics), portland and masonry cement, and wood.

One reason for public concern about periodic harvesting of forests is that forest harvest activity occurs over relatively large land areas – on the order of 0.75% of the area of commercially available forests (0.5% of the area of all forests) in the United States is harvested each year. However, environmental impacts associated with harvesting are relatively short term, even though visual impacts immediately following harvest are sometimes dramatic (Figures 17.4 and 17.5). Another reason is the continued disassociation of an urbanized society from its historical knowledge of where things come from. A common agriculturally related anecdote is that of children who know fully that milk comes from a cooler at a supermarket but nothing of cows and farms except in a very abstract sense. Wood products and forests are often cast into a similar lot in which the general population demands high-performance and cost-effective housing and products but knows little of the forestry practices that are required to sustain their lifestyle.

Despite public perceptions to the contrary, the area covered in the United States forests is slowly increasing, and standing timber volume is increasing each year. To the extent that forestland has been lost in recent decades, losses have been due almost exclusively to urban expansion and associated development: the widening of highways; the construction of housing developments, shopping centers, and industrial complexes; the establishment of power line corridors; and the creation of reservoirs.

Mining activity, including the mining of metals, cements, and fuel resources, impacts a much smaller land area than does forest harvesting. Although 0.1% or less of United States land area is affected by mining in any one year, mining activity shifts very slowly from location to location, meaning that more-or-less the same locations tend to be impacted year after year. Because the United States is a significant net importer of metals and some nonmetallic minerals, such as mica, fluorspar, and gypsum, additional land areas outside the US borders are impacted as well.

An effective means of assessing the relative environmental impact of a material is to examine them over the life cycle of the material from raw materials extraction, through

(a)

(b)

FIGURE 17.5. (a) Timber harvesting feller buncher equipment used in a southern pine plantation. (b) High-production grapple skidder used in a southern pine plantation.

processing and conversion, and ultimate use. Examination of energy use is particularly revealing, because a number of serious environmental problems are related to consumption of energy, including acid deposition, oil spills, air pollution (SO_2, NO_x), and increasing concentrations of atmospheric carbon dioxide.

TABLE 17.4.   Energy required in the manufacture of various wall elements.

Type of wall	Energy to manufacture 100 m² of wall (thousand MJ oil equivalent)[a,b]	
Plywood siding, no sheathing, × 4 frame	2.255	(1.988)
MDF siding, plywood sheathing, × 4 frame	2.883	(2.541)
Concrete building block, no insulation	19.385	(17.087)
Aluminum siding, plywood, insulation board, over × 4 frame	5.619	(4.953)
MDF siding, plywood sheathing, steel studs	5.792	(5.106)
Brick veneer over sheathing	20.291	(17.887)

[a] These figures include consideration of energy consumed in extraction or harvesting, transportation, processing, and construction.
[b] Figures in parentheses are million BTU oil equivalent per 100 ft² of wall.
Source: National Research Council (1976).

In the mid-1970s the National Academy of Sciences, through its Committee on Renewable Resources for Industrial Materials (CORRIM) conclusively established that wood had a substantial advantage in relation to other materials in terms of energy consumption per unit of finished products (National Research Council 1976). Although technologies have changed significantly in all industries since 1976, recent studies have confirmed the advantages of wood (Buchanan 1991; Honey and Buchanan 1992; Marcea and Lau 1992; Meil 1993; Perez-Garcia et al. 2005).

The mid-1970s CORRIM effort examined the energy required to build wall systems for residential homes. Energy use associated with raw material gathering (harvesting or mining), transport, manufacturing, and building construction was considered. Wood-frame construction was found to require the use of far less energy than steel, aluminum, concrete block, or brick (Table 17.4). Because of concerns about the potential for global warming, recent studies of environmental impacts associated with raw materials processing have examined carbon dioxide emissions as well as energy consumption. A 1992 Canadian assessment of alternative materials for use in constructing a 9300 m² (110 000 ft²) building showed all-wood construction on a concrete foundation to require only 35% as much energy as steel construction on a concrete foundation. Furthermore, the liberation of carbon dioxide associated with building the steel structure was over 3.1 times that when building with wood. In a New Zealand study, Honey and Buchanan (1992) found office and industrial buildings constructed of timber to require only 55% as much energy as steel construction and ~66–72% as much energy as concrete construction. When residential buildings were considered, wood-frame construction with wood-framed windows and wood fiberboard cladding was found to require only 42% as much energy as a brick-clad, steel-framed dwelling built on a concrete slab and fitted with aluminum-framed windows. Accordingly, large differences in carbon dioxide emission were noted (Figure 17.6). Similar differences were found by Perez-Garcia et al. (2005). In another comparison of wood and steel-frame construction for light-frame commercial structures, which examined a wide range of factors in addition to energy, Meil (1993) again showed low environmental impacts of wood construction relative to steel (Table 17.5). Very similar results were reported by the Consortium for Research on Renewable Industrial Materials (Perez-Garcia et al. 2005).

(a)

(b)

FIGURE 17.6.   Forest renewal over a 30 year period following clear-cutting. Source: Courtesy of Weyerhaeuser Company Archives.

(c)

(d)

FIGURE 17.6. *Continued.*

(e)

(f)

FIGURE 17.6.   *Continued.*

TABLE 17.5. Comparative energy use, air emissions, waterborne effluents, and solid
wastes for steel and wood wall construction.

	Wood	Steel[a]
Energy consumption (GJ)	3.6	11.4
Air emissions		
Carbon dioxide (kg)	310.0	980.0
CO (g)	2600.0	11900.0
SOx (g)	400.0	3700.0
NOx (g)	100.0	1600.0
Particulates (g)	200.0	500.0
VOCs (s)	350.0	1600.0
CH_4 (g)	neg.	100.0
Water and effluents		
Water use (l)	2200.0	51000.0
Suspended solids (g)	12180.0	495.640.0
Nonferrous metals (mg)	62.0	2532.0
Cyanide (mg)	099.0	4051.0
Phenols (mg)	17715.0	725994.0
Ammonia and ammunition (mg)	1310.0	53665.0
Halogenated organics (mg)	507.0	20758.0
Oil and grease (mg)	1421.0	58222.0
Sulfides (mg)	13.0	507.0
Iron (mg)	507.0	20758.0
Solid wastes (kg)	125.0	95.0

Figures include resource extraction, processing and manufacturing, transportation, and
construction of a structural assembly. Based on construction of non-load-bearing walls 3 m
high by 30 m long (or ~10 × 100 ft).
[a] One-hundred percent virgin steel (nonrecycled content).
VOCs = volatile organic compounds.
Source: Meil (1993).

The values shown in Table 17.5 are dramatic, and they show that although
wood construction clearly has environmental impacts, these impacts are minuscule
compared with those of steel. When the use of recycled steel is considered, the differences between wood and steel narrow, but wood retains a significant advantage.
As part of the wood versus steel wall comparison, Meil examined load-bearing wood
and steel-framed walls in which the steel contained 50% recycled steel content. In
this case the steel-framed wall was found to be "some four times as energy intensive,
and correspondingly ... at least that much more environmentally damaging, despite
its recycled steel content." The point here is not that wood should be used to the
exclusion of all other materials, but rather that production and use of all materials
have environmental impacts that must be considered when formulating environmental policies. In the future, it can be expected that development of building design
and construction technology will seek to take maximum advantage of the properties
of each raw material, thereby designing buildings so as to minimize the total
environmental impact. Wood will undoubtedly play an important role in buildings
of the future.

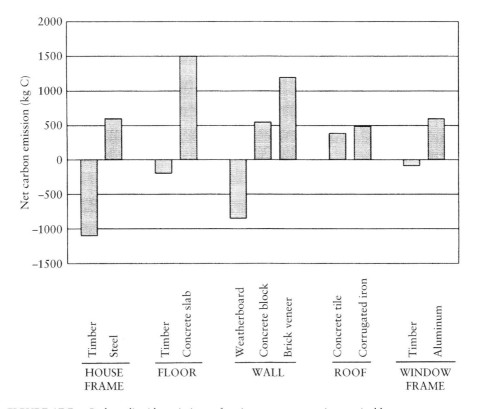

FIGURE 17.7. Carbon dioxide emissions of various components in a typical house.

References and Supplemental Reading

Bingham, C.W. (1975). The keynote. *For. Prod. J.* 25 (9): 9–14.

Bowyer, J.L. (1992). How wood dependent are you? *The Minnesota Volunteer*. March/April 24–25.

Bowyer, J.L. (1993). Wood and other raw materials for the 21st century: where will they come from? *For. Prod. J.* 45 (2): 17–24.

Bowyer, J. and Stockmann, V. (2001). Agricultural residues – an exciting bio-based raw material for the global panels industry. *For. Prod. J.* 51 (1): 10–21.

Buchanan, A. (1991). Building materials and the greenhouse effect. *NZ J. Timber Constr.* 7 (1): 6–10.

FAO (2015a). *Global Production and Trade in Forest Products, 2015*. Rome: Food and Agriculture Organization of the United Nations http://www.fao.org/forestry/statistics/80938/en.

FAO (2015b). *Global Forest Resources Assessment*. Rome: Food and Agriculture Organization of the United Nations.

FAO (2017). *FAOStat. Forestry Statistical Database*. Rome: Food and Agriculture Organization of the United Nations http://www.fao.org/faostat/en/#home.

Frederick, K.D. and Sedjo, R.A. (eds.) (1991). *America's Renewable Resources: Historical Trends and Current Challenges*. Washington, DC: Resources For The Future.

Honey, B.G., and Buchanan, A.H. (1992). Environmental impacts of the New Zealand building industry. Research Report 92-2. Dept of Civil Engineering, Univ. of Canterbury–Christchurch, Canterbury, New Zealand.

Howard, J.L. and K.C. Jones. (2016). U.S. Timber Production, Trade, Consumption, and Price Statistics 1965–2013. USDA For. Serv., For. Prod. Lab. FPL-RP-679.

Ince, P., Fedkiw, J., Dickerhoof, E., and Kaiser, F. (1989). National Measures of Forest Productivity for Timber. USDA For. Serv., Gen. Tech. Rep. FPL-GTR-61.

Isenberg, I. (1962). Fibrous raw materials and wood structure. In: *Pulp and Paper Science and Technology – Volume I* (ed. C. Libby), 20–53. New York: McGraw-Hill.

MacCleery, D.W. (1993). *American Forests: A History of Resiliency and Recovery*, Forest History Society Issues Series. Washington, DC: USDA.

Marcea, R.L. and Lau, K.K. (1992). Carbon dioxide implications of building materials. *J. For. Eng.* 3 (2): 37–43.

Mather, A.S. (1990). *Global Forest Resources*, 13. Portland, OR: Timber Press.

McNutt, J. and Rennel, J. (1997). The future of fiber in tomorrow's world. *Pulp Paper Int.* 39 (1): 34–36.

Meil, J.K. (1993). Environmental measures as substitution criteria for wood and nonwood building products. In The Globalization of Wood: Supply, Processes, Products, and Markets. For. Prod. Soc. Proc. 7319, pp. 53–60.

Miller, D. (1965). Kenaf – a potential papermaking raw material. *TAPPI* 48 (8): 455–459.

Moore, G. (1996). *Non-wood Fibre Applications in Papermaking*. UK: Pira Intl.

National Research Council (1976). *Renewable Resources For Industrial Materials*. Washington, DC: National Academy of Sciences.

Oswalt, S.N., W.B. Smith, P.D. Miles, S.A. Pugh, Miles, P., Vissage, J., and Pugh, S. (2014). Forest Resources of the United States, 2012. USDA For. Serv., Gen. Tech. Rep. WO-91.

Paavilainen, L. (1998). European prospects for using nonwood fiber. *Pulp Paper Int.* 40 (6): 61–66.

Pande, H. (1998). Non-wood fibre and global fibre supply. *Unasylva* 49 (193): 44–50.

Perez-Garcia, J., Lippke, B., Briggs, D. et al. (2005). The environmental performance of renewable building materials in the context of residential construction. *Wood Fiber Sci.* 37, CORRIM Special Issue: 3.

Perlack, R., Wright, L., Turhollow, A., Graham, R., Stokes, B., and Erbach, D. (2005). Biomass as Feedstock for a Bioenergy and Bioproducts Industry: The Technical Feasibility of a Billion-ton Annual Supply. US Department of Energy, Oak Ridge National Laboratory/US Department of Agriculture. http://www.eere.energy.gov/biomass/pdfs/final_billionton_vision_report2.pdf.

Postel, S., and Heise, L. (1988). Reforesting the earth. Worldwatch Paper 83, Worldwatch Institute, Washington, DC.

Powell, D., Faulkner, J., Darr, D., Zhu, Z., and MacCleery, D. (1993). Forest Statistics of the United States, 1992. USDA For. Serv., Gen. Tech. Rep. RM-234.

Ross, M., Larson, E.D., and Williams, R.H. (1987). Energy demand and material flows in the economy. *Energy* 12 (10/11): 953–967.

Schultz, H. (1993). The development of wood utilization in the 19th, 20th, and 21st centuries. *For. Chron.* 69 (4): 413–418.

Sedjo, R.A. and Lyon, K.S. (1990). *The Long Term Adequacy of World Timber Supply*. Washington, DC: Resources For The Future.

US House of Representatives. (1993). Global warming (Part 2). Committee on Energy and Commerce, Subcommittee on Energy and Power. Report of Hearings, Serial no. 103–192.

Winandy, J., Stark, N., and Clemons, C. (2004). Considerations in recycling of wood – Plastic composites. Proc. 5th Global Wood and Natl Fibre Comp. Symp. Kassel, Germany, pp. A6-1-A6-9.

Winjum, J.K., Meganck, R.A., and Dixon, R.K. (1993). Expanding global forest management: an "easy first" proposal. *J. For.* 91 (4): 38–42.

World Bank (2017). *Open Indicator Data*. Washington, DC: World Bank Group https://data.worldbank.org/indicator/AG.LND.FRST.K2.

Wright, L.L., Graham, R.L., Turhollow, A.F., and English, B.C. 1992. Growing short-rotation woody crops for energy production. In *Forests and Global Change*, Vol. 1. Opportunities for Increasing Forest Cover. R.N. Sampson and D. Hair, eds. Washington, DC: American Forestry Association, pp. 123–156.

INDEX

Page numbers in *italics* refer to Figures; those in **bold** to Tables.

Forest Products and Wood Science: An Introduction, Seventh Edition. Rubin Shmulsky and P. David Jones.
© 2019 John Wiley & Sons Ltd. Published 2019 by John Wiley & Sons Ltd.
Companion website: www.wiley.com/go/shmulsky

Printed and bound by CPI Group (UK) Ltd, Croydon, CR0 4YY

27/10/2024

14580161-0001